Better Journey Time – Better Business

Conference Organizing Committee

R Gostling (Chair)
British Rail Research

D Barney
English, Welsh, and Scottish Railways

B Clementson
Railtest

C Griffiths
Office of the Rail Regulator

R Muttram
Railtrack

N Ogilvie
Railtrack

D Razdan
ADtranz ABB Daimler-Benz Transportation

R Smith
European Passenger Services

R Smith
University of Sheffield

D Swan
Advanced Railway Research Centre

IMechE
Conference Transactions

I MECH E
150th Anniversary
1847 – 1997

International Railway Conference on

Better Journey Time – Better Business

Selected papers presented at

24–26 September 1996

Organized by the Railway Division of the
Institution of Mechanical Engineers (IMechE)

Co-sponsored by

The Institution of Electrical Engineers
The Japan Society of Mechanical Engineers
The International Union of Public Transport
The Institution of Railway Signal Engineers
The Railway Industry Association

IMechE Conference Transaction 1996 – 8

Published by Mechanical Engineering Publications Limited for
The Institution of Mechanical Engineers, Bury St Edmunds and London.

First Published 1996

© The Institution of Mechanical Engineers 1996

ISSN 1356-1448
ISBN 0 85298 997 0

A CIP catalogue record for this book is available from the British Library.
Printed by The Ipswich Book Company, Suffolk, UK.

Contents

Related Titles of Interest

Title	Author	ISBN
Railway Engineering, Systems, and Safety	IMechE Seminar 1996–16	1 86058 015 7
Railway Rolling Stock	IMechE Seminar 1996–17	1 86058 016 5
Design, Reliability, and Maintenance for Rail Transport	IMechE Seminar 1996–18	1 86058 017 3
Rail Traction and Braking	IMechE Seminar 1996–19	1 86058 018 1
Implementing Rail Projects	IMechE Conference 1995–1	0 85298 948 2
Railways for Tomorrow's Passengers	IMechE Conference 1993–7	0 85298 859 1
Railway Safety	ILS Sourcebook 1994	0 85298 941 5

For the full range of titles published by MEP contact:

Sales Department
Mechanical Engineering Publications Limited
Northgate Avenue
Bury St Edmunds
Suffolk
IP32 6BW
England

Tel: 01284 763277
Fax: 01284 704006

The International Scene

C514/085/96

A general review of speed-up in Japan

H SOEJIMA BEng
Central Japan Railway Company, Tokyo, Japan

SYNOPSIS

1. History of increasing speeds in Japan
Since the opening of the first railway in Japan in 1872, maximum operating speed has increased to 275 km/h introduced in 1990. This was made possible by the various technical innovations implemented in order to cut travel times.

2. Technical measures for increasing speeds
In order to shorten travel time, maximum speeds and average speeds have to be increased. For that purpose, various technical measures must be implemented.

3. Hurdles to be overcome for speed-up
For a successful speed-up, noise and vibration issues are the most important hurdles to be cleared. Compared to the Shinkansen, conventional lines face other factors such as level crossings and distance between signals.

4. Management effects of speed-up
Speed-up has a greater impact on society for railway lines with higher traffic volumes. As a result, speed-up allows an achievement of increased revenue. The 'Nozomi' Shinkansen using Series 300 is one of the example of this effect.

5. Increasing the added value of time spent inside trains
Speed-up creates extra time by shortening travel time. Methods of adding value to the travel time must be considered for customer satisfaction in the future.

1. HISTORY OF INCREASING SPEEDS IN JAPAN

Since the start of the first train operation in Japan between Shinbashi and Yokohama in 1872, various technical innovations have been implemented in order to shorten travel times.

There are two types of tracks in Japan: standard gauge for the Shinkansen and narrow gauge for conventional lines. Accordingly, speeds have been increased under different circumstances for each type of track.

For narrow gauge conventional lines, electrification, track reinforcement and improvement of signal facilities have been promoted. Particularly for passenger trains, speeds have been increased by a wide margin by adopting automatic couplers and compressed air brakes. As the result of these efforts, the maximum operating speed reached 110 km/h in 1958.

During this period, high-speed running tests were performed from 1948 to 1957 using prototypes of new trains, etc., and the highest speed in the world for narrow gauge track of 175 km/h was recorded in 1960. (Maximum test speed)

Following that, the maximum speed for service operation was raised to 120 km/h in 1968.

Furthermore, efforts were made to implement efficient speed-up by first reviewing regulations limiting speed increases for conventional lines and by increasing speeds through curves and turnouts. In addition, technical development was promoted to allow significant increases in maximum speeds and speeds through curves. As a result of these efforts, the maximum speed of 179 km/h was recorded on the Kosei line using a Series 381 pendulum train in 1985.

Following the privatization of the Japanese National Railways in 1987, the newly established JR companies increased speeds in order to vitalize conventional lines. Service operation with a maximum speed of 130 km/h was started on the Joban, Hokuriku and Kosei lines in 1989.

On the other hand, for the Shinkansen, the speed of 256 km/h was achieved over a model section of the Tokaido Shinkansen in 1963, and the speed of 286 km/h was recorded in 1972 on the Sanyo Shinkansen using a Series 951 experimental train. After that, the maximum speed of 319 km/h was recorded with a Series 961 experimental train on the Oyama test line of the Tohoku and Joetsu Shinkansen in 1979.

Following the privatization of JNR, the Japanese railway industry was vitalized when Central Japan Railway Company rewrote the record book for speed after 12 years using the Series 300 'Nozomi' in 1991.

East Japan Railway Company achieved a maximum test speed of 425.0 km/h in 1993 for the Shinkansen, while Central Japan Railway Company is currently conducting high-speed running tests using its 300X Shinkansen trains.

In terms of operating speeds, the maximum operating speed in 1964, at the time of the Shinkansen's inauguration, was 210 km/h. However, the maximum operating speed was increased to 220 km/h in 1986 and 275 km/h in 1990.

The maximum speed forms the image for the means of transport in question, and is easily applied as a measure of technical level. However, whereas maximum test speeds are limited by the combination of rolling stock and ground facility performance, maximum operating speeds are mainly determined in consideration of the cost performance of speed-up. Therefore, even though maximum test and operating speeds face different limiting factors here in Japan, efforts are being made both to raise the technical level and at the same time increase speeds in consideration of management aspects.

2. TECHNICAL MEASURES FOR INCREASING SPEEDS

The ultimate purpose of increasing speeds is to shorten travel times. Generally, maximum speeds are increased in order to shorten travel times on service lines. However, in consideration of Japan's topography, it is necessary to understand the various limiting factors which prevent travel times from being shortened. In other words, rather than simply increasing maximum speeds, it is also effective to increase average speeds by increasing speeds through curves and turnouts as well as acceleration and deceleration speeds. Therefore, it is necessary to ascertain which of these measures will produce the greatest effect and to implement measures for increasing speeds which are the most appropriate for the line in question.

Power must be increased in order to increase maximum speeds, and high-speed cars require large acceleration power when running at high speeds. However, the coefficient of adhesion drops at high speeds making it necessary to improve control technology. Also, since large amounts of energy are generated when braking from high speeds, regenerative brakes should be used as much as possible.

In this manner, both the drive and brake power increase together with the speed, causing train dimensions and weights to increase accordingly. Therefore, efforts to reduce weights and

4

provide a low centre of gravity are essential. In Japan, new technologies are being adopted one after another to allow power to be increased without increasing the size. These include the conversion of d.c. motor drive to a.c. motors, and the evolution of rheostatic brakes into regenerative brakes. High output devices which are both lightweight and easily maintained are also being widely adopted with recent developments in electric power semiconductors.

In addition, advances in simulation technology have resulted in improved aerodynamic properties, current collecting performance has been improved by increasing trolley wire tension, and track maintenance levels have also been improved.

Increasing average speeds is also extremely important for shortening travel times, and is a particularly major factor for lines with large numbers of sharp curves such as conventional lines in Japan.

Factors limiting speeds through curves can be broadly classified as the risk of overturning, track failure due to lateral force, and a worsened riding quality. Various measures have been taken to overcome these factors. These measures include moving equipment situated on car roofs to under floor positions to provide a low centre of gravity, and using a pendulum control system which smoothes car body inclination by controlling car bodies using computers which have been programmed with track data such as curve positions and shapes. In addition, self-steering bogies have also been put to practical use in recent years. These bogies vary the axle facing along the rails to reduce the lateral force.

In addition to these hardware limiting factors, soft aspects such as the number of stopping stations, reducing stopping times and increasing train frequencies must also be considered when creating train schedules.

3. HURDLES TO BE OVERCOME FOR SPEED-UP

Since trains run through densely populated areas in Japan, clearing noise and vibration problems is the most important issue for implementing speed-up for service trains. Railway noise is broadly classified by noise source into rolling noise generated by wheels rolling over rails, structural noise generated by the vibration of structures, current collection system noise produced by the pantograph and trolley wire system, and car body aerodynamic noise produced by the car body cutting through the wind.

Countermeasures for these noises are divided into noise source countermeasures which aim to lower the noise source emissive power and soundproofing countermeasures which aim to block noise.

Countermeasures for rolling and structural noise include smoothing rail profiles, ballast mats, slab mats, anti-vibration sleepers, sound absorbing and blocking work such as steel girders, lowering the spring coefficient of rail fastenings, installing noise barrier walls, and reducing car weights, etc.

Countermeasures for current collection system noise (spark, sliding and aerodynamic noise) are divided into ground facility and car-related countermeasures. For ground facilities, these countermeasures include increasing trolley wire tension and adopting trolley wire oilers. Car-related countermeasures include reducing the number of pantographs, laying special high-pressure bus lines, introducing fine adjustment contact strips, improving the lubrication performance of contact strips and attaching pantograph covers, etc.

A number of countermeasures are being implemented for aerodynamic noise including streamlining car noses and smoothing car body surfaces (smoothing windows and doors, under floor equipment covers, coupling covers), etc. In addition, measures to prevent TV reception interference and noise for houses along the line are also being implemented.

The Shinkansen is not simply operated at high speeds on existing lines. Rather, a completely new set of tracks have been constructed, thereby freeing tracks, signals, trolley wires, safety devices and all other facilities from the various limitations of conventional tracks. Thus, the Shinkansen have been constructed using entirely new technology and standards, and the tracks are completely separated from roads. However, for speed-up on conventional lines, close attention must be given to the issue of level crossings and the distance between signals.

Moreover, when maximum speeds are increased significantly with the current signal system, trains are unable to decelerate to the specified speeds making it necessary to alter the signal system. In particular, since specifications for level crossing safety devices are determined in accordance with the fastest trains when train speeds differ, crossing warning times become longer causing automobile traffic congestion and other problems. Thus, problems exist which must be resolved in the future.

In addition, distance intervals with preceding trains increase together with maximum speeds. These distance intervals are governed by the blocking length, which is in turn determined by the brake performance, car length and other factors. Lengthening this interval restricts the number of trains which can be operated. Therefore, signal and other facilities must be improved in order to reconcile the two conflicting goals of increasing speeds and achieving mass transport.

Note that since speed-up also increases costs, cost performance must of course be carefully ascertained.

Cost increases produced by increasing speeds include higher motive power costs, higher car and ground facility maintenance costs and additional investment for car and ground facilities.

On the other hand, increasing speeds is also thought to reduce costs by improving car and employee turn-around rates. Accordingly, for increasing speeds to be economically viable, the sum of the increased revenue (or the avoidance of drops in revenue if the situation is left unchanged) produced by increasing speeds and cost reductions due to increased efficiency must exceed the cost increases. In short, cost increases produced by increasing speeds must be kept as low as possible. The proportion of operating costs accounted for by motive power costs increases together with speeds, and is therefore the area which must be emphasized most in terms of cost increases. In order to reduce train running resistance which is closely related to motive power costs, efforts must first be made to reduce train weights and air resistance as much as possible.

Moreover, regenerative systems produce merits only for electric railways among means of transportation, and adopting these systems is also an important factor in working to reduce motive power costs.

Regarding car maintenance costs, increasing speeds also increases maintenance costs for worn portions such as wheel flanges, pantograph contact strips and mechanical braking devices, and fatigued portions such as running gear and driving devices. However, it is thought that most of these problems can be solved by adopting special wheel profiles, improving performance through curves by optimizing the axle box support rigidity, and optimizing the cant and slack.

Increasing speeds may lead to increased track failure. However, maintenance costs can be lowered by reducing car weights and unsprung mass, reinforcing weak portions of tracks and efficiently controlling track irregularity.

Increasing speeds are also expected to increase trolley wire wear. Therefore, integrating trolley wires and pantographs and other measures for limiting cost increases must be considered.

4. MANAGEMENT EFFECTS OF SPEED-UP

Then, why has speed-up been implemented in Japan? Or in other words, what are the effects of speed-up?

Generally, speed-up is said to produce both external and internal effects. External effects are those merits received by the national and regional societies, and internal effects are those merits received by the railway company.

Since time spent moving on trains generally tends to become lost time, there is a great need to reduce these time losses through speed-up. Reducing time losses produces ripple effects which extend over an extremely wide range of fields including alleviating labour shortages and creating leisure time, etc.

Given this background, speed-up has a greater impact on society for railway lines with higher traffic volumes. In other words, speed-up will undoubtedly pay off if implemented in locations with high cost performance.

6

As a result, speed-up will also allow railway companies to achieve increased revenue.

Taking the Tokaido Shinkansen as an example, Shinkansen transport volume has tended to increase uniformly with the rise in GNP. This is an example of a beneficial circle where Shinkansen transport volume rises due to economic expansion and economic development is in turn spurred by activated transport.

Speed-up also had a large effect on cities along the line. The inauguration of Tokaido Shinkansen service between Tokyo and Osaka cut the travel time by more than half from 6 hours and 30 minutes to approximately 3 hours. This allowed the Kanto, Chubu and Kansai regions which form Japan's three major commercial regions to be included within the same sphere of social activity.

For example, viewed from the Tokyo metropolitan area, this placed Western Japan within day-trip range. This greatly activated exchanges with Western Japan and produced major social effects.

In other words, the effects of speed-up include regional development effects such as promoting industrial and corporate location and promoting industry, and speed-up also leads to population increases and increased production capacity along the lines, etc.

Transport results for the Tokaido Shinkansen exceeded 3 billion passengers in the 31 year period following the start of operations. In terms of market share, the Tokaido Shinkansen accounts for 85% of all passengers between Tokyo and Osaka, indicating an overwhelming strength for intermediate distances of up to about 500 km compared to other modes of transport.

Annual passenger volume is currently 130 million passengers. The Tokaido Shinkansen carries approximately 350 000 passengers daily, and has become an indispensable means of transport for Japanese society.

Construction costs for the Tokaido Shinkansen were 380 billion yen, and this investment was recovered in about seven years. As mentioned above, annual regional economic effects of 50 to 60 billion yen are also anticipated, making the Shinkansen an example of management success.

Nozomi operation using Series 300 Shinkansen trains shortened the travel time between Tokyo and Osaka by 30 minutes, thereby attracting passengers from other modes of transport and shifting passengers to the Nozomi. This has resulted in an annual increase in revenues of approximately 15 billion yen.

On the other hand, Nozomi related investments totalled approximately 120 billion yen, making this an example of successful results obtained by shortening travel times.

5. INCREASING THE ADDED VALUE OF TIME SPENT INSIDE TRAINS

Speed-up generates the direct effect of creating time by shortening travel times. However, methods of adding value to travelling time must be considered as a future theme.

In other words, railways should aim to create new spaces inside cars including a business style in order to allow the effective use of travelling time.

As a part of these efforts, we have worked to ensure that passengers are not cut off from information by providing telop and radio news inside cars and to secure means of communication by providing the ability to receive external calls at public telephones using leaky coaxial cable facilities.

In addition, we have also worked to create comfortable travelling spaces by providing music services and installing private compartments in 1st class cars, etc.

In addition to speed-up, measures to ensure safety and improve service and comfort, etc. are indispensable for railways to continue developing in the future.

These wide-ranging policies will serve to satisfy passengers, thereby further increasing the effects of speed-up and constructing railways which will be selected by passengers in the 21st century.

JR Group Route Map

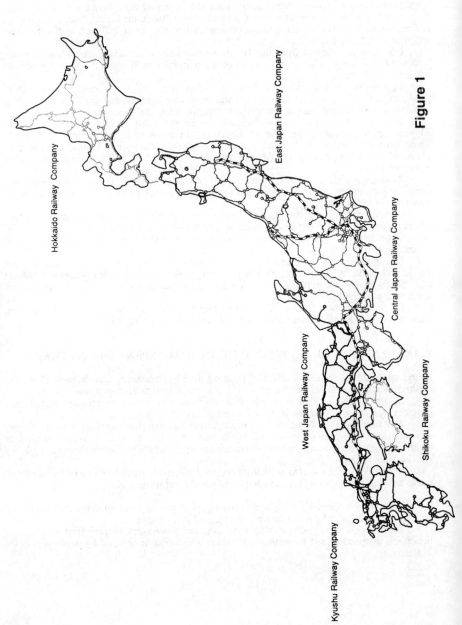

Hokkaido Railway Company

East Japan Railway Company

Central Japan Railway Company

West Japan Railway Company

Shikoku Railway Company

Kyushu Railway Company

Figure 1

Network of Tokaido, Sanyo and Other Shinkansen Lines

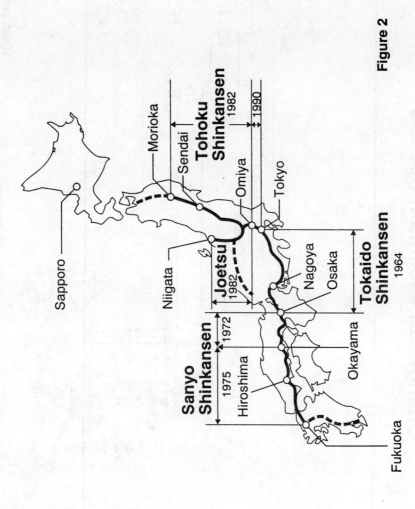

Figure 2

Changes in Maximum Speeds

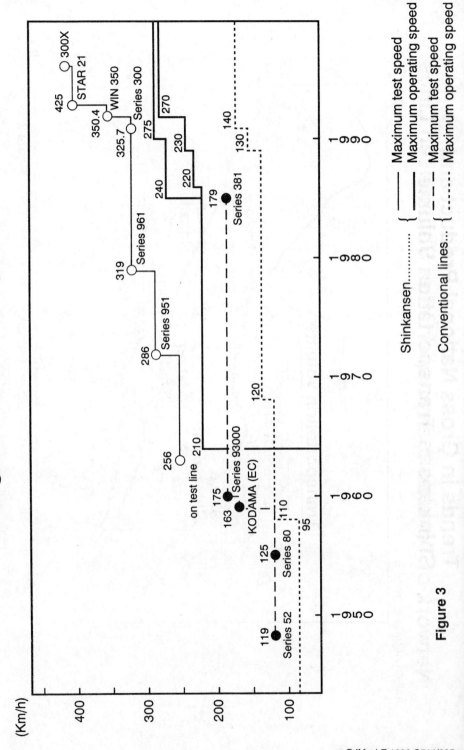

Figure 3

Trends in Gross National Product and Shinkansen Transportation Volume

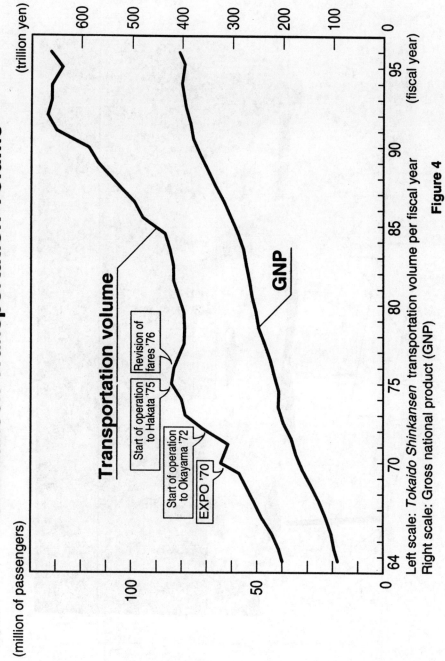

(million of passengers)

(trillion yen)

Transportation volume

Revision of fares '76

Start of operation to Hakata '75

Start of operation to Okayama '72

EXPO '70

GNP

64   70   75   80   85   90   95

(fiscal year)

Left scale: *Tokaido Shinkansen* transportation volume per fiscal year
Right scale: Gross national product (GNP)

Figure 4

From right to left
300X SHINKANSEN, Series 300, 100, 0

Figure 5

The British Scene

C514/054/96

Transport and energy: environmental issues

R A SMITH MA, PhD, FIMechE, FIM and N R HARRIS BEng
Advanced Railway Research Centre, University of Sheffield, UK

SYNOPSIS

It is accepted that economic activity world-wide has become large enough to effect global weather patterns. In the foreseeable future, further expansion of the global economy will be hampered by shortages of resources, particularly energy. Reviews of current transport problems should therefore be made with the longer term in mind. Although, in general rail has significant environmental advantages over other modes, automobiles will improve their performance over a short timescale. Rail 'speed-up' should involve the door-to-door journey, which implies that transport systems of the future must have a high degree of integration.

1 GLOBAL EFFECTS OF ECONOMIC ACTIVITY

Traditionally economic performance has been measured in monitory units. However, the basic activities of any economy are concerned with winning raw materials from the earth, the sea and the air and turning them into desirable products, either material goods or food. These activities consume energy and produce waste products; thermodynamic and mass conservation laws are more fundamental considerations than cost, although costs are more often a useful short term measure of small parts of these activities.

Only in the last decade or so has it been recognised that the scale of our economic activities world-wide has become sufficiently large to displace the equilibrium condition of the globe. It has been established that the average surface temperature of the earth has risen by about $1^{\circ}C$ in the last 135 years and the rate of change has increased in recent years. Recent reports (US Energy Administration; May 1996) suggest that world emissions of carbon dioxide, the main cause of global warming, are expected to increase by 54% above 1990 levels by 2015. This 'global warming' alters the earth's hydrological cycle, affecting the way water is transported around the planet to alter the distribution of rainfall and heat. This in turn produces climatic flips that have become increasingly common in recent years: driest

British summer in two centuries, heavy rain May to August in Korea, heatwaves up to 50°C in Chicago and so on.

Although the warning signs are now clear, because people are not too directly influenced in their daily lives, our leaders still propagate economic ideas based on the unconstrained expansion of our system: for example that already developed economies will expand by 3 to 4 % per annum, and that a higher still expansion of 7 to 8 % in developing economies will tend to equalise economic levels world-wide in the next 40 to 50 years. China's economy is doubling every ten years and its population is six times that of the USA.

However, to the extent that material consumption is tied to economic growth and given that growth rates of 3 to 4 % will double the size of an economy in about 25 years, this period will see material consumption equal to the total material consumption in the history of mankind and at the end of the period human beings will be commandeering virtually the entire output of nature [1].

If this statement seems alarmist, it is because the consequences of the runaway effect of exponential growth are not well enough understood. For this reason a detailed explanation is given in Appendix 1.

The conclusion is stark: our plans for increasing economic activity are unsustainable because shortages of material and energy are just around the next corner. It should therefore be against this longer term outlook that our current reviews of environmental issues of transport are conducted.

2 GROWTH IN TRANSPORT AND CHANGES IN LIFESTYLES

Transport, over the last forty years, has been dominated by the huge increase in automobiles and consequent switch from public to private transport mode. Air transport has also enjoyed very significant increases. Continental and inter-continental travel is now possible, in real terms, cheaper than ever before. As a result, in developed countries lifestyles have changed to match travel possibilities. The distance travelled for work has increased hugely, with a resultant increase in energy usage and production of atmospheric pollution. Cities have spread because people are able to travel large distances to work: for example, in Tokyo salaried workers looking to buy a 100 square meter house at, say ¥40m - 5 times their average annual salary - need to look some 50 to 60 km from the capital. In the UK and elsewhere, shopping centres are increasingly situated 'out-of-town' and shopping without a car has become difficult.

There are signs that the unsustainability of similar growth is becoming recognised such as the recent issue of a government Green Paper [2] and review of indicators of sustainability [3]. Recent reports in the UK suggest that one third of motorways and trunk roads will be congested within five years [4]. In Singapore planners have decided that £9bn - over £30k per head over 20 years - will be spent on virtually eliminating the car by building an ultra-modern public transport system [5]. Seoul authorities have recently announced plans following Singapore's example, by imposing peak-flow traffic tolls, increases in fuel prices and incentives to public transport operators. Clearly, public transport will play an increasingly important role in the future, both to ease congestion and to reduce pollution levels. In the UK, although the problems have been recognised, the recent Green Paper (Transport, The Way Forward [2]) avoided making any firm decisions on most of the questions raised by the recent debate on transport. The Paper was reviewed as a damp squib by interest groups as diverse as environmentalists, the roads lobby and the CBI. An RAC spokesman commented, "This is paralysis by analysis. This report could have been written by an undergraduate: it contains virtually no policy."

The next sections introduce some quantitative argument about traffic levels, environmental effects of transport and associated issues.

3 QUANTITATIVE DISCUSSION OF ENVIRONMENTAL ISSUES

3.1 Travelling further and by different modes

The following Table illustrates both the increasing distance travelled per person per year and the change in transport modes over the last 20 years in the UK.

Table 1. Changing mode of travel in the UK [6]

	1975/76	1985/86	1992/94
Total distance travelled per person	4710 miles	5317 miles	6439 miles
% of this distance travelled by car	72 %	76 %	81 %
% of this distance travelled by bus	10 %	7.6 %	5.6 %
% of this distance travelled by train	6 %	5.5 %	4.6 %

* The remainder of the percentage of total travel unaccounted for in the above is by bicycle, moped and aircraft.

3.2 Projected growth of automobile traffic

The growth and forecasts of future trends of passenger transport by car are shown in Figure 1 Congestion, pollution, noise, fuel shortage, cost, health and accidents will inevitably combine to intervene with this forecast trend. Planning to maintain control over any such intervention would seem prudent. Delaying tackling such issues until market forces dictate we can no longer travel cost effectively, is not a viable option, although politicians would like to avoid taking what would be perceived in the short term as unpopular measures.

3.3 Pollution

As a result of our transport requirements we are creating a huge quantity of pollution. Some modes pollute more than others as can be seen from Figure 2. Similar data exists for freight transport. The result of this pollution can be uncovered in a number of unseen costs discussed later in this paper. The modal shift toward road clearly produces an unhealthy environment.

All power generation, with the exception of nuclear (in theory) and some renewable sources, produce some emission. Wind, hydro and solar power appear only to have the capacity to meet local demand. Nuclear power seems unlikely to be exploited in the UK in the foreseeable future so until plentiful, clean power is developed we must aim to use the finite fossil fuel energy reserves we have as cleanly and efficiently as possible. Increasing the cabin (load) factor of transport modes would make a useful contribution towards this aim.

In Figure 3 (derived from Figure 2) maximum cabin factor is assumed. These figures give some idea as to the improvements that can be made by simply filling vehicles. This requires no research, no new technology - only increased willingness to travel with others.

Similar improvements in freight transport emissions can be obtained by ensuring higher utility of container space. In this case there may be a need to develop logistical software to facilitate system optimisation in this respect.

3.4 Using resources

Transport is based on the ability to convert an easily stored energy source into kinetic energy. The number and volume of suitable energy sources is diminishing rapidly. Oil, gas, coal and nuclear power are among the most useful in the UK. Figure 4 shows remaining global resources, while Figure 5 shows the proportion of current, reliably recoverable reserves used in a year.

Even accounting for tapping new reserves, it can be estimated that at current and forecast rates of consumption oil, gas and coal will run out in approximately 40, 60 and 250 years respectively. It is increasingly important that the most appropriate fuel, based on long term strategy, rather than present cost, is used for each particular purpose. For example, coal is more appropriate in power stations rather than lighter, easily stored gas. Recently in the UK nuclear power has suffered set backs as Government support has appeared to dwindle in the light of environmental pressure and lack of technical strategy. This may well prove to be a mistake. Control of nuclear waste and non-polluting power generation could be achievable with perseverance and well directed research. Reserves of fuel for nuclear power stations are not likely to run out for hundreds of years.

3.5 Creating noise and vibration

Without exception, all forms of transport create some form of noise through their operation. This noise comes in many different forms dependant on the mode of transport in question. Individual sources may be comparatively quiet but wide spread as in the case of the motor car. Sources may also be loud but concentrated as in the case of mass transport systems such as rail and air which have fewer, noisier sources and, in general, cause fewer noise objections than road vehicles.

More stringent legislation concerning noise is appearing. The Royal Commission report on Transport and the Environment suggested maximum noise levels at external walls of 65dB $_{Aeq\ 16h}$ and 59 dB $_{Aeq\ 8h}$ for day and night time respectively [10]. As progress and innovation is made in fields of technology such as Active Noise Control, sound proof materials, combustion and related control processes, further, even tighter noise restrictions will be possible. Successful work has already been reported where a diesel generator was fitted with an active exhaust pipe significantly reducing noise emission without effecting fuel consumption [11]. However, control and minimisation of noise is still a significant limit to speed-up of trains, particularly in urban areas.

3.6 Hidden costs of transport

Besides the obvious costs of transport, there are a number of other costs such as accidents damage and health costs that are significant enough to be included into the comparison. In 1993 the cost of road accidents was £8.3 billion in the UK [12].

Figure 6 illustrates the risk of injury or death in an accident by road and by rail. Road accidents are especially difficult to reduce as there are a huge number of vehicles each with an infinite number of degrees of freedom, controlled by individual persons each with their own infinite number of responses and actions. Road is improving its position despite these inherent difficulties, partly because congestion is reducing speeds. Rail, with fewer vehicles, is a one dimensional problem - trains can only travel along the rails. Although braking distances are longer, train drivers cannot swerve, cross the white, line hit the motorway barrier etc. Despite this, a number of accidents still occur. The application of, for example, current global satellite positioning technology (or terrestrial equivalent) and modern control theory to this comparatively simple problem could significantly reduce rail collisions.

As well as accidents there is the issue of both physical and mental health. Pollution and noise from transport modes effects a great number of people in the UK. Recently there has been concern in the press about air pollution, particulate matter and asthma. Transport related pollution is apparently giving rise to additional costs borne by the National Health Service. Research has acknowledged that driving causes stress and there is evidence connecting traffic nuisance with increased levels of stress in nearby residents. These matters inevitably give rise to health costs borne by the taxpayer.

Air pollution, largely acidic fumes from vehicle exhausts, is also quietly eroding many buildings and structures. Many of these require costly repair on a regular basis. Here again transport related air pollution appears to be giving rise to costs borne by individuals and the public purse.

4. ENVIRONMENT AND THE PACE OF LIFE

4.1 A day's travel

It is now possible to travel return from London to New York in a day, spending over four hours in America (and over £4600 to get there!). More common travel requirements within the UK have, however, become taken for granted. Travel has become a tremendous benefit to the economy. It is difficult to know if transport stimulates GDP or vice-versa, but they both go hand-in-hand. The ability to travel further in a short time means that time effectively slows down for business, more business can be conducted over a longer distance in a shorter time.

The ideal size of a country might be one in which return from the capital to all parts is achieved in a day. As countries become politically closer the desirable country might be replaced by the desirable continent and so on. Hence the pressure to increase transportation speed which is well recognised in Continental Europe.

4.2 Faster travel

To travel faster we must use more energy. (An IC electric uses 480 kJ/pass-km at 160 kmh and 650 kJ/pass-km at 200 kmh [13]) Increased energy use reduces the time we have to find sustainable energy sources to replace diminishing fossil fuels. This increases the likelihood that we shall loose our ability to travel as freely as we do, and suffer the economical consequences that such a loss will bring. This is a long term view; very difficult to bring to the forefront of short term politics.

In addition to higher energy consumption (and associated higher emission rates) faster transport is noisier. Concord is noisier than ordinary jet aircraft. High speed trains are noisier than ordinary trains. It must be recognised that faster travel is more environmentally detrimental. The 18th Royal Commission Report [14] recommends that no proposals be taken forward for high speed rail travel (>300kmh) unless the environmental benefits from reducing air traffic can be comprehensively demonstrated. Additional noise and energy consumption are cited as reasons for this recommendation.

The pressure for faster travel must inevitably produce a faster rate of damage to the planet. Already there are grave doubts that we can sustain the current rate. Furthermore, expensive efforts to improve the speed of one stage of a journey will not be productive unless the connecting and feeder modes are also speeded up and integration and ease of connection are also streamlined.

4.3 Congestion and the city environment

Increasingly the road network will not be able to cope with the increased demand for fast transport. Figure 7 shows that the car is much slower than we think for short journeys. Already in cities this mode of transport is beginning to fall behind the pace of life which people desire. These figures have serious implications for city environments.

The rise in the motor car has meant that cities have been planned for road traffic. City streets are now at maximum capacity. Buildings are being attacked by acidic fumes from roads. Residents are suffering increased incidents of respiratory disease. The result of this has been a trend toward light rail systems as in, for example, Docklands, Sheffield and Manchester. These are an improvement only if they provide advantages not offered by cars. Fast rush hour travel through off road or pedestrianised sections is required.

Such rail systems are only environmentally beneficial if they carry a high number of people. A tram carrying one passenger is many time more polluting than a car with one driver.

The solution to urban transport related environmental damage is fast, frequent rush hour transport provided by electric rail systems with high cabin factors, separated from other traffic. Future city plans should include where ever possible reduced access for cars and increased access for mass transport i.e. tram, bus and access roads only. Planners should devise convenient, safe and reliable park and ride schemes stopping traffic from outside the city at the outskirts providing a cheaper, easier alternative to parking in the centre.

Although this solution simply moves the pollution from one source to another, less pollution is generated by virtue of use of more efficient systems in the appropriate places. In the case of electric traction the pollution is also generated away from centres of population and in bulk allowing economies of scale in treatment and control of polluting waste gases.

5 CONCLUDING REMARKS

In the short term, time and environmental considerations will direct us to the minimisation of door-to-door journey time, perhaps at the expense of the highest speed reached on one portion of the journey. This implies the need for proper integration of appropriate transport modes. Train travel, which offers many environmental advantages over its competitors, needs to be used effectively at high load factors. In the longer term it may be that electronic communication will reduce the need to physically travel. In any case longer term alternatives need to be sought because current projected expansions of travel are unsustainable based on current known energy reserves.

REFERENCES

1. Rees, W. E., Reducing Our Ecological Footprints. Siemens Review, 1995 2/95 pp30-35.
2. Transport - The Way Forward - The Government's Response to the Transport Debate. April 1996   CM3234 London HMSO.
3. Indicators of Sustainable Development for the United Kingdom   March 1996 London HMSO.
4. D.o.T. Projections, reported in the Daily Telegraph, 2 March 1996.
5. The Independent, 4 January 1996.
6. Transport Statistics Great Britain 1995. 1995 London HMSO.
7. Transport Statistics Great Britain 1995. 1995 London HMSO. & National Road Traffic Forecasts 1989 London HMSO.
8. Carpenter TG The Environmental Impact of Railways p173 Wiley 1994
9. World Energy Council (WEC), Madrid 1992 & German Federal Institute for Earth Sciences and Raw Materials, Hanover, 1990.
10. 18th Report of the Royal Commission on Environmental Pollution - "Transport and the Environment", Objective H   p237 Oct. 1994 London HMSO (Cited as 18th Royal Commission Report)
11. Shipps, JC, Miller, S K, & Shaver, J Active exhaust silencers quiet diesel generator sets' Power Engineering Aug 1994 vol. 98 p 42-45.
12. Transport Statistics Great Britain 1978-89. 1989 London HMSO.
13. Farrington, J. 'Transport, energy and environment', in Hoyle, B.S. and Knowles, R.D. 'Modern Transport Geography' 1992 Belhaven London.
14. 18th Royal Commission Report 1994 Recommendation 73 p246.
15. National Travel Survey 1992/94. p57 1995 London HMSO.

APPENDIX 1
THE CHARACTERISTICS OF EXPONENTIAL GROWTH
The rate of growth of many activities and the rate of consumption of many resources is often proportional to the current size of the activity or consumption, this is:

$$dQ/dt = RQ$$

which on integrating gives:

$$Q = Q_o e^{Rt}$$

where Q, the current level occurs at the time t after a base level Q_o, R is the constant of proportionality and e is the base of natural logs. The time for the activity to double, t_D, is easily shown to be, $t_D = \ln 2/R$, thus a 1% annual growth rate ($R = 0.01$) doubles the activity in 70 years, 3% in 23 years, 5% in 14 years and at 10% in 7 years. These short doubling times for relatively modest growth rates are an important characteristic of exponential growth.
Figure A1 plots the exponential function in multiples of doubling time.

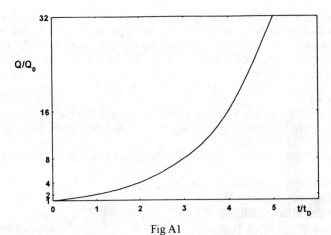

Fig A1

The total activity, or consumption of a resource, can be found by integrating this function between appropriate time intervals. In particular, total consumption from time 0 to a large time T is given, with a slight approximation, by Q_o/Re^{RT}. In a similar manner the consumption from time T to $T+t_D$ i.e. over the next doubling period, can be shown to be the same. Thus the important exponential characteristic, **the total resource consumed in a doubling period is equal to the consumption in all the previous doubling periods combined.**

Road traffic growth and 1989 forecasts

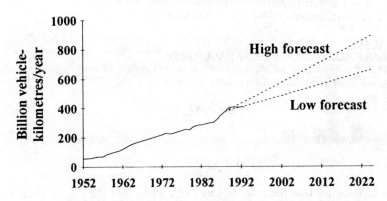

Fig 1. Past and future trends in the growth of the motor car in the UK [7]

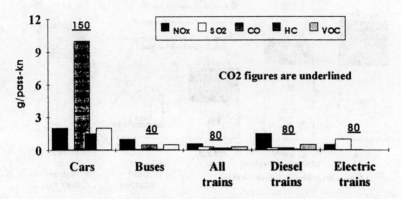

Fig 2. Passenger transport emission by mode [8]

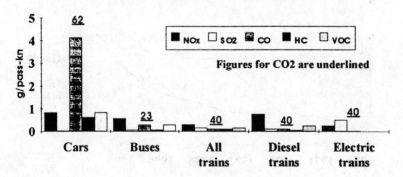

Fig 3. Improved emissions per pass-km from increased cabin factor.
[Adapted from 8]

World reserves of primary energy

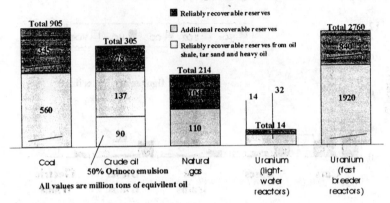

Fig 4. Energy reserves around the world [9].

Production of fuel as a proportion of reserves

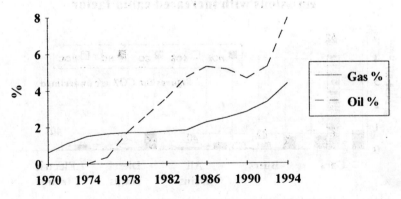

Fig 5. Proportion of reserves consumed each year [3].

Car, bus and train casualty rates 1983-1993

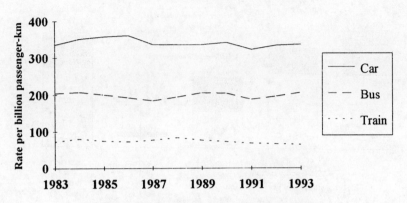

Fig 6. Comparative casualty rates of road and rail [6].

Average speed of 3 to 5 mile journeys

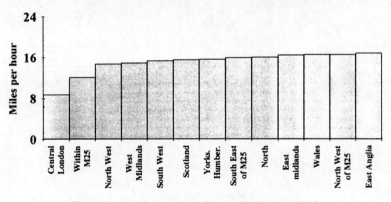

Fig 7. Door to door speeds of 3 - 5 mile journeys [15].

C514/062/96

Regulation, the changing industry and journey time

C J GRIFFITHS BSc, CEng, FIMechE, MCIT
Office of the Rail Regulator, London, UK

SYNOPSIS

The Railways Act was passed in 1993 and has brought about radical restructuring of the industry in separating ownership and management of the railway infrastructure from the operation of trains, the ownership of rolling stock and the provision of other railway services. The restructuring has introduced a new and complex set of contractual, and in some cases competitive relationships between the newly formed companies. Many of the new companies are now privatised. The restructuring, however, in itself, does not guarantee benefits for the users and customers. Those benefits depend on how the new players work together within the new contractual matrix. The Act requires the Regulator to exercise certain duties and functions which are designed to protect users and to promote the use and development of the network.

In the restructured industry partnerships between Railtrack, the train operators, rolling stock owners and manufacturers and funding bodies such as the Franchising Director and PTE's will be vital. This paper addresses the role of the Regulator in working with the industry to achieve a better railway: which includes better journey time when this is in the public interest. It describes the Regulator's duties, general features of the new industry structure and highlights the mechanisms that should bring about improved journey time both within the existing industry arrangements and by new investment.

1.   INTRODUCTION

Over the last three years, the railway industry has undergone an intensive process of reconstruction and privatisation. The industry has moved away from being a vertically integrated organisation where all assets and operations were under the management of the British Railways Board, to an arrangement where the owners and operators of the infrastructure, the train operating companies, the providers of rolling stock, the infrastructure maintenance companies and providers of other railway services are now separate commercial entities. Many are now in the private sector and the remainder are planned to transfer to the private sector over the next year.

Obligations and rights between the various industry players have been formally defined by licences and negotiated agreements and the resulting contractual matrix is almost in place. The structure of the new industry appears complex and effective relationships between the new constituent companies are essential to the efficient operation of the Railway. The new private sector owners are now setting about making this new rule based structure work in their commercial interests. The Regulator's job is to make sure it also works in the public interest. Figure 1 shows in diagram and summary form the various relationships between the industry players.

Previously, British Rail's investment policy was constrained by the amount of the External Finance Limit made available by Government and the absence of long term financial commitments by the State. Privatisation means that, for the first time, substantial private finance is available and that finance from the State and local authorities has been *secured* into the next century. New investment is promised by franchisees - very often they have bound themselves contractually to deliver it through the terms of their franchise agreements.

However, as long as fare-box revenues do not cover the costs of providing services and unless there is a radical change away from today's service timetable, the industry will continue to require subsidy from the State: This is provided by the Franchising Director and in some major conurbations outside London, by the Passenger Transport Executives. The passenger train operating companies being sold by the Franchise Director required in the order of 1.8 billion pounds in subsidy when they were in public ownership in 1995-6. Franchise agreements require passenger train operating companies to commit to service and performance levels: Whilst franchise commitments to improved services and operating efficiency will reduce the level of subsidy over time, there will be a continuing need for significant State subsidy.

Restructuring and privatisation, however, may not necessarily create benefits for customers and users unless the agreements between the undertakings allow development and innovation which work in the public interest. The restructuring of the industry will create new competition which will stimulate innovation. But Railtrack is the monopoly supplier and manager of the railway infrastructure; and certain other suppliers of railway services have either a monopoly or near monopoly in their own supply. The Regulator aims to promote the public interest as defined by his duties under the Railways Act.

© IMechE 1996 C514/062

He is required to protect users, to promote the use and development of the network, to promote economy and efficiency, to promote network benefits and to have regard to safety and the needs of the disabled. He has specific duties to impose minimum restrictions on operators and to enable service providers to plan their business with a reasonable degree of assurance and to not make it unduly difficult for Railtrack to finance its activities. He must also have regard to the financial position of the Franchising Director. In other words, he must balance regulation for the public interest with allowing reasonable commercial freedom for the railway undertakings.

The Regulator, however, can only exercise these duties through the exercise of his statutory functions: which are the approval of access agreements, the granting, monitoring and enforcement of licences and the approvals of closures. He has additional powers under general competition law. His aim is to work with the industry to produce a better railway; facilitating the industry partnerships within the set of rules that are in place, to produce improvement.

2.   BETTER JOURNEY TIMES

Better journey times can be brought about through improved performance and operation on the current railway, modifying the existing railway assets or by major investment schemes to renew and enhance the infrastructure, or purchase new rolling stock.

2.1   Operating Existing Assets

Good ' housekeeping' and operating arrangements on today's railway can, in many cases, bring about better and more reliable journey times for the travelling public. The benefits at the margin will flow through into operators' profits and provide them with commercial incentives.

Train operating companies must negotiate with Railtrack for rights to run trains on Railtrack's infrastructure. These are expressed in terms of the quantum of services, train types, frequency, intervals, patterns, time periods in which the trains should operate, departure, calling and destination points and journey times.

These rights are part of the track access agreements that must be approved by the Regulator. The Regulator in approving the access agreements assures himself that the access rights do not prevent efficient use of track capacity and do not unduly constrain other users. Similarly, Railtrack needs to reserve a quantum of track access rights for carrying out necessary works to the infrastructure.

All track access contracts incorporate the Track Access Conditions. These conditions, approved by the Regulator, are a multilateral set of 'rules' that define mechanisms for Railtrack and the operators on the network to work together and for promoting change. The Access Conditions include the procedures by which Railtrack makes bids for engineering possessions and train operating companies make bids for timetable slots. Bids for possessions or timetable slots must be compatible with the track access rights secured by Railtrack or the train operating company in the relevant access agreement.

Newly privatised train operating companies are already seeking, through this bidding process, to improve their services, and in some cases negotiating with Railtrack to improve their track access entitlement. The Regulator has published criteria which set out the public interest considerations he will take into account in considering access agreements - or modifications - for approval.

Both passenger and freight track access agreements contain performance regimes. The regimes offer financial incentives for both Railtrack and the operators to improve train running performance in terms of minimising delays and avoiding cancellations; and penalise them for a deterioration in performance. As a result of the new contractual structure, there are commercial incentives for Railtrack to operate trains on the network efficiently and to maintain the infrastructure; and for train operating companies to ensure their rolling stock and crewing arrangements perform according to the timetable. Similarly, if Railtrack exceeds or overruns its possession allowance, it will be penalised for the disruption it causes to train services. The aim of these commercial arrangements is to promote more trains running 'on time' with fewer delays for the customer.

The access agreements are binding agreements between the parties which the Regulator approves, but cannot enforce. The Regulator, however, must satisfy himself that Railtrack does not abuse its monopoly position, for example, by cutting costs on maintenance of the infrastructure for short term gains. He cannot rely solely on the incentives in the access agreements to secure this.

2.2   Condition of the Infrastructure

The Regulator published his policy document, in January 1995 on the track access charges that would apply for franchised passenger services from 1st April 1995. His policy allows Railtrack sufficient income from access charges to maintain the existing capability of the rail network. Additionally, he has allowed £3.5 billion (at 1995-6 prices) for investment in the renewal of the network in 'modern equivalent' form over the next six years to 2001. This has allowed Railtrack unprecedented certainty of funds for maintenance and renewal.

The Regulator will want to be satisfied that the amounts allowed in his review for bringing the network up to standard and for maintenance and renewal are properly spent for the benefit of the users: In other words, the users are 'getting what they paid for'. A key element in ensuring this will be the contractual commitments of Railtrack to the train operating companies, including commitments to journey times and sharing of investment plans.

The Regulator needs to be able to monitor Railtrack's efficiency and achievement. Information he will monitor and review includes:

- Railtrack's spending plans and its actual spend on infrastructure renewal and maintenance;

- Railtrack's achievement of its Network Management Statement: Railtrack is required to produce such a statement by its Licence. The first such statement was published in December 1995; and the Regulator is discussing with Railtrack the form of future statements. The aim is that this information should assist the Regulator, Railtrack's customers and the wider public to monitor Railtrack's effectiveness in maintenance and renewal of the network; and to promote a shared understanding of plans for the management and development of the network. The Regulator has already indicated that he wishes to see a greater degree of detail in the next statement;

- Railtrack's Key Performance Indicators: Railtrack has agreed with the Regulator it will publish key performance indicators which reflect the capability and reliability of the network. The first publication will be in September 1996;

- Accounting Information: Railtrack has agreed to produce information that will enable clear identification of monies spent on maintenance and renewals.

2.3   Securing Network Benefits

On a unified and integrated railway, a passenger would normally expect to be able to make enquiries regarding the timetable for services anywhere on the network; to be able to book tickets at stations or by telephone; to travel the network making changes between trains and to take alternative trains when necessary.

To safeguard these network services in the new industry structure, it has been necessary to ensure commitments of co-operation between industry players, who could be competitors. All operators are required under the terms of their Licences to enter into and honour the Ticketing and Settlement Agreement. This agreement sets out rules which require the operators to subscribe to through ticketing arrangements and to provide impartial information and ticketing services relating to other operators.

Under the access agreements train operating companies can bid for connections with other operators' trains and Railtrack must as far as possible accommodate such bids, in accordance with criteria approved by the Regulator which include 'maintaining and improving connections between railway passenger services'. Access agreements do not, however, refer to connections with non-rail services such as buses and ferries, except in a few, limited and special cases.

The Passenger Train Operating Franchise Agreements include requirements to co-operate on timetabling plans for defined connections and also include performance incentive regimes for specified connections.

The Regulator is keeping these mechanisms under review and can, if necessary, make changes to the Track Access Conditions to improve timetabling of connections.

2.4   Investment

In some cases, it will be investment that brings about the significant savings in journey times. The new railway structure must be able to finance both network enhancements and large scale new build projects if the network and use of the network is to develop. It must encourage and facilitate co-operation between train operators, Railtrack, Passenger Transport Authorities and Executives, local authorities and other funding bodies. Network users are dependent on Railtrack; Railtrack is dependent on the success of its customers; investment partnerships should develop.

The Track Access Conditions provide mechanisms for proposing changes to the network and to vehicles operating on the network. Both processes require consultation with affected parties; and compensation to be paid to any party adversely affected. In both cases, there is an appeals procedure; the ultimate appeal being to the Regulator.

The Regulator published his policy document 'Investment in the Enhancement of the Network' in March this year, which outlines his standpoint on investment. His policy is designed to encourage partnership in the development of investment schemes. He expects Railtrack to be proactive in developing investment proposals. Where Railtrack is able to negotiate charging arrangements for a project which enable it to recover project costs (including fair renumeration for investment), and receives a fair share of the benefits, then the Regulator expects Railtrack should carry out the scheme, unless it is contrary to the public interest.

The access charges created by such investment projects would fund the enhancement to the network. The Regulator would expect the level of these charges to cover avoidable costs. The charges would reflect sharing of any net benefits which should reflect the relative share of risk carried by Railtrack and other parties involved. Where the prospect of future reviews of access charging levels by the Regulator introduces material risk to the investment, the Regulator has indicated he would be willing to consider approving special arrangements for identified projects which would give protection and certainty at specified charging reviews.

Under normal circumstances, the Regulator would expect that any new track capacity produced by a new or enhanced facility should be shared between train operators: The Regulator recognises, however, in order to offer sufficient commercial incentives for the investment it may be necessary, in certain circumstances to allow a degree of exclusivity or moderation of competition. He is also prepared, if appropriate and providing certain public interest safeguards are in place, to consider the possibility of vertical integration; e.g. allowing the infrastructure operator or a rolling stock owner to have an interest in train operations.

3.   CURRENT INVESTMENT PROJECTS

At the time of this paper going to press, three major investment projects likely to lead to significant journey time improvements were at differing stages of development and demonstrate different approaches to investment. These projects are the Channel Tunnel Rail Link, Thameslink 2000 and the upgrading of the West Coast Main Line.

3.1   Channel Tunnel Rail Link (CTRL)

The CTRL is a proposed 68 mile new high speed rail link between St. Pancras station, London and the Channel Tunnel at Folkestone, Kent. The link connects into Railtrack's network both in Kent and the London area.

Following the government's competitive tendering exercise, a development agreement has been signed between the government and London and Continental Railways, a private sector consortium. Under this agreement, London and Continental will finance, build and operate the link. The government has transferred to London and Continental a number of public sector assets associated with the link (including land, accumulated knowledge and plans, and the existing European Passenger Services operation). London and Continental will operate high speed international services on the link, but are committed to provide to the government capacity for high speed domestic services between Kent and London. The operation of these domestic services will be franchised out to the private sector.

The CTRL is a partnership between the public and private sectors with the private sector taking on much of the risk. Powers to build the CTRL will be obtained under the Channel Tunnel Rail Link Bill currently before parliament. The CTRL will essentially be exempt from the domestic regulatory framework by virtue of the Bill. The government's interest and the interests of users, are protected by the development agreement, which includes the arrangements under which capacity is allocated to domestic services. The Regulator still has jurisdiction over operations on Railtrack's network by London and Continental and other operators in connection with the link. However, his duties will be amended to require him to take account of the needs of London and Continental; in particular, it is proposed that the Regulator will have an overriding duty not to impede the performance of any development agreement between London and Continental and the Secretary of State for Transport.

The government has chosen to transfer much of the risk of this relatively high risk project to the private sector, by means of a competition to build and operate. A conscious decision was made by the government that the most effective way of promoting the necessary partnership involved exempting it from the regulatory regime. This approach is unlikely to be practical or desirable for projects on the existing rail network.

3.2   Thameslink 2000

The Thameslink 2000 project consists of new rail infrastructure linking existing routes in London which should produce significant rail journey time reductions and increased track capacity for journeys into and across London from a number of locations both north and south of the city. The project will also provide an interchange at St. Pancras with international and domestic services on the CTRL and at various locations with the services of other train operators including

London Underground. Railtrack and the Franchising Director have entered into a partnership, formalised by an agreement for the construction of the project: Under the agreement, Railtrack has committed to construct the project and deliver specified train running capacity and journey times; the Franchising Director has undertaken to purchase Thameslink 2000 train paths for an agreed access charge and for a period of fourteen years.

During the development of this agreement, the Regulator's office gave informal advice as to the sort of contractual and charging arrangements the Regulator would be likely to approve in an access agreement. Before committing to the project, Railtrack needed some security regarding the approach the Regulator would take to access charging, and in particular the treatment of the charges at the periodic reviews, and this was given. As the project develops further the Regulator will be called upon to approve the access arrangements for the project. He will wish to be satisfied that capacity on the network is allocated in accordance with the public interest and that the interests of all affected operators and potential operators are taken into account.

3.3 West Coast Main Line (WCML)

The West Coast Main Line links London Euston with Birmingham, Manchester, Liverpool and Glasgow and is used intensively both by passenger and freight traffic. The southern part of the route was last fully modernised in the 1960s; the northern part in the 1970s. Much of the infrastructure is approaching life expiry. To accord with the assumption made regarding Railtrack's track access charges, the network, where appropriate, will need to be renewed in modern equivalent form: Railtrack has stated it is committed to the renewal of the West Coast infrastructure. This will include the proposed development of a new transmission based signalling control system. Renewal of the route's infrastructure in modern form should enable significant improvements in the reliability of operations.

The renewal of the route infrastructure presents an opportunity to also upgrade the route's capability at the same time. The Franchising Director and Railtrack are therefore, negotiating an arrangement whereby in return for increased track access charges, journey times on the route are significantly improved. This raises complex issues, especially the balance between how much should be invested into the capability of new rolling stock and how much into new infrastructure. When an upgrade has been agreed, the Franchising Director intends to franchise the route to a private sector operator who will need to purchase the rolling stock to an agreed specification. Again, therefore, this should develop as a partnership between Railtrack, OPRAF as the public sector funding body and the train operator.

If OPRAF and Railtrack are unable to agree terms for the upgrade, it will be open to OPRAF or a train operator to ask the Regulator to direct Railtrack to enter into an access agreement on terms determined by the Regulator. However, it is to be hoped that an agreement will be reached and the Regulator will want to be satisfied the agreement is likely to operate in the public interest and affords adequate consideration to the interests of other operators on the route. To facilitate the development of the project, the Regulator is giving guidance to OPRAF and Railtrack on the approach he would be likely to take on the proposed arrangements.

The Regulator may be called upon by operators on the route to exercise his appeal function regarding changes to the network or to Group Standards necessitated by the renewal or the upgrade, or the timetable changes necessitated by the project. In this, he will again aim to find a solution which operates in the public interest.

This will be the first major project committed by Railtrack since it has moved into the private sector. It is a key test of the new system and one to which all those involved, including the Regulator, must attach great importance.

4.   CONCLUSION

It is the duty of the Regulator, in exercising his functions, to protect the interests of users and customers and to promote the development and use of the network. The Regulator has striven to strike the correct balance between regulation to protect the public interest and allowing the new commercial structure to develop. He has put in place mechanisms designed to ensure maintenance of the network and approved arrangements that will offer commercial incentives both to Railtrack and the train operators to improve the reliability of train journey times and network benefits. He expects Railtrack to undertake investment schemes where this does not commercially disadvantage Railtrack and where it is in the public interest. His policies are designed to encourage and facilitate the development of partnerships in investment schemes where these are in line with his duties.

This text is written in the light of information available in mid-July 1996, but as a result of subsequent developments in the industry, may require revision for the purposes of presentation at the S'Tech Conference in September 1996.

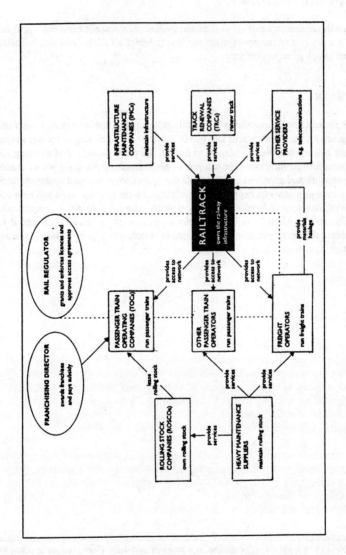

Figure 1. *Source : Railtrack Pathfinder Prospectus, 15 April 1996*

C514/083/96

Why a tram system?

P J ARMSTRONG BSc, MA, MSc, MCIT
Department of Planning and Economic Development, Nottinghamshire County Council, UK

SYNOPSIS

Transport thinking is changing rapidly in response to widespread concern about the environmental impacts of traffic. With urban areas experiencing increasing problems of traffic congestion, effecting the economic well being and the quality of life of inhabitants, local authorities are developing more sustainable transportation strategies; involving a greater role for modes other than the car.

Using the example of Nottingham, this paper considers the role of the tram, which, through its unrivalled efficiency as a people mover and its ability to combine speed in the suburbs with effective penetration of urban centres, may provide a particularly important element of strategy.

1. INTRODUCTION

Transport thinking is probably now changing more rapidly than at any time in the last thirty years. There are real concerns about the environmental impacts of the continuing growth in road transport. We are becoming more car dependent while at the same time access into and around our cities is being eroded by congestion. There is clear evidence in most cities that peak traffic is spreading and bus patronage is declining. With increasing congestion and reducing accessibility comes economic decline for central activities and associated employment and the worsening of environmental conditions and quality of life.

Faced with these problems the task of local planning and transportation authorities is to develop their future transport systems and infrastructure with the aim of balancing the often competing demands of the economy and the environment. While acknowledging the continuing importance of the car, this must include developing measures and consensus for managing travel demand and promoting alternatives to the car.

Following its successful revival in many European and North American cities the modern tram or Light Rapid Transit (LRT) has joined the bus, train and bicycle as possible alternatives to the car being considered by city authorities here in the UK. While capital costs and funding difficulties have moderated the aspirations of many smaller towns and cities it is believed that a tram or LRT system can fulfil a role in meeting the transport needs of larger cities and conurbations where other modes are likely to be less effective. Indeed, as part of a consistent package of measures, LRT demonstrates significant advantages. It is particularly good at meeting the main requirements of a modern transport system, including:

- to be environmentally sound;
- cost effective and commercially viable;
- accessible to all;
- having long term reliability;
- and to be sustainable.

2. SUSTAINABILITY IN TRANSPORT

Sustainable development means development which meets the needs of the present, maintaining and enhancing the quality of human life, without depleting non-renewable resources. For transport this will include:

- creating a closer relationship between homes and jobs;
- strengthening city and town centres;
- reducing energy consumption;
- introducing traffic and travel demand management measures; and
- promoting environmentally sound modes of travel.

Present trends in transport are compromising the needs of our children and grandchildren. But, while the love affair with the car is still a powerful advertising image, there is increasing realisation that the honeymoon is over.

Road building is not the answer to the urban transport problem. There is not the resource, nor the environmental capacity nor is there the political will to meet the demand. Even the present Government has taken a step away from Margaret Thatcher's "Great Car Economy" and indicated an acceptance of these issues.

Perhaps most significant in this respect was the March 1994 publication by the Departments of Environment and Transport of 'Planning Policy Guidance Note on Transport' (PPG 13) which provides advice on how local authorities should integrate transport and land-use planning. The key aim of the guidance is to ensure that land use policies and transport programmes help to:

- reduce growth in the length and number of motorised journeys;
- encourage alternative means of travel which have less environmental impact; and hence
- reduce reliance on the private car.

Recognising these issues in the Greater Nottingham conurbation, Nottinghamshire County Council has been moving towards more sustainable policies for some years and has developed a transportation strategy which provides a balance between transport provision and demand management and seeks a much greater role for public transport.

The overall emphasis of the strategy is to provide for good transport **accessibility** rather than unrestrained private vehicle **mobility**.

3. TRANSPORT AND THE URBAN ECONOMY

Greater Nottingham is the principal conurbation of the East Midlands and is considered the main engine of the region's economy. About 750,000 people look to Nottingham as the main job provider and commercial centre and are therefore dependent on its transport system. With a broad base of retail, tourism, leisure and cultural activities, the city centre alone provides about 50,000 jobs and has great importance in attracting new investment and maintaining the region's economic vitality.

However, unemployment is a significant factor in the local economy. A dozen inner city wards have unemployment levels of between 20% and 44% and similar high levels exist in parts of the Nottinghamshire Coalfield to the north of the conurbation where recent decline has been particularly dramatic. The decline in mining and other traditional employment activities is in contrast to an increase in retail, commercial and administrative employment, mainly in the city centre.

These changes in the economic structure of the area are increasing the number of people that commute to central Nottingham for employment. Accessibility to the city centre with its high concentration of job opportunities is crucial. But, with declining bus patronage and an increasing reliance on the private car, accessibility, and environmental conditions, are increasingly affected by congestion. While demand for journeys into the city is increasing, the ability of the city's transport infrastructure to accommodate this growth is fast running out.

This growing shortfall between projected demand and the capacity of the road and public transport network to handle it can be termed the Transport Gap (Figure 1). There is an economic cost of not bridging this Transport Gap and an environmental cost if it is bridged in a way which is not energy efficient.

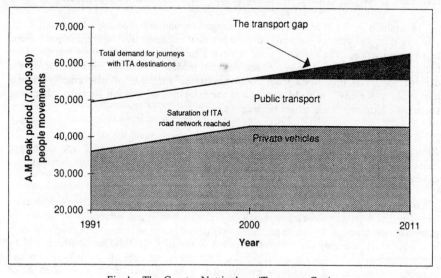

Fig 1. The Greater Nottingham 'Transport Gap'

The most obvious demonstration of these problems is worsening congestion illustrated by declining traffic speeds (Table 1).

Table 1   Morning peak journey speeds on radial routes into Nottingham.

Route	Corridor	Average Speed (mph) 1992	1993
A610	NW	14.9	13.6
A611	NW	15.0	11.8
B682	NW	14.8	11.8
A60	N	14.2	13.9
B684	N	18.8	17.1
A612	NE	17.7	15.2
A52	E	22.8	17.4
A606	SE	22.0	16.9
A60	S	23.4	18.8
A453	SW	25.3	25.0
A6005	SW	19.2	16.1
A609	W	15.2	16.8

Year by year journeys are getting slower, particularly in the northwestern corridor, and most major roads will soon be overloaded. They already are at peak times and the peak is spreading bringing congested traffic conditions to much of the day, reducing the efficiency with which the commercial and other activities in the city are serviced. Even assuming very low traffic growth, it is estimated that the conurbation's road network could reach 'saturation' level early in the new century, worsening already severe environmental problems.

In addition to the need to address the congestion and pollution aspects of traffic, an effective transport strategy must consider the potential economic development impacts from transport provision. The objective is to achieve a balanced transport system that provides accessibility but also gives people choice between modes. Failure to meet the increasing demand for travel into Nottingham could hinder the attraction of new employers and businesses to the city. It could also lead to the relocation of existing firms to other, less sustainable locations, more reliant on access by car.

A good transportation system allows a focus for development where it needs to be. In particular, the perceived permanence of a fixed track system provides for developer confidence and attracts economic opportunities. For example, there is much international experience of increased development activity around rail stations.

By adopting an integrated land use/transportation strategy for the conurbation it is intended that most new development will take place where it can be served effectively by public transport modes. In this respect, relatively isolated and low density developments built in a way which is highly car dependent are not sustainable. Instead, development will be concentrated within built-up areas, at suitable locations on the edge of the built-up area and at nodes within or beyond the green belt surrounding the conurbation along specified public transport corridors. Within and adjacent to the urban area public transport access can readily be provided by existing and extended bus routes but within the extended corridors it is envisaged that there will be a choice of public transport modes; bus and heavy rail or bus and light rail, linked with a network of park and ride sites.

This policy of development along transport choice corridors should encourage travel by public transport as well as improving its viability through increasing the potential catchment population.

In addition, demand management techniques, including parking and traffic management, the encouragement of walking and cycle use, car sharing and working hours flexibility, etc., are already being employed or investigated. Together with land use policies, these will minimise the Transport Gap. Nonetheless, a gap will develop and because of the city's central importance in commercial and employment provision, the regional economy will be difficult to sustain if the gap is not bridged.

4. ENCOURAGING PUBLIC TRANSPORT USE

Changing travel behaviour to bridge the gap will not be easy. The private car dominates existing demand, accounting for 71% of peak journeys. Congestion from too many cars is an everyday experience for most people travelling to work. In contrast, bus patronage is declining; nationally from 5500 million passenger journeys in 1982 to 4400 million in 1993/4, with a similar pattern locally.

This is mainly a result of buses getting caught up in congestion caused by too many cars - the declining spiral of public transport use.

To attract people in the proposed development corridors onto one of the public transport modes, these must have some comparative advantage over the car, for example by improving public transport's reliability and reducing its journey time by avoiding congestion.

Bus priority schemes are obvious measures, but the most effective way of achieving speed and reliability is through segregation. A major part of the County Council's transportation strategy therefore is the development of light and heavy rail services. Rail based transport, by combining comfort, reliability and speed, has proved especially attractive to car users.

For heavy rail, our aim is to build on the success of the Robin Hood Line; the County Council promoted rail reinstatement project serving towns and villages in the west of the County between Nottingham and Mansfield Woodhouse and, hopefully, up to Worksop in 1997 if Central Government finance can be secured. Further proposals involve increasing the number of stations and services on the under used rail network throughout the Nottingham Travel to Work Area. However, it is worth noting that the effects of privatisation on operating costs means urban heavy rail services are almost impossible to sustain without subsidy and current Government funding criteria severely limits their viability.

Thus, the high cost of heavy rail provision, together with its relative inflexibility and poor penetration of urban centres, means it has only limited application. By contrast, light rail, or light rapid transit, may prove the single most effective mode for bridging the Transport Gap and reducing the environmental cost of travel.

5. GREATER NOTTINGHAM RAPID TRANSIT (GNRT)

As an urban transportation system light rail is particularly attractive:

- it combines speed, reliability, comfort and modern, 'up-market' design;
- it is comparatively low cost and is energy and land use efficient;
- it has unrivalled efficiency as a people mover, with high vehicle capacity allowing for low operational costs;
- it can be given high priority over other traffic without causing much delay to that traffic; for the same movement of passengers, buses given priority will hold up traffic at signals three times (3 separate bus arrivals) for one LRV;
- its main advantage over heavy rail is its ability to run on street and negotiate tight radii and substantial gradients;
- it can therefore combine speed in the suburbs with effective penetration of the city centre;
- it provides the desired perception of permanence and quality needed to focus development and encourage economic activity;
- its power source, electricity, can be generated from a variety of fuels, so reducing dependency on any one source of fuel, and allows for relatively easy implementation of pollution-reduction technology at the power plant;
- it is free from noxious fumes and emissions at the point of use;
- it is accessible for people with impaired mobility through the use of low floor design;
- it is quiet and can fit comparatively unobtrusively into environmentally sensitive areas such as parks and the street scene.

This latter point is a sensitive issue. The Manchester system has been the subject of severe criticism concerning the heavily engineered and excessive poles supporting the overhead power lines. This was due largely to the procurement circumstances which led to the design and construction process being rushed and aesthetics being afforded little consideration. This is by no means inevitable as was shown by the more sympathetic on-street design used in Grenoble and, indeed, in Sheffield .

And it is not just the finished product that must be considered carefully. Building a railway through the city will inevitably cause short term traffic and environmental problems. Careful planning and control of contractors must be a high priority and it is vital for new promoters to learn from other's experiences.

Returning to Nottingham, the first LRT will run 14km northwest from the city centre to Cinderhill and Hucknall and for part of this it will share trackway with the Robin Hood Line, integrating the county-wide and suburban systems and improving access to the heart of the city (Figure 2).

The route links the majority of primary trip attractions in the City Centre, including the Broadmarsh shopping centre, the railway station, the main business and tourism areas around the Lace Market and Old Market Square, the Theatre Royal and Nottingham Trent University and then continues through the inner city and suburban areas of Hyson Green, Basford and Bulwell to Hucknall at the edge of the Coalfield area, with a branch towards the M1 motorway at Cinderhill. Five Park & Ride sites will be provided or converted from bus use and it is hoped also to encourage bike & ride and bus feeder services.

Ultimately, it is envisaged that there will be a conurbation-wide network of public transit priority routes, for bus, heavy rail and LRT, meeting the transport needs of the conurbation and serving new development along the public transport choice corridors described earlier.

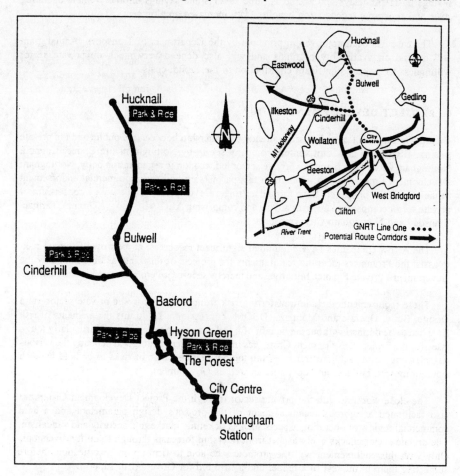

Fig 2. Proposed route for Line One
(Inset - future network)

Operating as the NET - Nottingham Express Transit, Line One will cost in the region of £85 million. Peak frequencies will be 5 or 6 minutes and the journey between Nottingham and Hucknall will take less than 30 minutes and to Babbington 23 minutes. The system is expected to attract between 8 and 10 million trips a year and remove over 2 million car journeys from the city's roads.

An Act of Parliament authorising the implementation of a light rapid transit route in Nottingham received Royal Assent in July 1994. The financial and economic cost benefit analysis, required to justify Government grant under Section 56 of the 1968 Transport Act, indicates that Line One will bring overall net benefits valued at nearly £30 million, including savings in traffic delays and accidents, time savings and savings in other vehicle operating costs and economic regeneration benefits from increased employment.

The case has been approved in principal by the Department of Transport. Actual grant will depend on availability of funds and will also depend very much on private sector willingness to make a significant contribution to the capital costs.

6. PROJECT DEVELOPMENT GROUP

Greater Nottingham Rapid Transit is already a partnership between the public and the private sectors. GNRT Ltd is a joint venture company set up by Nottinghamshire County Council, Nottingham City Council and the private sector led, economic regeneration body, Nottingham Development Enterprise. To strengthen this partnership and seek commercial endorsement of the project the promoters have formed a Project Development Group with a consortium of private sector companies known as Arrow and comprising AEG/ABB (now ADtranz), Tarmac, Transdev and Nottingham City Transport.

The Arrow companies have committed significant resources, at their own financial risk, to assist the Promoters meet the requirements for projects of this kind, as determined by the Government's Private Finance Initiative, and thereby secure Government support and funding.

These requirements include transferring risk from the public to the private sector via a Design, Build, Operate and Maintain (DBOM) concession. To do this the elements of risk have had to be defined without the benefit of a full design process. Thus, a particularly tough issue for the Project Development Group has involved understanding and debating the various risk elements in the scheme and assessing the optimum balance of risks to be held by each side: promoters, builders and operators as well as Government.

The close working with the private sector through the Project Development Group has also facilitated a rigorous technical audit of the project's design parameters and a hard commercial review of operating aspects such as ticketing, marketing, security and vandalism. The private sector checks every aspect from ridership forecasts through to utility diversions. Only with this endorsement will the promoters be able to firm up on specifications before subjecting them to the test of a tender and a firm bid for Government support.

7. CONCLUSIONS

Typically of all large urban areas, Greater Nottingham is experiencing increasing problems of traffic congestion and pollution. In line with PPG 13, Nottinghamshire has adopted a balanced transportation strategy which aims to reduce the need for longer journeys and encourage the use of energy efficient modes. This will be done in the longer term through control of land use and development and in the shorter term through a package of traffic and demand management measures. A choice of transport modes will be employed to provide the requisite levels of accessibility and the aim will be to integrate the modes into a complementary network in so far as the competitive market will allow. Because of its particular advantages in environmental impact and of speed, reliability, comfort and modern image and its important role in assisting in the attraction of economic investment and urban renewal, Light Rapid Transit will be a fundamental element of the balanced transportation strategy.

If support from Government is forthcoming contracts for construction of the LRT system can be let in 1998 and Nottingham Express Transit (Figure 3) will be providing an alternative to car congestion and pollution, helping to improve the environment of Nottingham early in the new century.

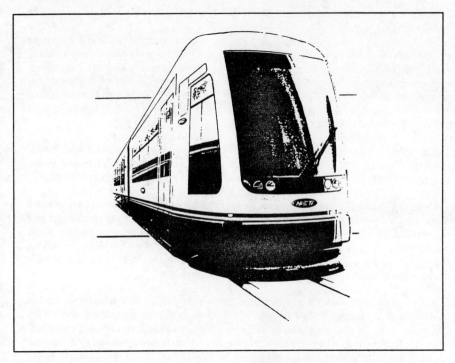

Fig 3. Artist's impression of a Nottingham Express Transit vehicle

C514/059/96

Total journey experience – connections and the customer

C E PERRY BSc, FIMechE, MCIT
M D Thameslink, London, UK

Connections are part of the customer love / hate relationship with public transport. They facilitate wide choice, yet they may slow down the journey and create anxiety.

The paper considers their impact in a suburban context and uses Thameslink as a case study from the customers viewpoint.

INTRODUCTION

This paper describes the Total Journey Experience for the rail passenger, and the ways in which connections and speed are important to it. Using a comparison between competitive modes, it then suggests important parameters in the design of the rail product as the Total Journey Experience.

MOTIVATION

Travel is a derived demand, and in general people will only travel if they have a reason to go somewhere. The reason behind the journey will carry with it a variety of conditions that must be satisfied to give the customer a positive perception of their experience, to the extent that they are prepared to generate further travel by recommendation or by a repeat journey.

This variety gives the rail product designer a significant challenge. Rail is not by tradition a server of a single niche market. On the same journey there can be those who are on a business journey and could possibly have used a chauffeur driven limousine or have flown; sitting next to them may be someone whose real alternative was to use a cheap coach or the cheapest method yet invented - scrounging a lift or hitch! Their different needs have to be recognised.

Journeys to work have their own set of requirements. There will be strict limits to the level of fatigue they are seen to involve, as they have to fit on top of a normal working day. This restricts the total time and the number of changes that the customer will tolerate.

There are also those journeys which generate the need for luggage handling. Going on holiday is the worst case, but shopping trips and visiting family and friends often involve the carriage of personal goods.

CHOICE OF MODE

With the spreading profile of car ownership in the UK, the real competition for passenger travel is between car and public transport, with the coach Vs rail choice made on a secondary set of criteria, cost and convenience being the most prominent. This choice is intensely personal, will be defended and rationalised to a surprising degree, but will not necessarily be made after a logical decision process. The decision may have been made in principle a long time before the actual journey, when the customer was standing in the protected environment of the car showroom. Habit and personal preference may be the determining factors ('I still go by ferry - the crossing makes me feel on holiday'), while the key option of public transport may never be consciously considered. If it is, the final choice of rail may put the mode through a series of hoops before the decision is made.

- Can I go there by train?
- Is there a train at the time I want to go?
- Does it get there when I want to arrive?
- Are there enough trains to give me a choice?
- Do I have to change?
- Can I afford it?
- How do I get to the station?
- Will the car be there in one piece when I get back?

If all these occur to the customer, there is another question which may stop the process - can I find out easily? At this stage the customer may even get as far as the engaged tone at the Enquiry bureau, if not, it may be easier to get the car out anyway!

TOTAL JOURNEY

According to the industry, Thameslink is a Train Operating Company. If the TOCs had the view that this alone was their main contribution to rail, we would collectively fail. The role of the TOC is to be close to the customer, understand his needs and package them around the train journey. We all have to be aware that use of the mode is about more than the train trip. The customer has to be attracted to the mode and informed before getting themselves to the station. Even at the station the train has to be located and reached. The in-vehicle time is, of course, the key to the whole experience, but a high percentage of journeys actually involve a change of train. The end of the trip nearly always includes a change of mode, possibly the only exception being those who work on railway stations. In London, the importance of choice is clearly important in managing this last change, and it is possible to trace how the existence of the connectional possibilities offered by Thameslink adds value to the public transport scene.

In summary, every Total Journey Experience includes at least one more connection than it does train trip. This establishes their first importance to the customer.

By way of contrast, the car trip is perceived as simplicity itself, a walk to the garage, drive to the nearest parking spot close to the final destination and then a short walk.

In reality, some or each of these stages may have risk, delay and inconvenience associated with them, but these are not generally expected.

COMPETING ON THE TOTAL JOURNEY EXPERIENCE

Competing with this perception of the choice is part of the TOC's challenge, and historically is not the part at which we have been most successful. Considerable efforts are being put into the promotion of the advantages of rail, while some of the risks and disadvantages of the road mode are reaching public consciousness through the efforts of others, the Newbury By-pass protesters among them.

Congestion and delay are a growing part of the motorway journey expectation, while road rage and car related crime are seeping into public awareness. Against this the train really does 'take the strain' and the speed differential is contained in a current Thameslink slogan 'Really motor to London'. The complete absence of jams and those super photographs of trains speeding past motorway queues are real advantages that we must make count with the potential customer. The image of on-train catering, toilet cleanliness, and staff friendliness are all far more important to the user than we have traditionally considered.

It is, however, quite clear that the customer weighs up these factors and will make their final decision based on significant aspects of the choice-

- Does the train go where I want?
- Is it reliable and punctual?
- Is it 'convenient'?

The definition of convenience clearly involves a through service, with seamless transfers if required, frequent and hassle free.

THE TOTAL JOURNEY CHALLENGE

We must achieve 'The Informed Customer' if the full connectional possibilities and simplicity of the mode is to be exploited. Much work remains to be done in this connection and we have direct experience of the way in which poor information affects the sale of Thameslink. Nearly ten years after the opening of a direct through route for London, and spending many hundreds of thousands of pounds on advertising, it is not unusual to find booking office staff on a route offering direct services to (say) Gatwick, seemingly unaware that there is any option in crossing London to using the Underground! If this is the situation among those who should know, the position in the general population must be worse. The full information includes not only the destination, but the practical possibilities that go with it - advanced check-in for an airline journey; cross platform or over bridges? Availability of help at the connection points; all of these factors may influence the decision, or in the absence of the information, prevent it.

At this stage it is worth remembering that the heavy rail mode competes best in the moving of high volumes of people between significant nodes of activity. It is not our strength to be stopping at every country hamlet and small new development, the journey time implications are distinctly sub-optimal. In cities the speed will drop as full advantage is taken of connectional opportunities offered by more frequent stops at stations with good inter-modal choices. These connections allow the range of stations reached to be significantly increased and therefore offer a much higher number of possible destinations. The ease of connection at these stops is important in the overall journey, and much work is required in London to optimise some of the possibilities. For a Thameslink example, West Hampstead presents a super opportunity for the Northern Line, the North London line, and Chiltern. However, the walk in the rain may put you off! Such factors, combined with stairs as the cheap approach to changes of level, plus the lack of all the required information will always make connections a second best alternative to a through journey.

The importance of Thameslink 2000 and Crossrail to London has therefore to be seen in the light of the connectional opportunities that they offer, each of them opening up new routes to an unbeatable number of destinations.

SPEED

Through generations of operating and marketing experience the strong link between journey time and the use of rail is well established and understood well enough for predictions to be made that are robust enough to satisfy the most cynical of investment appraisers! It is this link that has driven the constant improvement in journey times through investment in faster trains and straighter track.

By way of contrast, the value which a traveller places on their connectional time is not so well researched. But the main factor is clearly the overall time taken for the Total Journey Experience.

Speed is important to the customer in a number of ways. Firstly they may place a high value on any 'wasted' time and regard all travel as wasted time. Secondly, the perception is that motorways and modern saloon cars have reduced journey times, so the competition are seen to be making improvements. In fact this may not be the case at the time you wish to travel, but finding out is not easy!

The impact of speed on Journey planning is clearly important. The great step forward in this country came with the advent of 125mph travel, which was seen to open up many new day trip possibilities and clearly these generated travel. The same factor can apply as we move into and around our cities, in which the average speed of journeys can be both slow and variable.

Good journey times allow for the customer to make choices on how to spend time, more time at home for the commuter, more time with that special person for the young lover! Speed can give more time to enjoy the primary target for the day's experience - time at an event or with the family/friend. These factors combine to give journey time improvements the generative, as well as value added impact on the economics of transport operations.

Clearly the amount that will be paid as a premium for good journey times will depend on a number of preconditions. These include the purpose of the journey, and who is paying for it, but the relationship between good connections and their impact on the price is not well researched, and yet is clearly important in the design of services and facilities.

Benefits of speed

From the operator's point of view, speed has a double benefit on the bottom line. Speeding up the journey to increases in demand for travel as it widens the potential market for rail and sharpens up the comparison with the opposition.

In addition to the revenue benefits, the reduction in journey time also allows the more efficient use of resources. More intensive use of expensive trains is possible, and more productive use of the time of our staff who crew them. Line speed and connection improvements that reduce the overall journey time are therefore going to be the constant objective of a Train Operating Company in touch with its customers, and these benefits should be borne in mind by service planners, infrastructure investors and train builders.

For the train operator, speed usually represents more revenue at lower cost - good business if you can get it!

Speed at stations

In the Total Journey Experience it is of course possible to improve the speed of the station experience. Ticket office queuing is a matter for architecture in terms of the station layout, plus, of course, the operator having an eye to the needs of his passenger flows at different times of the day. The right number of windows open, and the availability of machines are important criteria. Nevertheless the preference of the customer for a real person to sell the ticket is an important factor, as well as the generative impact that effective policing of the ticket line gives. In general terms, longer opening hours are good news for the operator as well as the customer.

Queuing can also be cut by a reduction in the transaction time. Here simplicity in the fare structure and well trained staff make a positive impact, and the ticketing technology is critical in this respect. Ease of training lowers the entry barrier for staff recruitment, and therefore opens up the prospect of cheaper transaction staffing costs, while the time consuming elements of impartial retailing involve holding up a queue of one's own customers to make a difficult and time consuming reservation for another operator.

Telephone booking and self service procedures are still at their infancy in railway retailing, but well developed elsewhere. The purchase of tickets at automatic machines with credit cards is being actively pursued and looks to be revenue generative. Once again the importance of people contact for train customers should not be overlooked, although the customer (particularly the regular traveller) may well respond to some education in this respect.

FREQUENCY

The impact of service frequency on the customers perception of a train service is often misunderstood and understated. Most existing operators have come from a management regime where consistent pressure on costs has produced a reduced train service. Now we face a completely different situation, and for the first time in many years the full costs of track and train ownership are transparent to the operator. This is clearly a good thing for service frequency as the operator finds that marginal costs are a fraction of the revenue generated. A greater imperative for improved service cannot be found, and it would be a tragedy if this were to be diluted on the grounds of political dogma and the concept of a 'free' infrastructure.

A reliable frequent service from all the operators in a connectional journey chain produces a number of benefits for the customer.

Firstly it minimises waiting times. There is good market evidence to show that the customer values his waiting time particularly highly, and roughly at twice the value of travel time. The thinking behind this is clear as waiting time is seen as 'wasted' time whereas the actual travel at least has the benefit of being useful overall. When considering the total journey time it is these unproductive periods that stick in the memory. Good frequency minimises waiting time, and it is clear that the collection of several separate route frequencies on a common track, as will happen with Thameslink 2000 will produce Underground type frequency through London.

A second feature of high frequency is that it simplifies the journey planning. At a threshold frequency (4 trains per hour in outer suburban areas?) there is no need to consult the timetable, and a 'turn up and go' service is created. This has significant convenience benefits for the customer, and produces advantages for the mode as it removes the need to look up the train times, but not the need to consult the road atlas!

Thirdly, frequency minimises overcrowding as it produces a spread of trains to accommodate the customer peaks, even to the extent that on the Underground the informed customer can wait for the train that meets his chosen environmental conditions. This reduces the loading time as well and gives further benefit for the operator.

Finally, the marginal trips represent good asset utilisation, and the potential for revenue generation in excess of costs, as well as the public service benefits that a credible timetable generates.

TYPES OF CONNECTION

In addition to the variety of passengers with their own specific requirements, so each type of connection also carries different needs. For the heavy rail operator the important modal connection types are Train to Train, Train to Plane, Train to Underground/Metro, and Train to Bus/Coach. Our history in organising these, (or lack of) create the customers expectation. It is not to the advantage of rail that through privatisation the constant need of the media and politicians to say how dreadful things are also creates in the mind of potential customers the image that none of these inter-operator and inter-modal opportunities are being exploited. This is far from the case. One could look for a long time to find another set of suppliers who provide an impartial ticketing service that quotes the lowest fare even when it is provided by the competition! In addition, Thameslink is very active in the generation of significant intermodal ticketing and promotional agreements.

Train to Train

The requirements for this type of connection relate to information on the right connection, minimising the waiting time by inter-operator and service planning, and the provision of a decent environment for the wait. Where possible, cross platform or same platform connections should be provided in the plan, and operation, of the railway.

Train to plane

The touchstone of this modal interchange is through ticketing, as it eliminates the hassle of re-booking and the anxiety of a stressful part of the journey. Quite clearly, baggage handling facilities must recognise the special needs of the longer distance traveller. This is so obvious as to be hardly worth saying, yet the interface between trolley and platform is not right at Gatwick, where the train service choice and frequency is probably the best of any airport in Europe, if not the world!

Train to Bus

Information is probably the key issue here, as both location and frequency may be suspect, while the multiplicity of liveries now prevalent in local bus service provision has created uncertainty in the user's mind. Thameslink's co-operation with the local authority at Elstree and the use of technology to assist are important ingredients in the future use of this kind of interchange.

Train to Underground

The major issue for this interchange is the sympathetic and informed guiding of the customer to the right line and best route for their journey. The methods include trained staff, good maps and efficient signage. For regular travellers the easiest connection with least walking and stairs may well determine their route to work, and choice of mode, with considerable sophistication being evident in the choice.

Getting the best out of intermodal connections is therefore a combination of careful design and reliable execution of a number of factors. The weight given to each factor must be carefully judged in relation to the requirements of each interchange, and the need of the customer, particularly those with the dominant journey purpose. The combination aims at seamlessness, minimising the hassle factor at each interface. Through ticket to the other mode, good information, good signing, and co-operation between operators are essential ingredients in the mix. Much remains to be done and some significant investment opportunities exist.

The love/hate relationship

People clearly hate connections for the hassle they create, and the unproductive time they take. Anxiety surrounds each change as all of the planning questions have to be addressed in terms of security, convenience and comfort. There is no doubt that our current performance on connections creates an 'interchange penalty' in the minds of real and potential customers.

The operators hate them too, as they represent an operational constraint and a complication of the information that could, and should, be given to the customer. Operationally they also contain the lose-lose choice between breaking a connection and losing customer goodwill, or holding it and running a late service to the satisfaction of no-one.

An alternative view is that people love connections. They open up many more destinations and within London allow for the swift escape from those places where few people really want to go. I include the London terminal stations of most train operators in this list! Key to the multiple choice that Thameslink offers in terms of London destinations is the flexibility where the chosen route and mode may even take the weather into account. For the intermodal journey, the connection is a necessary component and its design becomes even more important.

THAMESLINK 2000

No paper around this subject by a London operator (let alone Thameslink) could be complete without reference to the huge benefits that this project will bring to the London travel market. It will relieve congestion, both on-road and on the Northern line. It will provide a new tube-frequency service through the heart of the City. All of the non-user benefits will be there in abundance, pollution, safety etc. And not one of them has been used to justify the investment! The reduction in the costs of road infrastructure maintenance will be noticeable. Most of all, from a railway marketing viewpoint, it will open up to more routes the kind of connectional flexibility that has marked out the success of Thameslink over the last 8 years.

SUMMARY

Speed is the key to many service improvements for the customer, to maximise the potential rail market and to streamline the use of expensive resources.

Frequency improves the perceived convenience, and minimises stressful waiting time.

Connections widen the rail market and choice of destination, but introduce hassle, anxiety, and reduce speed.

The Train Operator's job is to understand the customers' needs and expectations, and relentlessly push for the improvements they rightly expect and that our competitors are delivering.

We look to the Engineer, the Architect and the Systems specialist to provide the innovation and cost-effective solutions in station design, passenger information, ride, speed and comfort at each stage.

We look to our marketing specialists to better understand the views of our customers to help us set informed design parameters which will best serve the contribution of rail in the public transport scene for the next millennium.

Train Control and the Railway System

C514/036/96

The British approach to competitive journey times

P J HOWARTH
TMG International, London, UK

THE BRITISH APPROACH TO COMPETITIVE JOURNEY TIMES
BY DR PETER J HOWARTH

1.   INTRODUCTION

When the entrepreneurs and engineers of the mid-19th century sank their capital and expertise into the development of the railways in Britain, most of them believed that not only was each scheme viable but also that each would produce a return in excess of that from any other source. Perhaps the criteria on which they based their appraisals were not as sophisticated as those applied nowadays but the fact remains that they believed they knew sufficient about the market, or potential market in many cases, to take the risk. Although a considerable number got it wrong, in some cases before the first sod was turned, but unfortunately not before some significant expenditure, the vast majority got it right and many individual personal fortunes were made. More particularly, there was vast step forward in the mobility of the population and an expansion of industrial activity exponentially greater than anything that had happened during the first industrial revolution of the 18th century.

Today we inherit a core network of railways in Britain which has successfully survived the ravages of two world wars, the development of the internal combustion and jet engines, the scrutiny of Dr. Beeching, a myriad of legislative changes and not forgetting the absence of clear support from either political or social groups. However, of even greater interest, and the prime purpose of this paper, is the robustness of the network to accept the changes of the last half century and to achieve a position of such strong viability. Immediately prior to the restructuring of April 1994, designed to meet the requirements of "privatisation," the British Rail group included an InterCity sector which turned in a profit, a Network SouthEast commuter sector which covered more than four fifths of its operational expenses and an expanding Regional Railways sector, providing extremely good value for the subsidy involved.

All this has been achieved against a background of limited investment, particularly when compared with some other major systems in Western Europe, and against a background of improving the established network rather than building major new routes. This paper will demonstrate how the process has evolved, how results have provided material for research and forecasting, how train planning and timetabling have become major sciences and how these factors have combined to present a range of marketable products.

2. EVOLUTION OF THE NETWORK

Although the very early railways were built for specific purposes, some of which were predominantly freight, and not forgetting the constraints imposed by the "man with the red flag" the major builders of the 1840's soon recognised the benefits which emanated from creating a straight, level and potentially fast permanent way wherever possible. Some did it extremely well, Brunel's Great Western route from Paddington westwards is a classic example, whilst others had to contend with the problems imposed by difficult landowners or topography. In this later category, the West Coast Main Line from Euston is a particularly good (or bad) example.

Train speeds rose gradually as the individual companies became established and entered into competition among themselves. As the growing market for rail travel emerged, attention turned to how this growth could be handled and encouraged. Single lines were doubled, double lines were quadrupled and some main lines were only developed at the end of the century, at a time when mile-a-minute running was rapidly becoming a feature of the crack trains on most of the main lines. It is particularly important to note that the Great Central which opened at this time was built to an alignment which even in the 1990's is seen as appropriate for the development of international services of continental gauge rolling stock.

By the 1930's steam technology had reached its peak and although affected by the Second World War, it was unlikely that any further improvements would be possible. In fact, prior to the Modernisation Plan of 1955, there were few examples of significant speed improvements compared with the pre-war period. However, the 1950's signalled the advent of competition in a form not previously experienced. The motor vehicle thus precipitated a gradual process of defence which ultimately evolved into offence on the part of the railways.

3. EARLY DEVELOPMENTS

There was, in hindsight, much criticism of the 1955 Modernisation Plan and particularly the extremely high number of different classes of diesel locomotives which were a part of the process. However, there did emerge a number of diesel locomotive types which brought about a step change in passenger train journey times. A prime example was the English Electric Type 4 (later Class 40) which revolutionised services on the two main lines to Scotland as well as the Great Eastern and trans Pennine among others. On other routes the Sulzer Class 45/46s, the Brush Class 47s together with the range of diesel hydraulic machines on the Great Western, all made similar contributions to improved timings.

© IMechE 1996 C514/036

What happened in this situation was a classic example of an initiative which was originally driven by economy - the replacement of the costly and inefficient steam engine - but which ultimately brought about a significant improvement in the quality of service which itself led to growth in demand and increased revenue. The total effect on the net revenue brought about by a significant reduction in the cost base together with a modest increase in income began to demonstrate that the decline in rail can be slowed, if not reversed, in certain segments of the passenger market.

By the mid-1960's, a strategy to progressively improve journey times was gathering momentum with the introduction of the Deltics to the East Coast Main Line and the completion of the West Coast Main Line electrification to the North West, and later the West Midlands. This was accompanied by modest route improvement schemes incorporating a progressive reduction in permanent speed restrictions (PSRs) and improvements in line speeds.

4. ESTABLISHING & TESTING THE MARKET

The mid-1960's also marked a significant change in the management of the railways in Britain. The establishment of the British Railways Board and the appointment of Dr Richard Beeching as Chairman led to the formation of the passenger marketing directorate and an upsurge in market research, particularly in the inter urban market. Early computer modelling included the development of forecasting techniques. Commercial managers began to learn about the value of time to the business travel market, about the value of price to the leisure travel market, about the value of journey times and service frequency and about the impact that customer service can have on the overall perception of the rail product.

More importantly, railway management generally was becoming aware of the value of working as a team to identify engineering and operational improvements and how they might grow the business so as to generate a return on the initial investment. Marketing plans began to emerge which included not only the aspirations of the commercial managers in terms of their service specifications and income forecasts but also the expenditure requirements of the train operators, rolling stock and infrastructure engineers. The principle of the all-embracing Route Improvement Scheme became established, the West Coast improvements being a notable first example.

However, the most significant feature of this period was the progress being achieved across the network from the incremental approach to improving the passenger product, a situation which continued to develop through succeeding decades and which will now be examined in some detail.

5. SYSTEMATIC EVOLUTION

5.1 Investment-led Schemes

Many of these schemes have been developed on a modular basis where a progressive payback is obtained. The technique has been applied very effectively with successive timetables benefitting from journey time improvements as physical works have been achieved and radical service revisions are introduced with step changes in resources, such as new traction. The

East Coast Main Line from London to Leeds, Newcastle and Edinburgh is a prime example of sustained progress over 30 years.

Year	Traction	Max. Speed	Line Improvements
Until 1962	Steam	80 mile/h	None
1962	Deltic diesel	90 mile/h	Gradual 1962-70
1970	Deltic diesel	100 mile/h	Gradual 1970-78
1978	HST diesel	125 mile/h	Gradual 1978-91
1991	Electric	125 mile/h	140 mile/h capacity

The particular strategy for this route was typically one of evolution:-

* Deltic diesel: more powerful than the steam locomotives and earlier diesels (which the Deltic replaced) with potential for 100 mile/h as line improvements were completed;

* High Speed Train (HST): the means of achieving 125 mile/h without waiting for the Advanced Passenger Train (APT) and its dependency on electrification;

* Electric: predominantly driven by operational cost savings but with potential for 140 mile/h when cab signalling introduced, a further betterment still to come.

The journey time between London and Edinburgh has fallen from seven hours to four hours between 1962 and 1991 with similar proportional improvements over the shorter distance important journeys between London, Leeds and Newcastle. Service frequency is at least hourly and often half-hourly, many intermediate stations have been successfully developed as inter-city railheads and market pricing has both smoothed out the peaks and troughs of demand and generated welcome additional income. In short, a conventional railway has earned a quick return on the technical, operational and business opportunities of the time, without necessarily being able to see the ultimate, or even the next step.

There are other good examples particularly those involving the introduction of HSTs on the Great Western, Midland and Cross Country routes and the West Coast Main Line showed similar results following electrification in the 1960s and 1970s until the failure of firstly the APT and then the InterCity 225 schemes to materialise. Even off the major inter-urban routes there have been successes with investment-led improvements - the most notable being the Thameslink and Chiltern total route improvement schemes of Network SouthEast and the Sprinter Express train service expansion on Regional Railways.

5.2   Route Upgrading And Resource Productivity

Major benefits in terms of cost reduction and income growth have been generated by relatively cheap initiatives, where the initial outlay has both been modest and has earned a quick payback, often in the first year. This has allowed the risk to be covered out of revenue funds at a time when investment funding has been scarce. A wide range of measures have proved extremely successful:-

* higher track speeds from simple slewing or re-canting undertaken as part of routine maintenance or low-cost renewals, to bring short sections up to route standards;

* running shorter, and therefore lighter, trains;

* tighter timing schedules, either by reducing recovery time or by taking out little-used station calls;

* shorter turn-round times at terminals giving the opportunity to operate a more frequent service from within the same fleet; push-pull operation brings even further economies;

* greater traincrew productivity through a variety of initiatives which, although often difficult to negotiate, generally benefit those staff remaining.

5.3   Train Speeds

Whilst much has been achieved from improvements in train speeds, this has been constrained in some cases by the characteristics of the individual routes - the West Coast Main Line being the most significant - and in others by the limitations imposed by the critically important signalling and braking considerations. For train speeds above 100 mile/h, 3 or 4 aspect signalling, which shows the state of the next signal ahead, is required. Additionally, conventional locomotives and coaches cannot exceed 110 mile/h because of the excessive stopping distance required for existing braking systems. This is the maximum speed agreed for a single driver and for these reasons is therefore a common train speed across the network.

Although some trains, notably the HSTs and Mark IVs, have brakes fit for 125 mile/h and signalling is provided for this speed, a second driver is required and for this additional cost reason speeds are often limited to 110 mile/h outside those periods when journey times are sensitive. Above 125 mile/h cab signalling is required because:

* it is not felt safe to rely on visual signals and basic AWS at these speeds.

* an extra signal section is needed for stopping.

As a means of breaking through the speed barrier above 125 mile/h, Automatic Train Protection (ATP) has been considered on British Rail as a form of surveillance, with cab signalling enhancing lineside signals. Being a selective upgrade from the universal Automatic Warning System (AWS) its objectives are:

* avoidance of signals being passed at danger.

* adherence to permanent and temporary speed restrictions.

* authority to run at over 125 mile/h and up to 155 mile/h, which is probably the economic limit on mixed traffic lines, and requiring only one driver even on sections over 110 mile/h.

The requirements surrounding the achievement of higher train speeds are summarised in the following table:-

Train Speed	Mult Aspect Signalling	Special Rolling Stock	Automatic Train Protection	Number of Drivers
Up to 100 mile/h	No	No	No	1
100 - 110 mile/h	Yes	No*	No	1
110 - 125 mile/h	Yes	Yes	No	2#
125 - 140 mile/h	Yes	Yes	Yes	1

*	Must be Mark III with nominated electric locomotives
#	One driver if ATP provided

5.4 Timetabling and Service Planning

In ensuring that the maximum possible benefit is derived from journey time improvements, it is also important to ensure that the running times are 'clean', i.e. that they are free of any inadvertent padding-out. On British Rail there was a rigorous scrutiny over the make up of a schedule, as follows:-

* pure point-to-point times derived from a computer performance model were rounded to whole or half minutes according to context then shown verbatim in the timetable;

* standard recovery times were added at specific locations and shown in the timing column, corresponding to the level of temporary speed restrictions allowed to the civil engineer, as agreed under the "rules of the route" each year;

* pathing time was added where necessary to avoid conflict between train paths and shown in the timing column with suitable notation;

* different advertised times were adopted where appropriate, for example to make a destination arrival time more dependable, also notated in the timetable.

Regular review of these elements to minimise the extent of additional time is essential in those circumstances where a revenue value can be ascribed to relatively small improvements in advertised journey times.

In converting these schedules into a marketable timetable, further key factors need to be incorporated in the equation. The importance and attractiveness of clock-face and regular interval departure times, particularly at main centres, and the issues of connectional requirements and margins must be addressed. Whilst some of these are judgmental, sufficient research and data should be available for the revenue risks to be assessed and for best and worst case scenarios to be simulated.

Finally, one should not overlook the production process improvements now being achieved by the use of computer-aided systems for timetable planning, resource programming and performance monitoring.

5.5   Business Review

By far the most fundamental approach to high speed rail developments in the UK during the last few decades has been the growing consciousness that each change and innovation must generate an improvement in the overall financial position of the railway industry. Sales statistics and income results are subject to considerable scrutiny to identify the explanation for changes in demand and utilisation. Operational and engineering performance, both physical and financial, is analysed to ensure that productivity gains are being achieved and costs controlled. Customers are also subject to research and questioning to ascertain their perceptions of service quality and delivery and to anticipate their emerging needs.

Forecasting employs the sophistication of MOIRA (Model of Income and Revenue Allocation), which has been developed and proven over a period of more than twenty years. The considerable fund of practical experience gained in this area has been encapsulated in the highly refined Passenger Demand Forecasting Handbook.

Taken together with proven expertise in the appraisal of investment opportunities, it has been possible to synthesise the full financial effects of a range of investment and train service options based on the incremental approaches identified in this paper. In particular, the incremental approach has generated quick returns at a time when significant investment funds have not been available and which, during the late 1980s, permitted a limited degree of internally funded investment to be sunk in further incremental improvements.

Finally, the transparency of many of the successful schemes delivered on British Rail during the last decade or so has encouraged outside agencies such as local authorities and Passenger Transport Executives, to invest in quick return incremental projects. There are particularly good examples now coming to fruition where new lines are opening or re-opening in the East Midlands, South Wales and other provincial centres as a cheaper alternative to more innovative rapid transit systems.

6.   FURTHER EVOLUTION

It is clear that the next stage of evolution of these incremental improvement principles should be directed to the West Coast Main Line where there has been no significant reduction in journey times since the mid-1970s. A major upgrading is now contemplated involving private sector expenditure in renewing infrastructure, probably followed by leasing of a new generation of trains. A consensus is now emerging that the only real solution to the tortuous geometry of the route is to adopt tilting trains.

Following the regrading of the route, the following journey time improvements, with or without tilting trains, could be realised:-

From London Euston	Current Journey Time Minutes	Upgraded Conventional Train Minutes	Upgraded Tilting Train Minutes
Birmingham (2 stops)	92	85	75
Manchester (3 stops)	145	133	118
Crewe (non-stop)	99	91	83
Liverpool (2 stops)	146	131	120

These journey time reductions of between 15% and 19% will considerably enhance the competitiveness of rail travel over the West Coast and earn substantially more income. Whether this will generate an adequate rate of return to the private sector investors is yet to be revealed but what is certain is that the experience of past decades will enable a range of options and sub-options to be appraised.

Developments on other principal routes will be driven by individual circumstances but electrification of the Great Western and Midland Main Lines and ATP-inspired higher speeds on all routes are possibilities.

An area of particular interest relates to the further development of passenger traffic through the Channel Tunnel. Clearly the principle of incremental improvement cannot be applied where there is such a step change in topography, technology and the demands of both the customers and the partner carriers. It was, of course, politico-environmental reasons which brought about the earlier hiatus in development of the new route between Channel Tunnel and London. However, the situation is reversed 'beyond' London where various interest groups are clamouring for improvements. Whether the incremental approach can be adopted to meet these needs is as yet unclear but the demands of the new European markets will continue to present this challenge to the existing network.

7. COMMENTARY AND CONCLUSIONS

In considering how Britain compares with other countries with advanced railway systems, it is perhaps appropriate to set out the major strengths and weaknesses of the incremental approach to providing competitive passenger services, as follows:-

Strengths

* low risk modular evolution with quick returns.

* market-oriented with immediate and improving revenue stream.

* disciplined timetabling, resourcing and innovation.

* immune from catastrophic failures to innovate (e.g. APT).

Weaknesses

* effect of signalling limitations on achievement of higher speeds.

* absence of visible long-term strategy and funding.

* slippage of British railway industry in terms of major technological advances.

The future privatised structure of Britain's railways presents opportunities for the incremental approach to continue, given that Railtrack and the Train Operating Companies work together. In fact there is already evidence that this is beginning to happen and with the expected injection of private capital there is every expectation that the incremental approach will continue to be successful.

What are the measurable benefits of the British approach when compared with other major railways in Europe? Considerable train service improvements have taken place during the last two decades in all the larger countries with France and Germany particularly taking the lead in the building of entirely new routes. Real comparisons are difficult because of the relatively uncommercial approach by the continental railways to passenger marketing and pricing. However, it is clear that Britain's railways, when compared with S.N.C.F. and D.B., are both more productive and more profitable, despite only receiving a fraction of the investment. So far as the taxpayer is concerned, the overall subsidies and debts are much lower in Britain and better value for money is clearly being obtained. Some of this advantage must surely emanate from the incremental approach adopted in Britain, coupled with the tighter financial management and more aggressive and productive marketing.

In summary therefore it is submitted that the British approach to competitive passenger services has much to commend the range of benefits which accrue from its application. These particularly include the encouragement of disciplines in research, planning, development, engineering and operations, together with the provision to investors of early and transparent results of individual schemes.

C514/020/96

Harnessing the whole team to reduce journey times

M WINTERBURN BSc, CEng, MIMechE and **H McCORMICK** BA, MSc
London Underground Limited, London, UK

Synopsis
This paper outlines the centrality of the value of journey time for estimating the social benefit of London Transport's services, and the use of journey time for the planning and managing of the Company. It describes how London Underground has adapted a Netherlands Railways method for measuring customer disruption based on journey time in order to drive organisational and commercial behaviour. The paper ends with a discussion about the wider social and commercial implications of harnessing the whole organisation to reduce journey time as a prime objective.

Introduction
Underground railways have recognised since their inception that journey time is the prime attribute of the service their businesses offer. In more recent times, London Underground has developed financial and social benefit arguments for improving journey times. This paper outlines the reasons for that position and goes on to describes how the Company's organisation and suppliers are being focused to deliver the benefits of shorter journey times.

Customer benefit - the decision criterion
To establish why journey time is important, we need to understand what the overall objective is. London Underground recognises that its product is not a conventional commercial product: the provision of urban public transport is essential to the efficient operation of a city, and the benefits generated go beyond the immediate customers. As well as its economic impact, the Underground system is also relatively environmentally friendly - an advantage whose importance is increasingly being recognised.

In view of the general benefits to society afforded by a good public transport system, it is important that the decisions made by those running it accord with society's interests. The existence of an extensive infrastructure gives the Underground a partial monopoly, which requires that it be regulated in the public interest. The prices it may charge are set by Government, and the 1984 Act requires that London Transport provide transport services having 'due regard to the transport needs ...'. These needs are not clearly defined, although the withdrawal of services is subject to a lengthy enquiry process.

Beyond the basic provision of services, at prices agreed with Government, London Underground is free to decide its own priorities, within the guidance given by the Department of Transport. This recognises that the services provided and investments made should not be targeted at profit, but should take account of the full benefits derived by users and non-users. To provide a clear framework for decision making, London Underground has adopted as its business objective, or decision criterion:-

> 'to maximise net social benefit within available funds and subject to a defined gross margin target'.

This definition allows for the external benefits resulting from transfer from road, as well as the benefits to customers. However since additional travellers on the Underground impose costs on existing users, through extra congestion, the net benefit from external impacts tends to be small in comparison with the direct benefit to existing users. Saving in journey time is a key element of this direct benefit.

Evaluating the customer benefits of journey time
Research carried out over the years by the Government, academics and London Transport has been used to demonstrate the validity of two key underlying concepts:-

-       that people's time is worth money, and
-       that people's behaviour demonstrates this.

As services improve, demand levels increase, in just the same way as they would in response to a fare reduction. The improvement in journey time or quality is therefore equivalent to a money saving for the traveller. The value of the saving can be established through revealed preference - where a traveller trades off, for example, a slow, but less crowded service, against a faster, more crowded one, or a more expensive express service against a cheaper, stopping service. Market research can also place values on all aspects of the journey through stated preference techniques, though care is needed to factor particular responses within an overall valuation of potential improvements.

Values of time used by London Underground are based on values agreed by the Department of Transport for use in assessment of road improvement schemes. For the basic journey, riding in the train, the current value is a little over £5 an hour (9p a minute). But the evidence also supports applying higher values for those elements of the journey that involve more "hassle" or "pain". The following weightings are applied:

Walking	times 2
Waiting	times 2
Climbing Stairs	times 4
Congestion	variable, depending on degree, up to times 2.5

The importance of journey time - a typical underground journey

Taking these values, and including also the fares that customers have to pay, the true "cost" of a typical Underground journey to our customers is as follows:

	pence
Fare	88
Waiting Time	56
In Vehicle	124
On-train congestion	15
Station congestion	7
Queuing, station entry & exit	62
Total "time" cost	352

Only about one quarter is the actual fare. The rest is the value of the time taken up (about 15 minutes), plus the "pain" factors. Underground travel is generally not undertaken for the pleasure of the journey, but as a means to some other end. Minimising the "pain" involved, including reducing the time actually consumed means that people can do something else with their time, deriving a value from that, or they may be prepared to make extra journeys, opening up the possibility of gaining benefit from new activities or areas made accessible.

The likely gain, in terms of London Underground's objective, has also been assessed for a range of potential improvements. In comparison with the total value of the time involved in making a journey, the value of improving various comfort and information features of the journey has been shown to be contained within a maximum potential value of 63p per trip. Since some of the improvement needed would be extremely costly, the maximum achievable improvement in value is substantially lower.

Measures to improve service regularity and reliability, through their impact on journey time and on crowding, have been shown to be the top priority. Understanding this has led to a focus on journey time for planning and managing the Company.

Harnessing the whole team

For some time, journey time has been used to direct investment and schedule train services through the business planning processes. This sets the overall criteria of performance, however the operational managers are measured on a day to day basis by a variety of operational performance indicators, known as Key Performance Indicators ("KPIs").

Although many of the KPIs support the goal of improved journey times, their relative impact is not easy to determine and there is overlap leading to a lack of focus and dissipated effort. The support managers, notably engineers, require other performance indicators which are difficult to relate directly to the Company's KPIs and journey time improvement. This situation clouds the business imperative and leads to a lack of overall cohesion and drive, reducing Company effectiveness.

In 1994, we became aware of the Netherlands railways "EVB" method of measuring customer disruption as a sole indicator for operational and engineering performance [i]. This provided a simple process for identifying the relative loss due to operational incidents linked to the business objective of improving journey times. It also had the potential for providing the required focus for business improvement which could be used by everyone in the customer service support chain to measure, and therefore direct their performance.

The Netherlands Railway EVB method
The Dutch method was based on measuring the relative disruption due to incidents - the source of customer delays. It is now known that these comprise only 60% of all loss of journey time, the other 40% comprising the cumulative impact (bunching effects) of small variations in run time which are not normally recorded. Nevertheless the method provided a basis for

- simplifying the measurement process
- providing a business comparison of operational and engineering problems
- customer focused measurement instead of measures of technical failure
- measurement linked directly to the principal Company objective
- measurement capable of being cascaded to all parties supporting service provision

Furthermore, London Underground could express the disruption in financial terms using business models already in place. The impact of providing all staff with a direct value for service failure has proved to be a powerful motivate for change and driving home the Company Goals.

Measuring Customer Delays on London Underground
The Dutch method has been developed and piloted on the Piccadilly Line train service. The customer disruption caused by each incident over two minutes is looked-up on a set of charts and recorded on the pre-existing Incident Notification Form. When the information from the forms is entered routinely into the incident database, the amount of customer disruption is entered at the same time. This information can then be used to analyse the operational and engineering problems.

The amount of customer disruption is based on the length of the incident, its location, the time and the day of the week. The calculation uses the standard business planning train service models. These calculate the cumulative length of delay for the customers on the trains affected by the incident, both directly and indirectly, that is, the total increase in journey time. Although the principle is simple, the service patterns for our central areas are quite complex and it has required 12 top specification PCs running continuously for 3 months to compute the look-up tables to cover all Lines.

Practical details

The calculation is obviously an estimate. This is reflected in the acronym used for the process, **Nachs** (pronounced Nax), standing for "**Nominal accumulated customer hours**" delay. At first there was some resistance to a new, seemingly outlandish unit of measurement. The concept however seems to have caught on, perhaps assisted by the observation by some staff that a thousand Nachs makes a k*nach*ered service.

It has been found that it is not worth measuring anything much below 100 Customer Hours disruption and this is represented by a unit of 1 Nax. In financial terms this is presently equivalent to a "cost" as described earlier of about £500 social benefit, or about £135 in additional revenue.

Simple "Nachs" look-up charts were produced for the Operational Mangers to use at the pilot stage. These are now being superseded by simple PC based look-up tables which automatically display the answer when supplied with the pertinent facts. Within the next two years, when a new "direct entry" incident recording system is introduced, it is intended that the result will be generated automatically.

Although the process has been piloted on the train service, it is equally applicable to station services. Customer disruption is now being calculated for train service delays, train cancellations, speed restrictions, signal delays, lift and escalator failures and station closures - measuring in effect customer disruption on the passenger journey from station entry to station exit, for all the railway, due to any cause.

Some types of signal failure and speed restrictions have the dual effect of delaying customers and reducing service flexibility. The loss of flexibility should technically be reflected in an overall increase in the estimate of disruption for all the other incidents during the imposition of the restriction. At this stage however a simplistic approach is being taken by approximating the losses simply to the direct customer disruption and making no allowance for the cumulative effect.

The benefits of using Nachs

The Nachs approach has produced for the first time a focused measure of business performance related directly to customer benefits. This draws the whole organisation and suppliers into the process of reducing journey times and provides a whole range of new opportunities and insights into managing the railway.

Common goals

For the first time there is the opportunity to compare losses occurring in different parts of the business and allocate resources to the best effect. This knowledge is available not just to the accountants and business managers but to everyone in the organisation and support services. Everyone is aware of the relative need.

What is perhaps more powerful is the harnessing of efforts to common objectives. The previous plethora of indicators and measures could not be all managed at the same time and tended to result in each manager, business unit and discipline targeting different areas of concern with a lack of drive or stamina for concerted change. The common *Nachs* objective has the added advantage of providing a customer focus rather than a indicator of the operator's difficulties, which may or may not be of relevance to the customer.

Figures 1 & 2 are some early results from the pilot on the Piccadilly Line. These show for the first time the comparison of all types of delay available on a routine basis for train services. When the station service Nachs calculations are included, the picture for the whole railway will be available for the first time in any form. As can be seen, even comparison of incident numbers is an unreliable indicator of relative customer disbenefit.

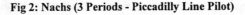

Fig 1: Incidents (3 Periods - Piccadilly Line Pilot)

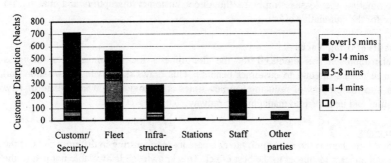

Fig 2: Nachs (3 Periods - Piccadilly Line Pilot)

Cascading the information

The Nachs process provides the means to cascade consistent performance information to all parts of the organisation and "attach" cost to the failure reports in support systems. A coding system is also used to allow the Nachs for each incident to be transferred to other databases, in-house or with Contractors, so that it can be used in fault tree analysis. For example it will be possible to know how much customer disruption has been lost through motor failures, or sticking doors, or even microswitches if these should become troublesome.

Previously you can imagine how impossible it was to count up train failures, lost mileage, delays over 15 minutes, lost headways etc in any meaningful way - and hence the impact of their failure was lost on the staff maintaining them. The Nachs process encourages empowerment - every member of staff is aware of the outcome of their actions.

Once experience and data has built up, it is intended that this process will also be used to inform designs, bringing home to those quite remote from the operational railway the business impact of their actions.

Financial awareness

Finally, the Nachs process provides a direct measure of financial loss due to delays. Ever tighter targets for traditional measures of operational performance are difficult to justify in absolute terms and tend to be seen as being imposed - for example where does a targeted 5% year on year improvement arise from ? On the other hand everyone can appreciate financial loss. It encourages teams to come up with innovative ideas about how to improve performance, especially if they can see that relatively small additional costs will be easily justified by the business benefits from their proposals.

A good example how Nachs has changed management practice is the proposal to fit fluorescent lamps with a polycarbonate coating. These lamps are more expensive but they avoid the need to withdraw a train immediately from service when a vandalised car has been discovered.

- The true impact of this problem was only noticed after the Nachs process had been introduced.
- The train service managers only had confidence to consider and put forward the solution because they knew it had an overwhelming business case. Previously a "why bother" could well have existed due to the difficulty of making a case.
- The fleet manager was willing to order and fit the tubes, because he was able to make the case for an increase his budget, rather than having to squeeze it out of the existing budget for little direct benefit to his personal business goals.

Treating the cause *and* the symptoms

The use of charts in the Piccadilly Line pilot scheme has illustrated another opportunity arising from the a single consolidated measure. Figure 3 shows the rise of customer disruption as the length of a train service incident increases. The curve is approximately exponential - at least for the first 15 - 20 minutes.

Figure 3 - Customer Delays - typical

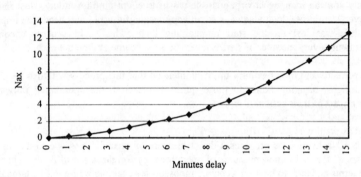

In practical terms if a delay extends from 6 to 9 minutes the customer disruption doubles ! This simple relationship has not been widely appreciated, although during peak hours when the effect is most apparent many managers instinctively react with urgency to incidents which are likely to be extended. Similarly, although 15 minute delays trigger customer charter claims, it is suggested that it has not generally been appreciated quite how disruptive such incidents can be until the Nachs process made this information more widely available.

Therefore as well as attacking the cause of incidents, there is also advantage to be had from tackling the symptoms and the recovery process. For example, incidents which require a driver to check the train due to faulty doors may be handled differently when the full costs are appreciated. For a 3 minute headway it is not long before it is better to tip out and get out of the way of trains that *do* work !

Journey time measurement in Contracts

The exponential rise in Nax with increasing delay also has implications for performance contracts. The level of customer disruption is affected as much by the operational handling of an incidents as by the initial fault itself. This situation has had the potential for acrimonious debate between operator and maintainer for many years - for example when dealing with delays over 15 minute. It will also lead to dispute if "Nachs" are directly written into the contract - the maintainer has only control over the number of faults, not how they are handled.

"Nachs" nevertheless provides the means to relate fault modes to mean customer disruption on a historical basis. Customer disruption information can then be used to relate the fault targets to the Operator's customer priorities, say on an annual basis. Performance contracts with train maintainers then will target fault levels by mode and number.

Regularity - the subtle losses
As mentioned earlier some 40% of all customer disruption (that is increase in unplanned journey time) is due to small perturbations leading to irregular service. These can only be measured routinely by their overall impact and are not "open" to cause and effect analysis in the same way as known incidents.

A working group has been set up in London Underground to understand the cultural and technical issues which determine staff and customer behaviour, and a vision of a "clockwork" railway has been set as an objective for the year 2000.

The best progress is probably being made on the Victoria Line. The service is already stretched and it is predicted that by the year 2000 the present service capacity will have been exhausted. In this situation, improving journey times also makes the most of the available capacity.

The Line has embarked on a process benchmarking exercise with Hong Kong MTRC, among other measures. This is breaking down the train service journey time into its constituent parts and determining best practice from the two railways. The exercise is involving the front-line operational and maintenance managers from both organisations to obtain lasting benefits.

The constraints of Journey Time and its relationship to business effectiveness
We would like to close this paper by putting forward some views which explore the limitations on the use of Journey Time as a driver for business improvement.

It is believed that Nachs will prove to be a powerful tool for driving a common agenda for the Company throughout the organisation. It is important to consider whether such a single minded objective is appropriate and will ultimately achieve its intended goal of maximising social benefit.

Journey Time does not determine how large the network should be, how well it is utilised nor whether it is self-sufficient. Reducing journey time does lead to increasing utilisation, but this is a by-product rather than an end in itself, and may not ensure the optimum cost recovery. Thus focusing the business only on journey time leads to some weaknesses and a tendency to ignore opportunities for obtaining greater gross margin.

Blind spots when focusing on journey time
First it is easy to ignore opportunities for market repositioning which could create additional revenue or reduce costs if the Company is too focused on journey time. Of course it is a much valued attribute of the railway that its services are difficult to change or withdraw - but to knowingly engrain this position into behaviour will turn away opportunities to increase cost effective marginal volume and improve gross margin.

Second, journey time fails to highlight where there is poor utilisation of facilities. Improving journey time may increase utilisation. It will not identify opportunities where a different approach to pricing, marketing or ambience may lead to increased demand, or where assets could be used more effectively to increase the potential for gross margin. Gross margin itself is too broad to drive out these crucial factors influencing Company success.

It is quite apparent that fundamental changes are taking place in the wider society's attitude to travel. The roads are becoming clogged, and increasingly the value of transport modes will be measured not just by their speed but by their availability, comfort and lack of hassle. Our measures will need a greater emphasis on the social value of freedom to travel and therefore of our passenger carrying capacity. Measurement of capacity as a desired social facility and its utilisation as a generator of gross margin will therefore become increasingly necessary.

For a Company in this context relying less and less on financial support from the Government, it will become more and more important that all sources of increased return on capital employed are exploited to the full. This situation is akin to that of the private operator for whom journey time is not just an end in itself but a means to an end, the end of filling seats. For all of us the use of a simple "Nachs" type measure, coupled with effective measurement of utilisation should provide complementary drivers for organisational success.

Conclusion
In spite of a final digression into the political, social and economic issues surrounding our business, journey time is, and will remain, a major driver of the success of a passenger rail transport business.

This paper has outlined how London Underground values that time, and the progress in driving that objective throughout its business. We are confident that given the continuation of the management development seen by London Underground in the last decade, the processes outlined will lead to continued improvements in service reliability and regularity. This will be achieved by harnessing the whole effort of the organisation and its suppliers to the goal of reduced journey times.

We trust this has stimulated your thinking about the drivers of good organisational performance and the opportunities for further progress in providing tomorrow's society with the freedom of movement it has come to expect, at a price it can afford.

Reference
i   ing. M.H. van der Werff:
     "Use of Failure Statistic Data to Improve the Availability of Infrastructure".
     IRSE Seminar on Railway Control and Signalling System Failure Analysis,
     London 15th April 1994

C514/044/96

A task-based interface for European train drivers

D P ROOKMAAKER, L W M VERHOEF, and J R VORDEREGGER
SE Arbo Ergonomics, Utrecht, The Netherlands

Summary
This paper describes how the ERRI-European Train Control System / Man-Machine Interface (ERRI-ETCS/MMI) has been developed. The main characteristic of this development was involvement of European drivers, European Railway experts and ergonomists. The result is a task based interface that can be accepted by any railway and any train driver. The experiences with the implementation so far indicate that users are very enthusiastic and the interface meets all their needs. Within the standard set there is sufficient freedom for customisation of the interface in order to meet specific needs.

1 ETCS/MMI Studies in the period 1991 - 1994

1.1 Orientation at railways (1991)

In 1991, the ETCS/MMI-group visited ten railways to establish their needs, concerns and opinions on the presentation of Automatic Train Protection / ATP and Automatic Train Control / ATC information in the future. ATP is in addition to line signals as ATC is not. The group interviewed 38 specialists from ten railways throughout Europe.

1.2 Workshop for feedback (1992)

With the orientation-phase completed, the MMI Group went on to draw up the basic presentation concepts; these were discussed in a workshop (1992) with 35 specialists from 13 European railways.

1.3 Elaboration of MMI Designs (1992 - 1993)

The most important information to present to the driver is:
- V_real: the actual speed of the train.
- V_permitted: a continuously changing speed which takes into account the train data (e.g. length, weight, brake performance) and track data (e.g. gradient, curves).
- V_target: The permitted speed at the target.
- D_target: The distance to the target.
This kind of information can be presented with one of the following basic designs.

Fig 1 A classical design

1) In a **classical** design, V_real, and V_permitted move on a circular line. Speed is presented on a round scale and pointers indicate the speed in the usual way.

2) In a **diagrammatic** design, V_real, V_permitted and D_target
move in a two-dimensional area (a graph).

3) In an **animated** design, the information moves in a central perspective (a 3-dimensional illusion). For each basic design two variants were elaborated.

Fig 2 A diagrammatic design

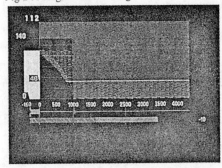

Fig 3 An animated design

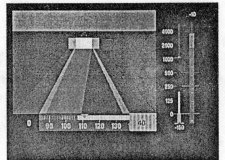

SE Arbo Ergonomics

1.4 Development of the Cab Display Test (1992 - 1993)

After this Cab Display Test (CDT) was developed, with the intention of evaluating six designs empirically and obtaining the opinions of drivers in a systematic way.

1.5 Test of designs (1993)

101 well-trained European train drivers from nine European Railways have tested three basic designs with the Cab Display Test. In this test European train drivers perform all kinds of driving tasks in several trains and in various situations (high speed, complex, dangerous and situations where information inside and outside were incompatible).

The drivers were matched in age and system experience (Automatic Train Control and Automatic Train Protection Systems). The effects of the design on safety, efficiency and driver comfort were established by measuring reaction times, number of errors and opinions with regard to:

1) **brake performance**: type of braking curve, number of warnings and system brakes, driving time, distance to the target;

2) **perceptual performance**: eye movements, readability and detectability of information inside (speeds, warnings, alarms) and outside (signals, boards, irregularities such as hot boxes, truck on level crossing);

3) **memory performance**: recall of speeds, learning curves;

4) **insight**: evaluation of complex situations (explaining what happens, evaluation of safe/ unsafe situations).

1.6 Data analysis (1993 - 1994)

The data collected with the Cab Display Test was analysed. In May 1994 a workshop was organized to discuss the results with railways delegates.

2 The proposed Eurodisplay (1994)

The empirical data from the experiments, ergonomic knowledge and the opinion of European drivers and railway experts has formed the basis of the design the ETCS/MMI-Group proposes now.

2.1 Hardware requirements

Ergonomic knowledge is used to specify the hardware requirements for hardware: touch screen technology, colours, luminance, screen polarity, graphical capabilities, reflection and physical size.
The proposed technology is completely in line with the actual development, state of the art and available hardware. The MMI-FRS will present more details.

2.2 Task-based screen areas

In this study the basic point of departure is that the design should assist drivers in carrying out their tasks. The proposed design is therefore task-based, i.e. the presentation is optimized for each part of the driver's task.

In general terms the driver's task is the following: to drive a train in a safe, comfortable and efficient manner. Other tasks, such as services to passengers and traffic control, finding remedies for failures, shunting and (un)coupling are not considered here.

For the Drive task, on which this study is focused, a distinction is made between:
1) **Speed Control**: Braking and traction actions for Maintaining Speed and Braking to a Target.
2) **Driving Mode Planning**: Planning all kinds of actions such as which speed to maintain and when to brake.
3) **Monitoring**: Detecting relevant information inside and outside (alarms, signals, irregularities)

Consequently, the proposed Eurodisplay has a structure that is compatible with those various tasks. Each task-part has a screen area that is optimal for that task.

2.2.1 Proposed design for the task-part: braking for a target (primary information)

For the driver the most important task is braking for a target. For this task he needs to know V_real, V_permitted, V_target and D_target.

This information is presented on a round scale. When the driver brakes to a target, the design shows him 'V_real minus V_permitted' as clearly as possible. The contribution of V_real and V_permitted to this difference should be made visible. This is possible on a fixed scale where both values move opposite each other. At the bottom of the round scale actual information concerning other tasks are presented, mostly by means of icons or abbreviations.

When confused or in distress, looking at the round scale is sufficient to ascertain that everything is under control and to establish which actions have to be taken immediately.

2.2.2 Proposed design for the task-part: braking for a target (secondary information)

The left section of the screen displays additional brake information. At the top, there is a visual indication zooming in to 'V_permitted minus V_real' when close to system brake. Next to that there is a graph indicating the distance to the stop target and a the distance to the predicted stop position of the train, assuming traction and brake remain unchanged. A prediction of the speed at a target > 0 is the next information presented.

The additional brake information in the left section is enabled or disabled depending on the time to system brake, train acceleration and the driver preference. Detailed brake information such as the speeds that are used to calculate V_permitted is presented at the driver's request.

2.2.3 Proposed design for the task-part: Driving Mode Planning

Driving Mode Planning information is presented in the right area of the screen (signals, gradients, static speed profile, length of authority, targets, pantograph instructions, stations, etc.). With this information, the driver can plan his Speed Control (when to brake, which speed to maintain).

On request of the driver this screen area displays detailed Route Information such as: the meaning of an icon or abbreviation or the identification and the position of elements and the characteristics of the platform.

2.2.4 Proposed design for the task-part: Maintain Speed

Results of the experiments made clear that for the scale problem and for the mental model of the driver, an indication of V_real readable in 1 km/h units should be recommended for that purpose. A round scale, a bar and a nonius are not recommended. For the 'Maintaining Speed', task the best presentation for V_real is digital.
There should be no other digits similar to or near the position of the V_real presentation.

Other actual speed control information is presented below (a digital presentation of V_real, actual Route Information items, advisory speeds).

3 After today

The ERRI-ETCS-MMI is being implemented now on the high speed line Vienna - Budapest. The MMI and its task based concept is now extended to other systems to be standardised in European cabs. The next system for that will be the cab radio. The open character of the ETCS-MMI and the task based structure proved to be a firm basis for standardisation, international acceptance and integration of several system.

Fig 4 The ERRI-ETCS-MII
Only the information that is relevant now is shown to the driver

References

ROOKMAAKER, D.P., VERHOEF, L.W.M. AND VORDEREGGER, J.R. (1991). **The** presentation of Train Control Information, Part I Orientation: analysis, criteria and presentation. November 1991, Netherlands Railways, SE ARBO/A&E/1991/1125-part1.

MAESSEN. N, VERHOEF, L.W.M., (1994). Development of a European Solution for the Man-Machine Interface (European train cabs). Draft Part II (Workshop Report): Evaluation and Suggested Presentation for a Euro Display. UIC/ERRI A 200/M.f4-945222-00.01-940509

SE Arbo Ergonomics

Improving Freight Performance

C514/082/96

Reduction of noise and ground vibration from freight trains

C J C JONES BSc, PhD, CPhys, MInstP
B R Research, Derby, UK

The effects of noise and vibration from freight trains represent an important part of the environmental impact from railways. Current trends in Europe to adopt new noise legislation and the concerns of railways over ground vibration from freight trains have led to considerable interest in research in these areas. This paper outlines the issues of noise and vibration as they particularly affect freight traffic. It then reviews the options for the reduction of these effects that have been the result of recent studies.

1   INTRODUCTION

A number of countries in Europe have introduced noise regulations for their railways in recent years. In the UK, regulations identifying entitlement to sound insulation for new and altered railways became effective in March 1996 (1). The regulation is stated in terms of an equivalent noise level (L_{eq}) 1m in front of the facade, from the total number of trains passing during an 18 hour day-time period and a 6 hour night-time period. The night time L_{eq} trigger level is set 5dB below that of the day time trigger level because it is thought that noise causes greater disturbance at night. The setting of different day and night time limits in terms of L_{eq} is typical of the noise legislation in other countries. Under these rules a small reduction of the noise level to meet the L_{eq} trigger level to avoid the high costs of noise insulation implies a significant change in the operation of trains - unless the trains can be made quieter. For instance to achieve a reduction of the L_{eq} by 3dB would require a halving of the number of trains. Although the highest absolute noise levels are generally associated with high speed passenger trains, because freight trains run more at night than during the day, they are under a tighter implied limitation than passenger stock.

In a growing number of European countries additional requirements are being put in place which require lower maximum noise levels during the passage of each train. This applies to new vehicles or railways but in some cases, for example, in the Italian regulations, there is a staged introduction of lower permissible levels for existing railways. This noise legislation has a wide implication for freight as this type of traffic is that most commonly operated internationally. For example a large fraction of the freight traffic leaving the UK through the Channel Tunnel is bound for Italy. There is therefore urgent Europe-wide pressure for the reduction of noise from freight trains. This has been recognised in the research that has been carried out under the funding of the European Railway Research Institute (ERRI), for instance in the OF WHAT (Optimised Freight

Wheel and Track) project (2), and also recently by the European Union (EU) in the form of a BRITE EURAM project, 'Silent Freight'.

The effects of ground vibration are, like those of air-borne noise, under increasing scrutiny and have become a major concern in the assessment of the environmental impact of new railways. A recent questionnaire conducted amongst the European railway administrations (3) has shown that increasing concern is expressed about all the effects of vibration from railways.

Ground vibration from trains is perceived in buildings adjacent to the track in two ways. Vibration of the walls and floors in the frequency range from about 30Hz to 200Hz radiates a low rumbling noise directly into the rooms. This effect, which is often known as 'ground-borne noise', is particularly associated with trains running in tunnels. However, in countries where very effective insulation for air-borne noise has been installed, ground-borne noise has been cited as a problem for surface railways (3).

Vibration in the frequency range between about 4Hz and 30Hz propagates along the surface due to the layered structure of the soil. This vibration is felt in houses near to the railway as whole body vibration. High amplitudes can be associated with heavy axle loads and with surface tracks. This vibration effect is therefore especially connected with freight traffic and for that reason is the main concern in the discussion of vibration later in this paper.

2   OPTIONS FOR THE REDUCTION OF ROLLING NOISE

Away from stations and depots on plain track rolling noise is the dominant source of noise from railway vehicles running at speeds below 300km/h. This is generated by the action of the combined surface roughness of the wheel tread and the rail head as the wheel moves over the rail. Remington (4) produced a scheme showing the process by which rolling noise is generated. This scheme, which is represented in Figure 1, also represents the process by which the system may be modelled. Such a model was produced by Thompson (5) as the BR Research software package 'Springboard'. Later extension of this by ERRI has led to the TWINS (6) package. This model has been extensively validated in full scale running tests conducted by ERRI and found to predict the noise levels close to the track to within ±2dB (6, 7). It is therefore a good tool for the development of low noise wheels and track.

Figure 1 shows that three possible sources of noise radiation are considered. These are the sleeper, the rail and the wheel. The process of the generation of rolling noise can be summarised in four stages as,

1   The forcing of vibrations in rolling due to the roughness of the wheel and rail surfaces.
2   The vibration response of sleeper, rail and wheel.
3   The radiation of noise from the sleeper, rail and wheel.
4   The propagation of the noise from each source and their summation at the observation position.

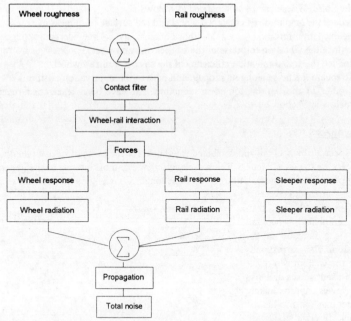

Fig 1 The process of rolling noise generation implemented in the TWINS model

Figure 2 shows a typical predicted one-third octave band sound power spectrum for each component. These sound powers are due to the action of a single 914mm diameter wheel rolling at 140km/h on 113A section rail on mono-bloc, concrete sleepers with a nominal 0.3m layer of ballast. The rail pad has been modelled as having a dynamic stiffness of $3.8 \times 10^8 \text{Nm}^{-1}$ and a loss factor of 0.15 (dimensionless). This may be taken to represent a rubber-bonded cork pad typical of those used in the UK.

In order to reduce the total sound power substantially the sound power from both the wheel and the track must be reduced. At high frequency the length of the contact patch acts as a mechanical filter averaging out the roughness of the wheel and rail at very short wavelengths. This effect is significant above 1000Hz and causes a frequency limit of the sound power (at this speed) of about 4000Hz.

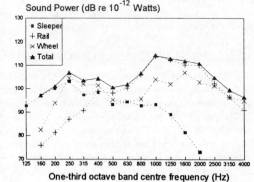

Re-examination of Figure 1 indicates the options that exist for Fig 2 Predicted sound power spectrum of rolling noise

reducing the noise. These can be summarised as follows.

1   Reduce the amplitude of combined wheel and rail surface roughness in order to reduce the excitation force.
2   Reduce the vibration response of the sleeper, rail and wheel.
3   Reduce the sound radiation efficiency of the sleeper, rail and wheel.
4   Attenuate the noise along its propagation path to the observation position.

A number of the options for noise reduction that have been researched recently are discussed below in Section 2.1 to 2.4.

2.1   Roughness

Measured spectra of wheel rail and combined roughness for the cases of tread-braked and disc-braked wheels are shown in Figure 3. The wavelength range of most interest to the generation of noise is approximately from 0.01m to 0.5m. (The longer wavelengths are relevant to ground-borne noise.) It can be seen that for wavelengths shorter than 0.5m the combined wheel/rail roughness is a function of the braking system. This corresponds directly to the well known result that tread-braked wheels give rise to noise levels approximately 10dB greater than for disc-braked rolling stock. Freight stock with disc brakes are used in the UK. However, current European standards for international freight traffic effectively discourage the adoption of disc braking. Additionally there are a very large number of existing tread-braked wagons which would be too

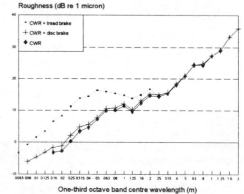

Fig 3   Spectra of combined roughness of wheels and continuously welded rail

expensive to modify with disc-braking systems. This predicates development of prototype solutions that can be retrofitted to tread-braked wagons.

2.2   Control of the vibration response of the rail and wheel

The relative vibrational responses of the rail and the sleeper can be controlled through the choice of the dynamic stiffness of the rail pad. The OF WHAT project (2) specified and commissioned the development of two acoustically optimised pads for UIC rail and bibloc sleepers. The running tests in August 1995 demonstrated a 4 to 5dB(A) reduction of noise compared to a softer pad in common use on the SNCF. The optimum stiffness value depends on the relative masses and radiation efficiencies of the rail section and the sleeper type and, to a lesser extent, on the speed and brake type (expected wheel roughness) of traffic at the site. Figure 4 shows the predicted sound powers for the same wheel, rail and sleeper as the prediction presented in Figure 2 (i.e. 113A rail on monobloc sleepers) but, in this case, the stiffness of the rail pad has been doubled. This stiffness may be taken to represent that of a typical EVA pad such as those used on the UK railway system. Comparison of the results of Figure 4 with those of Figure 2 shows that the rail component of noise has been reduced by 3dB but that the sleeper component is also increased by 3dB. The total track noise has been reduced about 1.5dB. This indicates that the pads most commonly in use in the UK are reasonably near the optimum. It should be noted that railways are generally moving to softer pads for reasons of sleeper protection from high impact forces, and that an acoustically optimum pad lies at the stiffer end of the range. A wide range of pad stiffnesses is currently in use and the reduction it might be possible to achieve by optimising the rail pads clearly depends on that starting point.

Once an optimum pad stiffness has been installed, the rail noise may be further reduced by introducing damping to increase the rate of decay of vibration amplitude along the rail and thereby to reduce its effective radiating length. The OF WHAT project developed and tested a tuned absorber device and work has been carried out at BR Research using a model for the evaluation of constrained layer damping. Because of the equal contributions of rail and sleeper for a track with already optimum rail pad stiffness, rail damping can only be expected to reduce the track noise by about 2dB.

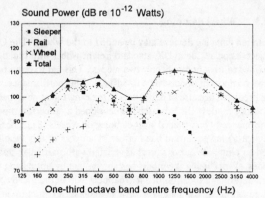

Fig 4 Predicted sound power spectra for stiffer rail pad

In order the reduce the vibration response of the wheel, a number of measures have been analysed theoretically and tested in experiments such as that of the OF WHAT project. The options that have been studied are described below.

2.2.1 Small wheels

Smaller wheels increase the modal frequencies of the wheel pushing some modes that are important noise radiators out of the range of excitation. An additional benefit of reducing the wheel size is a reduction in the radiation efficiency. A number of small wheels (down to 540mm) are already in use and an existing design of 640mm diameter wheel (with very thick straight web - see Section 2.2.2 below) was tested in the OF WHAT experiment. A reduction of the wheel component of noise of 18dB was achieved in the roughness-normalised results. However, in reducing the diameter of the wheel, the contact patch is also reduced in length and so too, therefore, is its filtering effect. This was found to limit the reduction in overall noise to about 1dB as it led to a 2dB increase in the noise from the rail, the dominant source in the test.

2.2.2 Wheel shape

The wheel cross-sectional shape may be optimised to modify the mode shapes. The aim is reduce to reduce the surface normal component of vibration on the wheel disc, i.e the acoustic source, for a small number of important modes. Such wheels have been manufactured and tested (2, 6,

7) and the wheels have been shown to reduce the noise by 1 or 2dB. The wheels that have been developed have had the constraint that the web must be curved to allow for thermal expansion during the tread braking cycle. A disc-braked wheel can be made with a straight web cross-section which would be more productive in reducing noise as it allows the decoupling of the axial and radial modes so that radial modes have negligible axial (i.e. surface normal) amplitude. It is difficult to reduce the axial vibration of the wheel disc for all possible running positions on the tread. This leads to the adoption in optimised wheel shapes of a thicker web.

2.2.3   Wheel damping

Vibration damping devices may be added to the wheel. The OF WHAT project tested a 920mm UIC standard freight wheel with tuned absorbers designed to treat primarily only two modes. This led to reductions of the wheel component of noise by 4 to 5dB (2). This performance was below that predicted because of difficulties in realising tuned absorbers for high frequency modes. BR Research has developed a numerical model for the performance of constrained layer damping on wheels (8). These dampers (Figure 5) have a broad band effect. Dampers of this type have been used to eliminate curve squeal associated with particular sets of wheel modes. Work using the model has shown that by optimising the design parameters and location of the treatment within practically acceptable bounds sufficient damping is achievable to obtain a reduction in the wheel component of rolling noise of 9dB using this technology.

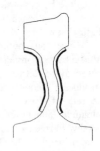

Fig 5 Wheel damper

2.3   Radiation efficiency

The radiation efficiency of a component is most simply reduced by reducing its radiating area. The benefit of small wheels in this respect has already been mentioned. Further ideas for the reduction of the radiation efficiency of the wheel including spoked wheels and wheels with holes in the disc are currently being pursued through the Silent Freight project.

Tests of a small cross-section rail were conducted in the early 1990's. The results were disappointing and the benefits did not merit further investigations at the time. The tests were set up before the Springboard model was sufficiently advanced for the role of the track design and the sleeper component of noise to have been recognised. The explanation for the poor performance of the track with the reduced rail section is that for the tested combination of rail, pad and sleeper, the sleeper noise dominated over the rail noise at the frequencies for which the rail's low radiation efficiency should have been effective. This demonstrates the need to analyse and design prototypes using sufficiently developed theoretical models before undertaking the expense of full scale track and vehicle tests.

2.4   Propagation control

Conventional noise barriers have been used extensively in European railways and are known to be effective. However, they can present a significant visual intrusion for line-side residents and also, by their presence, convey a negative message regarding the noise from the railway.

For these reasons, a combination of low, close track-mounted barriers and vehicle-mounted 'bogie shrouds'(Figure 6) has been tested by BR Research on passenger coaches (9). With the combined treatment a reduction of the total rolling noise approaching 10dB(A) has been demonstrated. With the bogie shroud alone a reduction of 5dB(A) has been achieved. The application of this technique to freight vehicles is being studied in the Silent Freight project.

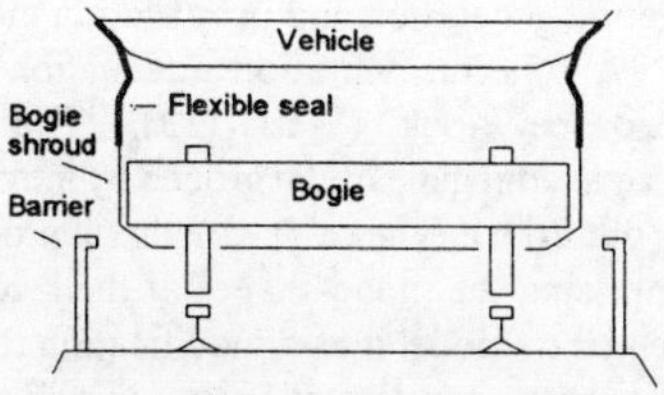

Fig 6 Bogie shroud and barrier

3 THE REDUCTION OF GROUND VIBRATION

Vibration in the frequency range relevant to ground-borne noise can be dealt with by incorporating vibration isolation elements into the design of the track. Modern tunnel designs tend to use concrete slab tracks for reasons of low maintenance and stable alignment. In this track form the vibration isolation that is afforded by a ballast layer is lost (10) and other resilient elements must be introduced. The principle of vibration isolation is to introduce a resonance into the track at low frequency. At frequencies much greater than the resonance, the force amplitude at the ground is reduced compared to that applied to the top of the track. The performance of the system depends on achieving as low a frequency of resonance as possible by reducing the rail support stiffness and increasing the mass above the resilience.

Vibration isolating track designs that have resonances down to about 20Hz are widespread in metropolitan railways for light passenger vehicles. These tracks incorporate soft baseplates or sleepers/slabs that are on elastomeric pads. For heavy axle loads the performance of these systems is limited by the allowable rail deflection that is imposed for reasons of rail and fastening component fatigue and to prevent gauge spreading. Nevertheless mainline tracks which incorporate resilient elements to reduce ground-borne noise have been implemented. One such example is in the tunnel under the Birmingham International Conference Centre (11). There, a ballasted track which has elastomeric pads under the soffit of each sleeper has been installed.

For the low frequencies associated with feelable vibration from surface-running freight trains it is impractical to engineer a track system with a low enough resonance. The track deflections would be too great.

Ground vibration from surface trains is the sum of dynamically induced vibration from the roughness of the track (wavelengths 0.1m to 5m) and the passage of the quasi-static deformation of the ground under each axle as the train passes a point on the track. The relative importance of the two mechanisms depends on the ground properties, the track structure, the axle load and spacing and the propagation characteristics of the ground at a particular site (12).

For the quasi-static vibration the major elements of the vibrating system within which the vibration is generated are the spring stiffness of the first few metres of soil and the mass of the entire track structure which may include an embankment (12). For these reasons, where this mechanism is significant, it cannot be expected that a change of track component such as the rail section, the rail pad or the sleeper should make a significant difference.

A small number of experiments to test various ideas for the attenuation of low frequency, surface propagating vibration have been carried out (11). These include trenches, masses placed on the ground and concrete slabs under the track. The results of experiments such as these are notoriously difficult to analyse conclusively given the inherent variation of the ground between different stretches of track and the differences in vibration levels generated by different trains. Again the need is clear for expensive experiments to be based on sound theoretical analysis of the vibration generation and propagation mechanisms.

A possible vibration attenuation device that has been studied recently is the 'Wave Impedance Block' (WIB) (13). The principle of the WIB is to modify the modal wave propagation regime of the ground by introducing a stiff layer under the track and extending some way out from it (Figure 7). Vibration propagates in parallel to the ground surface as wave fronts comprising the mode-shapes of the layered soil structure. To analyse the effect of a WIB it is essential to model it as a modification to the existing layered ground. A WIB could most cost-effectively be constructed using soil stiffening techniques such as high pressure injection grouting. With this technology is would be possible to install a stiffened layer of soil under the track without the need to remove and re-lay it.

The situation is difficult to analyse as the shear and compression wavelengths involved require large hybrid finite element models with wave-transmitting boundaries, or boundary element models to be produced. A WIB construction using the jet grouting technique, would be approximately 0.5m thick extending to about 6m from the centre-line of the track and lying at the bottom of the weathered top layer of soil. The conclusions of the work to date are that the performance of such a WIB would be dependent on the ground properties at a particular site. At some sites a worthwhile benefit (~10dB) is indicated for frequencies down to 8Hz.

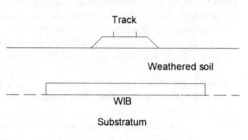

Fig 7 A wave impedance block

4   THE FUTURE

It has been shown that there is interest amongst the European railways in the reduction of the environmental noise and vibration impact of freight vehicles. A number of ideas for solutions have arisen in recent research and some have been tested with success. The high cost of testing prototype solutions has led to European cooperation through ERRI in the OF WHAT project and through the BRITE EURAM scheme with the current Silent Freight project. At the time of writing two further, relevant project proposals have been submitted to the EC. These are (i) the 'Silent Track' project which seeks to continue the track part of the work started in the OF WHAT project and so complement the Silent Freight project and (ii) the 'RENVIB' project which addresses the issues of ground vibration for both tunnels and surface railways, for mainline, freight and light passenger railways.

5   ACKNOWLEDGEMENT

The author is grateful to the Managing Director of BR Research for permission to publish this paper.

6   REFERENCES

1   Building and Buildings: The Noise Insulation (Railways and Other Guided Transport Systems) Regulations 1996, <u>Statutory Instruments, Department of Transport</u>, 1996 No. 428.

2   JONES, C.J.C. and EDWARDS J.W. Development and testing of wheels and track components for reduced rolling noise from freight trains, to be presented at Internoise '96, Liverpool, 1996.

3   A Questionnaire on the Concerns of European Railways Regarding the Effects of Ground Vibration, conducted by BR Research on behalf of ERRI Committee C163, Nov 1995.

4   REMINGTON, P.J. Wheel/Rail Noise, Part I: Characterisation of the Wheel/Rail Dynamical System, <u>Journal of Sound and Vibration</u>, 1976, vol 46, pages 359-380.

5   THOMPSON, D.J. Theoretical Modelling of Wheel-Rail Noise Generation, <u>Proc. Instn Mech. Engrs</u>, 1991, vol 205, pages 137-149.

6   THOMPSON, D.J., VINCENT, N. and HEMSWORTH, B. Experimental Validation of the TWINS Prediction Program, Part 1: Method, <u>Journal of Sound and Vibration</u>, 1996, vol 193, part 1.

7   THOMPSON, D.J., FODIMAN, P. and MAHE, H. Experimental Validation of the TWINS Prediction Program, Part 2: Results, <u>Journal of Sound and Vibration</u>, 1996, vol 193, part 1.

8   WANG, A., WILLIAMS, D.J. and JONES, C.J.C. Prediction of Damping in Railway Wheels with Constrained Viscoelastic Layers, Vehicle/Track Interface Dynamics, London, 1994, (ImechE).

9   JONES, C.J.C., HARDY, A.E.J., JONES R.R.K and WANG, A. Bogie Shrouds and Low Trackside Barriers for the Control of Railway Vehicle Rolling Noise, <u>Journal of Sound and Vibration</u>, 1996, vol 193, part 1.

10   JONES, C.J.C Ground-borne Noise From New Railway Tunnels, to be presented at Internoise '96, Liverpool, 1996.

11   JONES, C.J.C Use of Numerical Models to Determine the Effectiveness of Anti-vibration Systems for Railways, <u>Proc. Instn Civ. Engrs Transp.</u>, 1994, vol 105, pages 43 - 51.

12   JONES, C.J.C and BLOCK. J.R. Prediction of Ground Vibration from Freight Trains, <u>Journal of Sound and Vibration</u>, 1996, vol 193, part 1.

13   BLOCK, J.R and JONES, C.J.C. Stiffened Layers of Soil for Reducing the Surface Propagation of Vibration from Trains, BR Research contract report RR-DYN-95190 for Railtrack, Dec 1995.

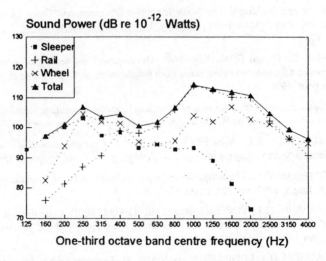

Fig 2 Predicted sound power spectrum of rolling noise

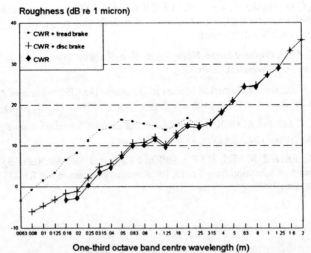

Fig 3   Spectra of combined roughness of wheels and continuously welded rail

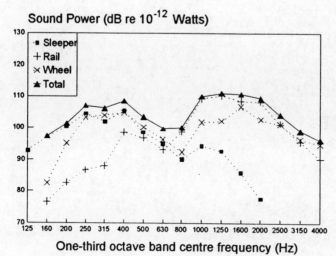

Fig 4 Predicted sound power spectra for stiffer rail pad

Piggyback concept for freight journey time

BERKELEY MA, MICE, FRSA
The Piggyback Consortium, London, UK

Synopsis

A pioneering piggyback service in the UK will start this year between Glasgow and London. A network of services within the UK and Western Europe is envisaged, following loading gauge height enhancement in the UK. The paper examines trends affecting piggyback and its road freight customers, and looks forward to non-stop trans-European services, unaffected by national railway or government preserved frontiers.

1. THE PIGGYBACK CONSORTIUM

The Piggyback Consortium is a group of 37 terminal, freight and train operators, business and local authority groups, infrastructure providers and other organisations. Their common objective is to get standard European lorry semi-trailers carried on rail wagons in Britain and through the Channel Tunnel. Membership covers Eire, Great Britain, France and European Institutions.

2. BACKGROUND

Lorry semi-trailers have been carried piggyback on rail wagons on the continent for many years, although there has often been more publicity given to carriage of complete lorry and trailer combinations with their drivers by rail through the Alps. More recently, piggyback in the United States has expanded as rail freight there revives to become a major freight player in a very exciting way.

In Britain, the small loading gauge of the railways was one of the reasons for the growth of the cross-Channel road traffic; total unitised traffic has doubled in the last 20 years, and MDS Transmodal, in a recent report for the Piggyback Consortium (1), calculated that 75% of this traffic was carried in lorry semi-trailers. Compare this with Sweden which, like Britain, relied on train ferry links to its main continental markets, and which moved eight times as much tonnage by rail despite a far smaller population.

The Channel Tunnel has created a major opportunity for rail freight, and the growth in cross-Channel rail freight traffic since the Tunnel replaced the train ferry two years ago has been steady, although concern has recently been expressed about the downward pressure on rates in competition with those for road freight using the ferries or le Shuttle.

However, there still remains the large proportion of road freight operators who have invested in tractor units, trailers and the relevant systems to offer a very efficient and cost effective service to customers. The challenge is to attract them to rail, offering a service to carry their trailers between freight terminals by rail, being delivered and collected by road to their ultimate destinations.

In the main, these trailers are standard 4 metre high in road-operating mode with air suspension. Those adapted for piggyback are already capable of being carried by rail on the continent on existing rolling stock, but the combination is too high and too wide for the smaller British loading gauge. Many studies were undertaken to see how much it would cost to widen the tracks and station platforms and increase the height by raising bridges or lowering tracks.

Widening was impossibly expensive. It would have involved reconstructing all station platforms and widening track beds, including structures, for the tracks to be laid further apart. In addition, passenger rolling stock on the routes upgraded would have to be modified so that passengers could get safely between the trains and the cut back station platforms. Unsurprisingly, this solution was costed at over £4 billion for main British Rail network.

However, if a wagon could be constructed to carry these trailers within the loading gauge width, then no changes to platforms or track bed would be required, and only the height would need increasing; the railways already had experience of this as part of electrification projects. Four international rolling stock manufacturers have now confirmed that such wagons can be built and one set of prototypes will be in operation by the end of 1996; these wagons are the key to piggyback being introduced in Britain.

3.   THE PIGGYBACK PROJECT

Piggyback in the UK is the carriage of standard 4 metre high lorry semi-trailers on rail wagons. These wagons can pass British station platforms and other trackside obstructions, but some increase in gauge height is required. On the continent, the gauge is larger and the piggyback trains will not require any gauge enhancement for the great majority of the continental network.

Standard lorry semi-trailers will need the addition of lifting points, since the trailers are lifted onto the rail wagons using the same equipment as used for swap bodies. Other changes may be required to make them suitable for both continental and UK piggyback but these are minor and well known on the continent. A piggyback trailer only costs 5% more than a standard trailer.

The aim is therefore to encourage the construction of these special rail wagons, the lorry semi-trailers, and to persuade Railtrack to create the necessary increase in gauge height to enable piggyback trains to operate on selected routes in the UK.

4.   THE ROUTE CORRIDOR

The   MDS Transmodal study (1) to assess the likely demand for these services, recommended that the first route to be upgraded should be Channel Tunnel-Glasgow with links to Irish Sea ports, initially connecting to existing rail freight terminals.

The Piggyback Consortium has been working closely with Railtrack over the gauge height increase and, with the West Coast Main Line development team within Railtrack, to ensure that any improvements undertaken are fully in line with long term development plans for that route.

The routes being studied are those recommended in the original consultants study, with some local variations:

- Channel Tunnel - Tonbridge, Redhill, Clapham Junction, Willesden.

- From Willesden, either up the West Coast Main Line or via the Great Western to Greenford, then Gerrards Cross, Princes Risborough, Aylesbury, Milton Keynes.

- The West Coast Main Line is then followed all the way to Mossend, near Glasgow, except that alternative routes through Stafford and Stoke on Trent will be studied south of Crewe.   There will be a branch to Trafford Park near Manchester.

- Links to Irish Sea ports comprise Glasgow to Stranraer, Heysham, Seaforth (Liverpool) and Crewe to Holyhead.

At the south end of the West Coast Main Line, the two alternative routings are still open, and the study will help Railtrack and the operators judge whether the operational advantages of keeping to the WCML rather than going via Aylesbury outweigh the possible additional cost of upgrading the tunnels at the Southern end of the WCML.   Similarly, the route through Stoke to Crewe, although not electrified, would avoid a difficult tunnel at Shugborough.

The work currently in progress is designed to identify the capital cost of upgrading the selected route.   It is due to be completed in the autumn of 1996.   The Consortium believes that this can be achieved on the spine route for under £100 million, and the upgrading work completed by 1998 to enable the service for full height trailers to operate.

5.   THE MARKET AND SERVICES

The potential market for cross-Channel piggyback services was studied by MDS Transmodal (1).   In addition to unaccompanied lorry semi-trailers, there is a significant growth in 9' 6" high containers which can be put on rail when the piggyback gauge is available.   There is also a domestic market for piggyback and 9' 6" containers within the UK, particularly on the longer routes such as London to Glasgow.

The objective of the MDS Transmodal study was to assess comparable journey times and costs between road and rail, considering the complete journeys from origin to destination, supplemented by in-depth interview with a number of operators.

Journey times for the rail services were assessed on the basis of present day operating practice between Glasgow, Manchester, Birmingham and London in the UK and Paris, Bordeaux, Lyon, Avignon, Perpignan, Aachen, Strasbourg, Basel, Milano, Frankfurt, Stuttgart, Munchen, Wien on the continent.

Out of the 20 - 25 trains per day in each direction, some examples of journey times between terminals are:

Glasgow - Paris	15 hours
Manchester - Perpignan	29 hours
Birmingham - Milano	32 hours
London - Lyon	19 hours

To these times must be added the road delivery time; the operators consulted felt that it was essential that the journey time to and from the terminal did not exceed half a day's legal driver's hours, perhaps 200 km, and they also stated that local haulage costs are likely to be the critical factors in deciding the competitiveness of the service. Operators confirmed that the rates and journey times used in the MDS Transmodal study were reasonable, competitive and attractive.

Since the time of the study, we have seen a reduction in real terms in road freight rates, in the UK, across the Channel and on the continent. Rail freight shippers, now grouped within the Alliance for Channel Tunnel Rail Freight (ACTR) are, in turn, putting pressure on the railway organisations to become more competitive and efficient, improve the reliability of the services and reduce rates. As regards service quality, they have reported a significant improvement since April 1996.

Railtrack has confirmed that the additional 20 - 25 piggyback trains in each direction can be accommodated within existing rail service patterns. Similarly, there are generally sufficient trains paths available through the Channel Tunnel. If traffic increases further, it must be assumed that revenue from additional business will enable improvements to remove any bottlenecks to be undertaken and financed out of revenue.

6.   THE LONGER TERM

Piggyback is designed to be attractive to the road freight market, which has been extremely successful in its costing, its flexibility and reliability. The whole logistics infrastructure of our towns, factories and distribution outlets is designed around road transport, which has developed its services in response to demands for increasing efficiency. In contrast, the railways have, over the last 30 years, failed to deliver a service that the customers wanted.

Road haulage in the UK is one of the most efficient in Europe and, partly as a consequence of this, the proportion of freight carried by rail in the UK is much less than in many continental countries.   For example, the Report of the Royal Commission on Environmental Pollution (RCEP) Report on Transport (2) quotes:

Share of Freight Transport - % of tonne-kilometres in 1992

	Road	Rail	Water	Ratio road/rail
Great Britain	65	8	28	8:1
France	64	31	5	2:1
Germany	48	37	16	1.3:1

7.   ROAD CONGESTION AND POLLUTION

There are now strong moves throughout the European Union to reduce the congestion and pollution caused by road traffic, the environmental and health damage done by the heavy lorry being seen as a major problem.   Forecasts of traffic doubling over the next 20 years conflict with recommendations by the RCEP (2) that CO_2 emissions should be reduced to those of 1990.   Similar target have been adopted by most EU governments.

With the present reliance on road freight for much of the Union's freight transport, it is clear that, if the environmental targets and political pressures are to be met, rail freight has a major and increasing role to play in the carriage of freight.   How can this be achieved?

8   FREIGHT BY RAIL

Over the last 30 years, the railways have become more regulated whereas controls on the road freight industry have been relaxed and often are poorly enforced.   The road freight industry would argue that it needs the flexibility to provide the services that their customers want but, conversely, the railways have become more inflexible to freight as that traffic has declined and passenger operations have become more intensive.

While long distance passenger train services, with many individual coaches going to different and far flung destinations, have declined, fixed formation trains using specific signalling and traction systems have, in many ways, made the railway network much less flexible for freight.   Only when new high speed lines have been opened has the existing railway become more 'freight friendly'.   EU Directive 91/440 on Open Access still requires to be implemented in a positive way on the continent.   However, the privatisation and greater accountability on the railways in Britain have provided a useful start in allowing open access, balanced by the enforcement of an uncertain and inflexible safety case regime which is causing many potential operators to shy away until the procedure and responsibilities are clarified.

9 HOW CAN PIGGYBACK MEET CUSTOMERS' NEEDS?

How can piggyback and other types of rail freight meet customer demands in the future? In summary, by providing a lower cost, faster, secure and more reliable service than the competition, along with the necessary customer service to make sure that the customer knows the status, temperature and location of his load at all times. In other words, we have to match the services offered by road haulage.

9.1 Tracking the load

These criteria may seem obvious but whereas tracking of rail wagons is now being developed, there will in the future be a need to track the load, whether it be on the road, rail or sea. Tracking of lorries is now extremely efficient, and incorporating this onto the rail legs of journeys, coupled with the necessary continuous monitoring of temperature is now being energetically addressed.

9.2 Temperature Control

Eurotunnel's lorry shuttles incorporate electric plugs for drivers to connect the refrigeration units, but these have not been widely used. Most drivers prefer to retain the use of the self contained unit instead. If we are to attract temperature controlled piggyback traffic, even if the self-contained unit continued to operate on the rail leg of the journey, there will be no lorry driver in attendance and so an alternative form of monitoring will be required.

9.3 Overall speed and reliability

There is much to be improved in overall speed and reliability of the rail section of the journey. It is clear that EWS Railway has identified the need to improve the reliability of their traction, so that schedules can be relied on as much as those for passenger trains. Similarly, there are delays on all parts of the British and continental system which probably reflect the low priority at present given to rail freight.

9.4 Rail frontiers

Finally, there are the frontiers between what are essentially national rail networks within the European Union. Each railway and member state jealously guards its own procedures, safety requirements, traction, customs and other rules, all of which help to delay freight trains at frontiers. With the single market in theory in place, one must ask why these organisations are not actively working towards common standards, interoperability, total removal of government customs and other checks, rather than playing lip service to the principle while holding tight to their own rules and regulations. There are many working parties, but when will we see the action, led by the European Commission?

Britain is no better and no worse than other member states. The open access here perhaps makes up for the muddled safety regime; the likely change to a more flexible rail freight security regime from cross-Channel freight in the North to South direction is welcome, but this has been achieved by pressure from the industry through ACTR against, until recently, strong government resistance.   Why was it imposed in the first place?   No other rail freight services are screened for bombs.

We now have to convince the French and other continental governments to change the regime which, only a few years ago, the British Government virtually imposed on them.

10   OUR 10-YEAR VISION

What is the vision of piggyback services in 10 years time?   I see a regular network of services around western Europe, daily on the little used routes, more frequently between more popular centres such as London, Birmingham, Manchester, Glasgow, as well as Yorkshire and Humberside, Bristol and South Wales along with Lille, Paris and Benelux.

Most important of all, these service would be non-stop between their terminals, working to timetables which are adhered to in the same way as for passenger trains.   A train from Glasgow to Milano would thus roll continuously, with no stops for meal breaks, frontiers, congestion or even waiting for a passenger train in the rush hour.   This requires one locomotive with two drivers, who will have food, drink, bunks etc to enable the journey to be done without stopping.   They would be trained for all the networks they would operate on, as Eurostar drivers are now, and would be paid accordingly.

Would the locomotives be electric or diesel?   Does it matter?   Why not either or both?   It is said that diesel locomotives cannot go through the Channel Tunnel, but it was built safely using mainly diesel construction locomotives, and there is a very effective ventilation system.

Piggyback then could average more than 80 kph from Glasgow to Milano.   That is the way to attract lorries of the roads!

For our lorry operator/customer, it would be an extension of their existing service, providing the long distance leg of their overall journey, at a price and a reliability that could no longer be matched by road all the way, due to congestion and other road pricing.

References

(1)   MDS Transmodal and Servant Transport Consultants; A Report for the Piggyback Consortium, April 1994, obtainable from Mick Sutch, Secretary to the Consortium, c/o Kent County Council, Springfield, Maidstone, Kent

(2)   Royal Commission on Environmental Pollution; Eighteenth Report - Transport and the Environment, October 1994; HMSO Cm 2674

20th July 1996, Ref p060721

Urban Systems

C514/035/96

Inter-operable urban rail transport

T GRIFFIN BTech, CEng, MIMechE
B R Research, Derby, UK

SYNOPSIS

Eliminating interchange can have a significant impact on reducing journey times and making public transport effective. This can be achieved in urban areas by making railway, light rail and tramway systems interoperable.

The city of Karlsruhe in Germany has shown what can be done. Urban trams now operate over the local railway network providing service direct from the city centre. Dramatic increases in patronage of rail services have occurred as a result.

A number of schemes to introduce interoperability exist in the UK. BR Research has developed technology to enable these to be implemented.

1. INTRODUCTION

National rail networks and transport systems in towns and cities have tended to be developed separately. They have been seen as serving different markets. Historically there were a few exceptions to this rule such as the Belgian "Vicinaux" national light railway system, but most of these earlier applications are now memories.

There is now more need than ever to make urban transport interoperable with inter-city railways and to eliminate interchange wherever possible. The private car has been seen as a great liberator, providing mobility for all, but the cost has been intolerable congestion, especially in urban areas, resulting in inefficiency, severe pollution and the decay of towns and cities. Public transport, including rail, has a vital role to play in reversing this trend. This cannot be done by compulsion, car users must be attracted by quality of service.

European passenger railways have been relatively successful in two markets i.e. inter city and commuting, but have been generally performing very poorly in the short distance regional and urban off-peak markets. This is despite the fact that most travel by car is for relatively short journeys and the streets of towns and cities exhibit a constant stream of traffic. Railway networks do not provide door-to-door convenience. Journeys by public transport usually require use of mix of buses and trains, interchange, long waits, uncertainty, and often quite a lot of walking.

The penalty of interchange can be understood by considering "generalised journey time", a term used in passenger demand models. If a rail passenger changes to a bus, metro or tram for the last part of his journey, such models add together the time taken to walk to the other service, the service interval, and the time that it then takes to reach his destination, and then double the total to give the additional equivalent journey time. So relatively short journeys of this type will always appear extremely unattractive. Old railway networks often do not correspond to the corridors along which people now wish to travel.

It is normally prohibitively expensive to build entirely new heavy rail systems, they can have an adverse effect on the environment and run into insurmountable problems from objectors. Underground railways are extremely expensive and are also usually difficult for people to access from street level. Light Rail systems have been developing because they are relatively cheap, minimise environmental impact and are easy to access. Their routes can match current demand and minimise interchange.

Urban interoperability might be achieved by interworking existing railways and tramway systems where these already exist, and incorporating light rail sections to provide a complete totally integrated network. The end result would be a system which might compete far more effectively with private transport and which could be achieved at less cost and with less pain than building new networks.

The particular attraction of using light rail or tramways as part of the system is that they can utilise street running in the city centre, so as to give direct access into shops, offices etc. On the other hand an existing railway will often provide better reserved track alignments to outer suburbs and faster journey times than could be achieved by a new extended suburban LRT line.

To achieve interoperability between heavy rail, light rail and tramway systems a number of potentially serious technical problems, need to be solved. Development has taken place, principally in Germany, this has recently accelerated as it has become clear that dramatic results can be achieved. The concept is now being seriously considered in the UK, and could provide the key to making our railways much more effective in the urban market.

2. BACKGROUND

Trains and trams have shared tracks ever since railways began but over the years each type of system has developed its own technology and standards. These are not always compatible.

Modern expectations that rail systems should provide a high level of safety are a prime consideration. Street operation requires good quality all round visibility for the driver, this means that it is not possible to design light rail vehicles with structures which are sufficiently strong to meet railway end strength requirements. If a train collides with a crowded tram there are likely to be many fatalities.

The current revival of interest in the concept began in Karlsruhe, Germany. There the City tramway had amalgamated with a local narrow gauge railway in 1957. The latter was converted to standard gauge and through working was commenced using conventional trams. In Germany there are different standards for railways and tramways, as in the UK, so Karlsruhe found ways of reconciling the two. Between 1979 and 1989 the route was progressively extended on the other side of the city, sharing track with a German Federal Railways freight line.

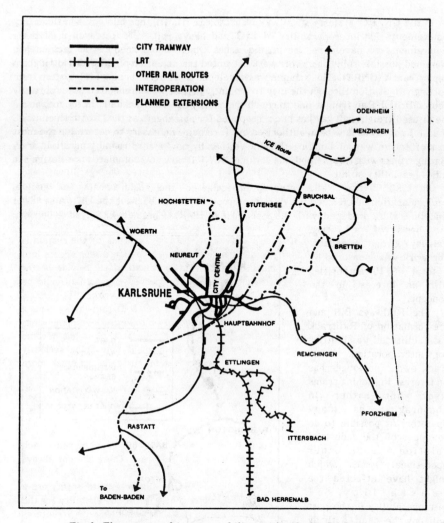

Fig 1 The extent of interoperability in the Karlsruhe Region

A much bolder use of the concept was introduced in 1992. Karlsruhe uses dual voltage trams which now operate over German Railways routes which are also used by regional passenger trains. Traffic has increased substantially compared with the local rail service which the trams replaced and which used the main station at the city edge. This has demonstrated the value of interoperability. Several extensions over other railway routes are planned. A number of other German cities have serious plans for regional rail/light rail shared track systems based on the "Karlsruhe" concept. One of these will link urban tram systems in three cities with a single route 150 km long.

In the UK, Her Majesty's Railway Inspectorate (HMRI), has laid down basic safety requirements for interoperability of LRT and heavy rail. BR Research used these requirements to develop a list of the issues which would need to be tackled and examined possible solutions. This work attracted the attention of Greater Nottingham Rapid Transit (GNRT). This scheme was developed as a street running LRT system from Nottingham Station through the city to a point where it was intended to use a partially disused British Rail freight line to reach the nearby town of Hucknall. In the meantime the disused freight railway has been reopened for passengers as the "Robin Hood line". The LRT route has now been authorised but it remains necessary to determine the most cost effective way of building the line, either by new tracks running parallel, or by sharing tracks with Robin Hood line services. GNRT have now completed pre-design and build feasibility studies.

In 1993 British Rail Research investigated the requirements for general implementation of LRT and heavy rail interoperability throughout the UK. A series of guidelines were produced, with the assistance of HMRI. These are expected to become the basis of a future Railway Group Standard. The work was sponsored by a total of 10 authorities with an interest in the concept.

The Railways Bill and the formation of Railtrack has facilitated possibilities for interoperation. The track owner no longer has an interest in running trains itself but rather in encouraging as many operators as possible to do so. Most of the industrial relations and other operational issues which might have affected the feasibility of interoperability in the past will now be dealt with by standard track access arrangements.

Apart from Nottingham other authorities have shown interest in the concept. Hampshire County Council has plans for a "Regional Metro" linking Southampton and Portsmouth. Kent County

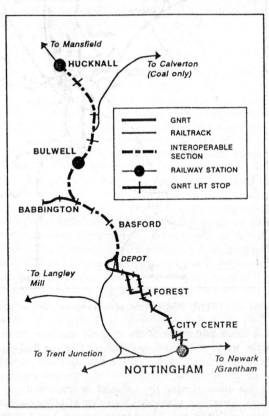

Fig 2 Greater Nottingham Rapid Transit Line 1

Council is planning to make use of the Strood to Paddock Wood line to link Maidstone with the Medway Towns, it also has plans for the Sheerness branch. The possibility of converting the entire Cardiff Valleys network into mixed LRT and heavy rail operation is being studied as are possibilities on other parts of the Railtrack system. In addition Tyne and Wear Metro is currently progressing its proposed Sunderland extension on the basis that its Metro cars will operate over Railtrack south of Pelaw.

3. TECHNICAL SOLUTIONS

3.1 Safety standards

Railway safety standards are normally devised to maximise safe interworking of trains throughout the system. Urban interoperability developments need not comply with generalised standards, operation can be restricted to pre-determined sections of route, over a relatively small part of the network. This makes the application of standards slightly easier. Solutions which involve an element of reconstruction of the existing railway are usually acceptable and any approval given can be route specific.

3.2 Structural strength

As already stated LRVs are much less crash resistant than heavy rail vehicles. BR Research considered whether it would be possible to build LRVs to meet railway crashworthiness standards but determined that it is not practicable at the present state of technology. Key factors are the requirement for the driver to have good all round vision in street traffic, and variations in floor heights and vehicle dimensions.

HMRI have accepted that it is not realistic to expect such vehicles to meet the normal crashworthiness standards for railway rolling stock and are prepared to tackle the problem from a risk assessment point of view. Projects involving interoperability of LRVs can be designed, operated and managed so that the risk to a passenger is not greater than anywhere else on the Railtrack network. The solution which BR Research developed in conjunction with HMRI involves a package of measures centred on a train-stop or simplified form of Automatic Train Protection (ATP).

Safety and system requirements have now been established for the proposed Nottingham application and a number of suitable train stop/ATP systems have been identified. Early indications are that the cost of these systems will not be prohibitive.

A similar approach was adopted in Karlsruhe where the Indussi ATP system is in use. Other factors which made the risk acceptable were the short braking distances and low maximum speeds of trams.

3.3 Wheel and rail profiles

The wheels used on trams and Light Rail Vehicles (LRVs) usually have narrower flanges than those used on railway vehicles. This is because the running groove in tramway rail is relatively narrow and shallow so as not to create too serious a hazard for other road users. LRV wheel diameters are comparatively small, typically in the range 500 - 750 mm (new), and the wheels of some of the modern novel low floor designs are smaller still (e.g 375 mm). If these vehicles operated over conventional railway track they would be liable to derail on turnouts or crossings, because the crossing nose gaps and check rails would be too widely spaced to guide the small flanges.

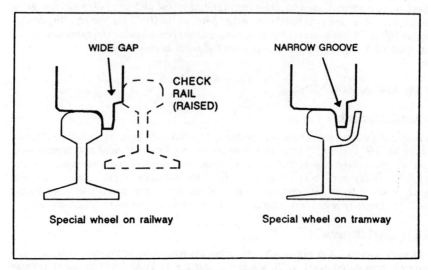

Fig 3 Special wheel profiles

Solutions to this problem have existed for many years. The Karlsruhe system initially used a special profile wheel profile in conjunction with raised checkrails. The principle is that the small flange meets normal street tramway standards but the back of the wheel, which is wider above street surface level, contacts check rails which are set back sufficiently to allow railway wheels to pass. Raised check rails have been accepted both in Germany and Britain.

The problem is also reduced by avoiding diamond crossings and restricting their angles. This may not cause any difficulties for route specific solutions.

The development of wheel profiles needs care because of detail differences between railway and tramway rail profiles and rail inclination, and the critical requirement to minimise wear and noise.

Research identified three basic solutions for use in the UK:

1. Systems without any street running can use conventional wheel profiles on LRVs.

2. Tramways which do not use flange tip running in street trackwork can use the BR P9 profile in conjunction with raised checkrails. This is the solution used by Manchester Metrolink, which does not interoperate but runs over conventional railway track for most of its route. It is likely to be the solution most widely adopted for large scale application.

3. If flange tip running is used then a wheel profile similar to that originally used in Karlsruhe, with raised check rails is suggested.

3.4 Structure gauge at platforms

The majority of new light rail systems in the UK will use vehicles with a floor height between 300 and 450 mm above rail, compared with railway coaches which have floor heights typically between 1150 and 1300 mm. This allows low platforms to be used in sensitive city centre areas where high platforms and long ramps would be obtrusive and obstructive. The standard railway platform height in Britain is 915 mm, which implies a step down from railway vehicles, but light rail systems will have platforms at virtually the same level as their floor, in order to meet new access standards.

LRVs are restricted to a maximum width of 2.65 m in the UK, in order to be compatible with traffic lanes, which is narrower than railway coaches. But the bodies of low floor LRVs extend well below railway platform level, so there is a possibility that the kinematic envelope of LRVs on tracks passing conventional railway platforms will not meet clearance requirements. It may be necessary to set existing platform edges back to provide clearance. In this case, step over distances to conventional trains will need to be checked to ensure that they meet HMRI requirements.

If low platforms are provided alongside existing railways for the use of LRVs then these will infringe the structure gauge for new railways laid down by HMRI. In reality however a dispensation will usually be possible. It is necessary to maintain a "passing clearance" of 50 mm between the platform edge and the kinematic envelope of trains. Platforms may have to be set further from the track centre on shared track than on normal light rail systems. A number of measures have been identified in order to overcome this problem which includes the fitting of sliding plates at the doorways of LRVs to close the gap. LRVs are often fitted with sliding plates of this type to improve access for the mobility impaired.

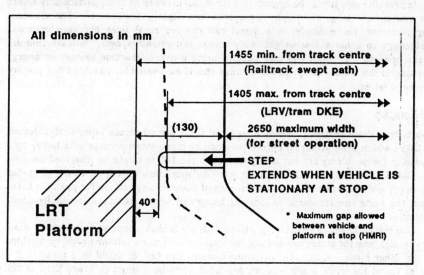

Fig 4 LRT platform and LRV step area on a main line railway.

3.5 Traction

LRT systems use electric traction both for efficiency and to make them environmentally acceptable, particularly in streets. In the UK voltages are now standardised at 750 and 1500 V, and all systems built to date have selected d.c. traction. Overhead power supply is the norm for street operation.

There are no significant technical problems in electrifying an existing non-electrified line for LRT operation. Experience in Nottingham has shown that it is possible to achieve the clearance requirements to allow heavy rail vehicles to pass underneath LRT wires without raising existing structures. In some situations however Railtrack would not approve LRT electrification, it might create a "barrier to entry" for other potential train operators who might wish to use the line in future.

In theory it should also be possible to install LRT electrification on a third rail (or fourth rail) electrified route. Care will be needed at the detail design stage to maintain clearances between LRVs and conductor rails.

LRT electrification is clearly incompatible with high voltage overhead. This problem has been solved in Karlsruhe by using dual system LRVs which operate on 750 V d.c. on the tramway system and 15 kV a.c. on DB. The centre section of the three unit articulated vehicles contains the transformers and associated equipment. These cars are about 3 tonnes heavier than the equivalent 750 V vehicles but have a higher seating density. About 15% of the vehicle capital cost is associated with the additional traction equipment. Voltage changeover takes place on the move while the vehicles coast through a short dead section. A similar solution ought to be feasible in Britain for application on 25 kV electrified routes.

In certain cases it may be appropriate for diesel LRVs to be used, particularly where the routes are long, relatively segregated and lightly trafficked. The Siemens RegioSprinter, for example, is a diesel rail car for rural lines but uses tramway technology to achieve low weight, easy access and attractiveness. Similar vehicles might operate into town and city centres using a hybrid traction system or energy storage, in due course clean power sources should be available, which do not require overhead wires.

3.6 Vehicles

Apart from those features already mentioned there are numerous other design details of LRVs which need attention before they can be made interoperable with heavy rail. However the problems are not as significant in the UK as might be imagined because the HMRI's "Railway Safety Principles and Guidance" now covers all forms of guided transport and a commonality and consistency of standards exists. So for example LRVs meet the same fire standards as those of heavy rail where the operating environment is similar.

One principle which HMRI have wished to apply is that wherever two systems exist on vehicles, one for street use and one for "railway use", then neither should be disabled in the other mode. Automatic switching systems can fail, it would be a disaster if a driver found his track brake was not available when he wanted to brake hard in the street. Occasional use of a railway horn might be unwelcome in a shopping centre but might save the life of someone who did not heed a quieter bell.

3.7 Other Issues

I have concentrated in this paper on issues with mechanical engineering interest but there are a number of other technical areas of varying importance which need to be considered before LRT and heavy rail interoperability can be implemented. I will mention them now briefly, for completeness.

The safety of passengers, and potential trespassers, on LRT low platforms alongside lines used by heavy rail trains is an issue in the UK. Solutions exist based on design of the platform, track treatment and warning systems.

There are a number of issues associated with signalling and control systems including track circuit operation, monitoring and diverting non-ATP fitted trains, control centres and communications.

Many of the other technical requirements have already been faced where light rail runs in close proximity to, or shares facilities with Railtrack, e.g. Manchester Metrolink with shared alignment, stations and level crossings near Altrincham.

4. IS IT WORTH DOING ?

It is clearly complicated and costly to integrate LRT systems with heavy rail or to add tramway sections to existing railways. The viability of extensive interoperability of urban systems remains to be determined but there are some pointers available from the experience of Karlsruhe.

It would be cheaper to create integrated public transport systems using buses. Buses perform, and will continue to perform, a valuable transport function but will they attract people from cars ? Comparitively slow journeys and a poor image deter car users. The evidence from Karlsruhe is that the percentage of passengers who are car drivers on buses is only 3%, on trams is 19% and on LRVs operating on DB is 40%. These figures are impressive considering that the three modes of transport are integrated with a common fares structure, publicity and timetables.

The attractiveness of interoperability is seen on the Bretten line. The DB used to carry 2000 passengers a day on this route in 28 trains. Ridership has increased to 14,000 per day now that the service operates into the city centre, has 11 more stops, and attractive, accessible vehicles. Weekend traffic, which used to be negligible in DB days, has become very important.

Karlsruhe has also saved operating costs by use of LRVs. DBAG bought four dual voltage LRVs and operates them on local services in the Karlsruhe area. They cost about 5.50 DM/train km, including staff costs, as against 12 to 17 DM/km for the local trains previously used. Another attraction for DBAG was the faster overall journey time of LRVs, e.g. a saving of 18% on a local service with the same stopping pattern.

Operating costs will be reduced if infrastructure is shared. However operators will be charged for the use of the track by its owner and if that organisation is commercially aware it will be charging the maximum the operator can bear, taking into account capital savings and increased revenue. The Karlsruhe Tramway company pays track charges to DBAG which add 8 DM/tram km to its average operating costs of 10 DM/tram km. The company hopes to re-negotiate a lower figure.

5. OVERVIEW

Interoperability can be achieved between railway and tramway or LRT systems so as to achieve dramatic results at less cost. The key element is the fact that street running can provide a "door to door" service without interchange which makes the effective journey shorter and more convenient than the private car.

Technical solutions have been developed which allow this form of interoperability to be achieved and meet safety requirements.

The concept could be used to greatly improve the quality of urban life in the 21st century. It could also increase the revenue which flows into national rail networks by providing better access both for passengers and additional rail services.

6. ACKNOWLEDGEMENT

The author wishes to thank the staff of BR Research who have contributed to work in this area, to clients who have sponsored it and to Karlsruher Verkehrsverbund GmbH and VCK, its consultancy organisation, for their cooperation and help.

C514/064/96

Developments in ultra light rail

P J WATKINS BEng and M J M PARRY BA
JPM Parry and Associates, West Midlands, UK

Congestion and pollution by private vehicles mean that public transport provision is increasingly
necessary. Examination of the roles which public transport is required to fill reveals that a wide
range of vehicles are required, from high speed trains to moving walkways. In this paper the
case for low cost, short distance guided transport, either to serve as a feeder to larger scale
networks or to circulate within urban areas is discussed and a new, economically and technically
feasible system, the Parry People Mover, is introduced.

1. INTRODUCTION

Society today depends both economically and socially on much higher levels of personal mobility
than were acceptable even forty years ago. Internal combustion engined private vehicles provide
most of this increased transport capacity, but it is now clearly apparent that mass movement by
car is unacceptable on grounds of energy consumption, pollution and sheer lack of road capacity.
The obvious solution is for many journeys to be transferred wholly or in part to public
transport, but it is difficult for any form of public vehicle to offer the flexibility and comfort that
car drivers take for granted.

2. GUIDED OR NON-GUIDED TRANSPORT?

Buses, which need no permanent infrastructure, are the simplest form of public transport to bring
into operation, but have little appeal to drivers of private cars, who see no reason to wait in the
rain just to stand in a crowded vehicle in the same traffic queue. Buses also have the
disadvantage of emitting pollution at their point of use. While a heavily laden bus emits much
less pollution per passenger mile than private cars carrying an equivalent number, for lower
loadings the reduction in pollution is not nearly so great. Buses with on-board non-polluting
power sources are rare and in the case of battery equipped systems have met only with limited
success. Trolley buses operating with overhead electrical connections to power lines are

similarly non-polluting and more technically successful, but are disadvantaged by sharing roads with and often being subject to the same delays as conventional road traffic.

Public transport operating on segregated alignments, such as trains and guided buses, can bypass congestion caused by other road vehicles and is more easy to power from continuous electrification or other power sources which do not pollute at their point of use. Although any system on a segregated alignment is more costly to bring into operation and less capable of giving flexible service, the advantages of fixed links can in many cases outweigh their disadvantages. Public perception of fixed link transport systems is that they are reliable and predictable and they are increasingly perceived as agents to regenerate urban areas.

3   CLASSIFICATION OF FIXED LINK TRANSPORT SYSTEMS

Fixed link transport systems for urban areas can very roughly be divided into the following classes. This hierarchy is decided on two grounds, firstly the distance over which the service operates and secondly the complexity and cost of the technology. The transport system can be viewed as a circulatory system, with major arteries leading into minor arteries which in turn lead into capillaries.

These are briefly summarised below:-

Heavy rail systems functioning within urban areas share much of the equipment used by interurban rail services. The defining characteristic is the area covered by and total length of the system, with suburban services typically serving a radius of up to 30 miles from the centre of conurbation. Suburban railways such as the electrified S-Bahn systems in many German cities provide a medium range commuter service from outer suburban areas to city centres, often sharing tracks with heavy rail long distance services.

Metro systems have a considerably shorter separation between stops, usually less than a kilometre. Some metro systems such as the London Underground use technology similar to electrified heavy rail, but other systems have distinct features such as rubber tyres and completely automatic operation. The term metro is usually applied to underground systems, but can also be applied to surface and elevated systems. The defining factors are taken to be the capacity (up to 80 000 passengers per direction per hour) and the stop spacing.

Light Rapid Transit, or LRT, is a term used to describe the evolved form of the traditional street running tramway. The technology is generally overhead DC electrical supply with earth return through the running rails, but unlike metro systems and heavy rail these systems are not confined to segregated routes but often street run on alignments shared with road traffic and pedestrians. Distances between stops are comparable to those for buses and there are no 'stations' as such, merely stopping points which in some cases have a low boarding platform. Systems can have passenger capacities between 2000 and 20 000 passengers per hour. The Sheffield Supertram and Manchester Metro are the first UK LRT systems, but similar networks are in operation world-wide.

Guided bus is a term under which a variety of technologies are grouped. The most obvious of these is conventional diesel engined buses steered by rollers acting on kerbs on either side of a concrete guideway (as in the extensive Adelaide O-Bahn system). Other possibilities include guidance by signals from buried cables or by central guide rails. The Guided Light Transit system developed by Bombardier in which vehicles are powered by a DC supply from an overhead cable and steered by a central guide rail falls midway between light rail and guided bus. Guided buses can operate on much narrower alignments than the roads which would be required for similar sized steered vehicles, and have the advantage over rail vehicles that they can also operate with conventional traffic in unguided mode.

People movers is the category of public transport on which this paper concentrates. Best summarised by the title 'horizontal lift', the class 'People Mover' can include as feeders to LRT and S-Bahn types of services, as well as systems enabling the circulation of visitors to exhibition sites, large shopping centres and airports. People Movers can provide remarkably high capacities despite their low speed and small vehicles. This results from the specific function and manner in which people movers are operated. Horizontal lift implies a frequent, reliable, untimetabled service which passengers use without pre-planning.

Figure 1 shows a comparison of the cost and performance (in terms of capacity) of the spectrum of public transport technologies.

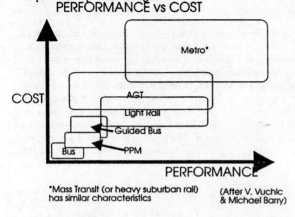

Figure 1

4. TYPICAL PEOPLE MOVER APPLICATIONS

In the past requirements for people movers have arisen at large leisure, retail or airport developments, either because increasing private car use leads to increasing demand for parking space at destinations, or because planning considerations mean that mass transit systems terminate at some distance from the development itself. It is not unusual at large exhibition or retail sites for parking spaces to be a quarter of a mile or more from the businesses they serve and at airports parking is even more distant from the terminal buildings. For such sites to be viable it is therefore necessary for transport to be provided between the parking area and the actual development. This can be provided by circulating buses, but is often more suitably achieved by a fixed link transport system. Similar systems are increasingly being built in new business and residential districts.

Now it is seen as desirable to reduce traffic in town centres, the distances between shopping centres and parking are being increased and 'People Mover' type systems have an application in the traditional town centre. They can serve both to bring people into the town centre from outlying car parks and as feeders to other, higher capacity transport systems. For example, a business park development in Rotterdam makes use of a people mover to link the district with the metro system.

A wide range of technologies are used for people mover systems, many of which are derived more from equipment for the bulk handling of materials in mining or industry than from 'conventional' rail technology. These include conveyors (moving walkways), cable hauled systems and automatic guided vehicles. Some technologies commonly used in people mover systems are:-

• Moving walkways

Safety considerations mean that moving walkways operate at speeds less than walking pace, and are therefore not a realistic way of transporting people more than a few hundred metres. The typical lengths of individual walkways are less than 100m, but systems have been devised which make use of several separate walkways (combined with escalators). An example of this is the Hillside or Mid-Level escalator opened in Hong Kong in 1993. This consists of a series of three covered moving walks and twenty escalators and is capable of carrying up to 6000 passengers per hour. The total cost was around $26 million dollars.

• Cable hauled

Probably the oldest and most conventional means of powering a people mover is cable haulage. In the early days of street running tramways numerous cabled hauled systems were constructed, of which the Great Orme Cable Tramway in Llandudno is now the only remaining in the UK. Cable hauled systems are now used for numerous people mover applications at airports and, of course, in ski lifts. The disadvantages of cable hauled systems are obvious, firstly the difficulty of expanding either the capacity of a system or its length and secondly the safety implications of the moving cable. It would not now be possible to bring a cable hauled system into operation on a route shared with other traffic or pedestrians. The 1.5 km long Laon system, installed on an existing viaduct, is a typical cable hauled people mover. This cost approximately 170 million French francs and has a capacity of 600 passengers per direction per hour. Cable hauled vehicles have been built which run on rubber tyres, are magnetically levitated or supported on air cushions.

• AC and DC electrification

Some of the most successful people movers in terms of reliability use electric propulsion from either AC or DC continuous electrification. Typically these consist of automatically controlled vehicles running on segregated, elevated tracks and powered by conductor rails, sometimes in the form of straddle monorails. Such systems vary widely in cost and capacity, for example the 3.0 km long Miami Metromover system cost $159 million in 1986 and has a passenger capacity of 7 000 passengers per direction per hour, while the 3.6 km long Sydney monorail cost 65 million Australian dollars and has a capacity of 3 000 passengers per hour.

• Innovative systems

People mover systems have often formed part of prestigious developments such as airports and theme parks. Leading edge prototype technology has been used for applications where simpler systems could perform equally well, often because the development is enhanced by serving as a showcase for an innovative system. High profile applications of new technologies have not always met with success, a case in point being the Maglev magnetically levitated system at Birmingham airport. While technically successful, the fact that this system used technology which never went into full scale production meant that it became impossible to maintain economically. If a people mover is to serve as a viable and cost effective means of transport its level of technology must be matched to the application.

5. PEOPLE MOVERS FOR URBAN APPLICATIONS

Elevated, segregated, automated, high cost people mover systems have been installed in several urban areas, but in many cases a short range transport system is required where neither a segregated alignment or the capital for a high cost system are available. Urban distributor systems, while having similar stop separations, speed and capacity requirements to airport, exhibition centre, or development site applications described above, cannot in the main be served by the same people mover technology. Systems operating on reserved elevated guideways are ruled out for almost all applications, due to the cost of the guideway and the visual impact of such constructions. The system must therefore be street running, compatible with both pedestrians and road traffic. There is no place for 'high risk' technologies incorporating one-off components or using unproved propulsion systems. A major justification for an urban people mover is to reduce pollution due to traffic and therefore the people mover itself should not emit pollution at its point of use. Automation at its current stage of development is only suited to vehicles operating on segregated alignments.

A specification for an urban People Mover can therefore be written as follows:-
- Stop separation of less than 500 m
- Low speed vehicles running at short headways
- Operation both outside and within covered areas
- Capable of operation on non-segregated track shared with pedestrians and road traffic.
- No pollutants emitted at point of use

Consideration of the above reveals that this specification was met by an existing technology, this being the traditional street running tramway. The Parry People Mover light rail system has been developed as a modern interpretation of this concept, without the drawback of the need for continuous electrification.

The carrying capacity of urban people movers such as those previously described ranges from as low as 500 passengers per direction per hour to 5,000 for more. The PPM system should therefore be capable of at least performing at the lower end of this range.

6. THE PARRY PEOPLE MOVER CONCEPT

6.1 Selection of guidance system

In order that a transport system can safely operate in a pedestrianised area, it is necessary that the vehicles follow a predictable path. Guidance can be achieved either by physical means or by some form of electronic control. Electronic control systems which follow buried cables or other marked tracks are being tested for operation in such applications, but the sophistication and unproved nature of the technology made it unsuitable for this application. Physical means of guidance include steering rollers running on concrete upstands and flanged wheels on rails. The low rolling resistance of steel wheels on steel rails as well as the fact that street running tramways have been in safe operation for well over a hundred years led to the selection of conventional tramway wheels and rails as the guidance system for PPM.

6.2 Selection of Propulsion Technology

An early decision in the development of the PPM was that in order to use simple infrastructure, the vehicles would have an on-board energy store, so that there would be no requirement for a continuous external power connection. The main options investigated were as follows:
- Conventional fossil fuel storage with internal combustion engine
- Chemical battery storage providing electrical power for traction motors
- Chemical fuel cell generating electricity for traction motors

- Flywheel kinetic energy store
- Other (compressed air, spring energy, stored steam energy etc.).

With the exception of hydrogen storage which was rejected on grounds of difficulty in handling and possible safety risks, internal combustion engines were deemed unsuitable for this application as they would not be zero-emission. Even cleaner burning fuels such as propane and natural gas were not favoured, the storage and transfer problems again being a disadvantage (Compressed Natural Gas powered vehicles require a compressor station).

Battery traction was experimented with at the turn of the century as a way of powering both road and rail vehicles, but except in specialised applications such as delivery vehicles (milk floats) has never been found to be economically viable. Although day-to-day power and maintenance costs for battery electric vehicles are low, this advantage is negated by the high cost of the battery replacements which are required every two or three years.

Fuel cells are a promising technology which is currently undergoing very rapid development, but at present are not developed to a stage where they are suitable for a public transport application.

Other energy storage technologies such as compressed air, steam, springs and flywheels have been used in vehicles in previous eras with varying degrees of success. Evaluation of these along with the possibilities described above led to the selection of a kinetic energy storing flywheel as the prime mover of the PPM system.

Flywheels have had a previous public transport application in the Oerlikon Gyrobuses which ran in Switzerland and Belgium in the 1950's and also, surprisingly, in Kinshasa, Zaire. These used a very large (1.5 tonne) flywheel operating at up to 3000 rpm. AC electricity was used to spin up the flywheel from an overhead pick-up at each bus stop. The flywheel then generated Ac which was rectified to DC to supply the vehicle's traction motors.

A flywheel was selected for the PPM for the following reasons:
- Simplicity (of the storage unit itself)
- Ability to rapidly absorb energy
- Ability to rapidly discharge energy (high power output)
- Indefinite life

It was, however, recognised that a number of issues had to be addressed before a flywheel could be used as an energy store onboard a vehicle.
- Cost - the manufacture of a reliable and safe flywheel is generally seen as being an expensive process requiring precision engineering and the use of high cost materials.
- Safety Concerns - unlike the storage of explosive fossil fuels and similarly the large volumes of acid contained in most battery packs, the immediate public perception of a rotating flywheel is that this is a potentially very dangerous device.
- Energy Density - the primary disadvantage of flywheels compared to other energy stores is that in terms of energy per unit weight the storage capacity of a flywheel will usually be very small.
- Means of energy transfer from and to flywheel. The instantaneous power output of a flywheel is governed by the capability of the transmission system, and, because the speed of the flywheel is constantly changing, the transmission must be capable of supplying a constant speed output from a widely varying input.

6.3 Design choices in flywheel spin-up and transmission

At the outset it was necessary to decide whether energy would be supplied to the flywheel by an onboard prime mover supplied from an external power source, or whether both prime mover and power supply would be external. Early investigations showed that it was possible to make a mechanical connection between the flywheel in the vehicle and an external motor, but this was rejected early on due to the complexity of such systems and the difficulty of establishing a design in which the charging station would be acceptable and safe in a public area. Furthermore, such a system would make it difficult to provide a back-up power system, as the spin-up motor is left behind at the station!

Having decided that the most appropriate means of increasing the flywheel rotational speed would be to use an on-board motor, it was possible to conceive of a small number of options for the transfer of energy to the vehicle from outside. This could be by means of a compressed air connection, or a hydraulic hose, but neither these or any other perceived method appeared superior to the obvious solution of an electrical connection. Thus virtually all PPM development has been based upon a flywheel connected to an electric motor with detachable connections to an outside electricity source. Due to the requirement that the vehicle be safe within a pedestrianised area, it was necessary to use only low voltage DC electricity, current regulations requiring that this be 70 volt or less. Additionally, it is recognised that it would be appropriate for the electrical supply to be live only whilst the vehicle is present at the charging station.

6.4 Flywheel design

In the design of the PPM flywheel energy storage system the above issues have been addressed and a practical system arrived at. Although several organisations are developing high speed composite flywheels, it was decided that a low speed unit using more conventional materials would be better suited to the PPM system. The cost of the flywheel unit is correspondingly low. The low energy density of a low speed flywheel is not a major drawback in a rail vehicle operating on the PPM system where energy is supplied to the vehicle at each passenger stop.

The current flywheel is a 500kg unit of 1m diameter constructed from several plates of high tensile steel. Within the limits acceptable for safety, i.e. a normal operating speed of 3600 rpm, the flywheel will store over 1 kWh of energy. The multiplate design leads a safe mode of failure in which a failing flywheel will break into large numbers of small fragments. The flywheel is enclosed within a heavy steel casing in which bearings are mounted. To reduce costs and for ease of maintenance conventional rolling element bearings were used.

A back up battery pack onboard the vehicle provides a power supply in the event of unscheduled delays making it impossible for the vehicle to complete its journey on the energy stored in the flywheel.

6.5 Selection of transmission system

It was not possible to develop the drive system concept in isolation to the flywheel spin-up system, because the rotational speed, range of speeds and other characteristics should suit both the energy input and output. General options for a drive system were as follows:
- Electrical - where the flywheel is connected to an electrical generator which produces an electricity supply for on-board traction motors.
- Hydraulic - where the flywheel powers a hydraulic pump which in turn supplies hydraulic motors to produce tractive effort.
- Mechanical - where the flywheel rotation is transferred by direct mechanical means to the vehicle wheels.

Of the above, both electrical and hydraulic systems would be capable of driving the vehicle at a range of different speeds independent of the flywheel rotational speed. In the case of an electrical drive this would be achieved by the use of transistor-based controllers, for example, and in a hydraulic drive by appropriate variable valving. A further feature that is very desirable in the PPM design is that it should be possible for energy to be recovered when the vehicle decelerates or maintains its speed on a descent. Hydraulic systems can do this without difficulty, although losses may be fairly high both in this and the conventional drive phases. Although not investigated in detail, an effective electrical regenerative brake would be practical.

Where an electrical drive is disadvantageous within the PPM concept is that a proprietary electrical generator would be unsuited both to the low speed of flywheel rotation, and the large range of speeds utilised (usually between 1000 and 3600 rpm). Furthermore, the electrical technology required is by no means conventional and would be beyond the scope of the intended operators of the PPM, it being intended that they be run by bus companies or the like. Hydraulic systems could be more within the scope of automotive technicians, but would still be a comparatively unfamiliar technology with anticipated high maintenance costs.

From the outset the PPM was designed around a mechanical drive. Whilst it was straightforward to connect the flywheel and driving wheels to give an appropriate cruising speed, it was not sufficient for the drive speed to be controlled simply by clutches. Early investigations showed that it was possible to produce a moderately usable vehicle with a stepped mechanical transmission, but this was soon supplemented by a wide ratio mechanical variator. Figure 2 is a schematic of the flywheel and transmission arrangement.

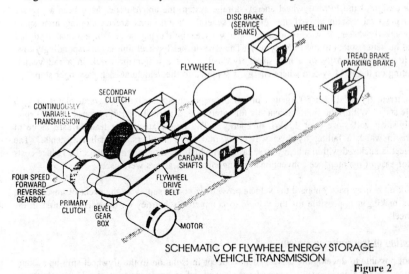

SCHEMATIC OF FLYWHEEL ENERGY STORAGE
VEHICLE TRANSMISSION

Figure 2

6.6 PPM System Performance

Through the construction of several test vehicles and trial running of these, acceleration, braking and other performance characteristics have proved satisfactory for on street running with conventional road traffic. Constant running speeds of 25 mph (32 km/h) have been achieved, but it is anticipated that the existing chassis design will be capable of over 30 mph without

© IMechE 1996 C514/064

difficulty. Flywheel spin up times now approach those necessary for a 20 - 30 second dwell time at passenger stops, this being appropriate for boarding and disembarking.

The overall dimensions and body design of the PPM vehicles enables around 35 passengers to be carried including 20 seated per car. A larger number of standees are possible but performance figures are based upon 4 persons per square metre. At present only single car vehicles have been constructed, but design work is under way on a two car set. Figure 3 shows an artists impression of a single car, modern styled vehicle

Figure 3

Capacity figures for PPM systems depend upon the frequency or interval between vehicles. In a loop system this could be from around 5 minutes down to a minimum of 2 minutes, giving a capacity of 500 to 1000 per direction per hour. Two car sets would enable capacities of 2000 per direction per hour to be achieved.

A typical single car vehicle could be supplied at a cost of £100 000. Suitable street running track would cost approximately £300 per metre and track laid on a segregated alignment approximatly £100 per metre. The target price for a two car set is £200 000. It is estimated that a system using PPM technology can give one third of the capacity of conventional LRT for one fifth of the cost.

7. PARRY PEOPLE MOVER APPLICATIONS

A large number of potential applications for the PPM system have been investigated, and in many cases detailed feasibility studies were carried out. These can be shown to cover most of the categories outlined previously for different public transport functions:

1 - A link between railway stations in two adjacent towns utilising a disused railway alignment, but extended into the town centre and passing adjacent to various sources of patronage such as schools and offices. Total route length around 7 miles, mostly segregated track with stops at up

to 2 km spacing and speeds of 50 km/h or more on segregated sections. Twin tracks installed with reversing loops at each terminus, bi-directional vehicles.

2 - A town centre tramway connecting shops, car parks and other public transport facilities. Low speed, 500m stop spacing, continuous loop of around 5 km, street running.

3 - Resort town tramway serving as a visitor attraction as well as a park & ride function. Set to run through pedestrianised areas for much of the route. 4.5 km loop.

4 - People mover connecting superstore with leisure centre. 400m shuttle service running segregated alongside car park.

5 - People mover connecting regional railway station with town centre. 1.2 km shuttle service on existing standard gauge railway track to be separated from regional railway system.

6 - People mover / light tramway connecting resort town centre and out-of-town shopping centre, but also extending through town centre to sea front. Loop system. Mostly segregated or running in pedestrianised areas but street running with traffic for around 30% of the route.

The requirements of these applications could not be met by public transport in any format other than a people mover, but sophisticated, segregated systems would neither be cost effective nor desirable on planning grounds. These possible applications demonstrate the case for systems in the format of the Parry People Mover light tramway.

8. CONCLUSIONS

Although People Mover applications have served as a showcase for new propulsion and automatic control technologies, the above discussion demonstrates the case for driver controlled vehicles and a simple propulsion technology not requiring a segregated track. The Parry People Mover represents one technology capable of meeting this requirement, although a multitude of other possibilities exist.

9. BIBLIOGRAPHY

Through the Cities - The Revolution in Light Rail - Michael Barry
A Planner's Guide to APMs - Trans21 - Transportation Systems for the Twenty First Century
Light Rail & Modern Tramway Magazine
Swindon Light Tramway - Presentation - Trevor Griffin - BR Research
The Parry People Mover - Paper presented at Emcee Conference 'Implementing Rail Projects'- F. Schmidt et al
Light Tramway Feasibility Studies (Various) - Parry Associates Consulting Engineers
Trials & Testing Programme Reports - JPM Parry & Associates
Draft Type Approval Application - Parry People Mover
Independent Appraisal of the Parry People Mover for London Transport Ltd. - David Catling

C514/071/96

People movers

A S ROBBINS BS, MS, PE
ABB Daimler-Benz Transportation (North America) Inc., Pennsylvania, USA

1. INTRODUCTION OF TECHNOLOGIES

Over 25 years ago, ABB Daimler-Benz Transportation (Adtranz), then known as Westinghouse Transportation, installed its first automated transit system in Pittsburgh, Pennsylvania in the United States, and demonstrated an innovative alternative to traditional transportation modes. In the years since, automated transit system technology has been refined through the use of microprocessors and solid-state electronics and extensive application experience.

In 1971, the first fully automated, driverless system went into operation at Tampa International Airport, Florida USA.

And in 1972, the Bay Area Rapid Transit in San Francisco, known as BART, began public service. It was the first totally new metropolitan rapid transit system to be built in the United States in almost 60 years and the first fully automated system.

Automated Transit Systems from Adtranz consist of attractive, comfortable vehicles operating on a dedicated guideway under a totally automatic control system and are available in differing system configurations.

The Automated People Mover Systems, designated as C45 and CX100, consist of electrically powered, rubber-tired vehicles with four pairs of guidewheels which lock onto and are steered by a steel guidebeam anchored to the guideway.

The CX100 is 11.89 meters long and features a spacious interior designed for passenger comfort and convenience.

The C45 is 8.5 meters in length and allows lighter, less expensive guideway structures for systems with lower passenger densities.

Both types can be used individually or coupled together to form multi-car trains, depending on system requirements.

The Adtranz Monorail technology is often referred to as a straddle-beam monorail. Individual cars are linked together into an articulated train unit with each car module supported by bogies placed between the individual cars, operating on pneumatic tires. The electrically propelled cars are steered by guidewheels along the elevated guideway providing total segregation from other traffic.

2. APPLICATIONS/CAPACITIES

The Adtranz family of Automated Transit Systems serve a wide variety of applications which benefit from this relatively new form of transportation system. The applications include activity centers such as amusement parks and fairgrounds, transportation and circulation of passengers within airports and the movement of large numbers of peoples in urban corridors.

Automated People Movers and Monorails may be characterized by smaller vehicles and shorter trains operating on dedicated guideways and running at frequent intervals to provide a high overall level of operational capabilities. Due to the nature of their design, including their ability to turn in small radiuses, they can be integrated within existing buildings, into new or existing airports and can be routed through buildings or to stations within buildings. Operation can also be placed in tunnels.

All of the automated transit systems operate with passenger carrying capacities typically above that found with buses and below those of heavy rail (full metro) rapid transit.

The Monorail is usually applied to special activity centers with system capacity requirements up to approximately 3,000 passengers per hour per direction (pphpd). Train lengths up to 8 cars are possible with each car carrying about 20 passengers. The monorail may run with top speeds of up to 40 km/hr.

The Automated People Movers may also be applied to activity centers including airports and as downtown circulators with passenger carrying capacities of up to approximately 18,000 pphpd. Trains lengths may vary between one to four cars in most applications with the C45 accommodating 39 passengers nominally and the CX100 accommodating 95 passengers nominally.

3. ROUTE CONFIGURATIONS

The increased use of automated transit systems, together with the variety of applications in which they are employed, has led to an assortment of possible network configurations. Planners have several basic configurations from which to create numerous routing concepts, and these decisions significantly affect system performance. Most automated transit systems in use today are either of the shuttle, loop (typically contra-rotating loop) and pinched loop (turnback) configuration.

Shuttle systems are ideal for connecting activity centers and for serving remote areas such as car parks. In airports, they have proven ideal for connecting "landside" functions such as bag claim and ticketing with "airside" gate areas. Shuttle systems generally fall within three basic types:

© IMechE 1996 C514/071

- Single lane, with one train shuttling back and forth

- Single lane with bypass, allowing dual-train operation

- Dual-lane, with trains operating on their own guideway

The single lane configuration is the least expensive and serves applications where wait time and capacity are not major concerns.

The addition of a bypass in a single-lane shuttle system increases the system capacity and decreases wait time. Trains are synchronized so that they cross every time in the bypass area.

The most flexible shuttle system is the dual lane system. While each lane is independent and responds individually to passenger demand, the trains are continually synchronized with the goal of crossing mid-distance for maximum efficiency. System capacity is maximized and passenger waiting times are minimized with this configuration..

System maintenance for shuttle systems is normally accomplished on-line, typically below one of the stations. An added benefit of the dual lane shuttle configuration is that maintenance may be performed on one lane while service continues on the other lane.

Shuttle systems may also operate on-demand, utilizing call buttons similar to lift systems.

Loop systems have been used in applications where service to multiple stations is required. Two or more trains travel in the same direction at safe following distance, serving stations at close headways thereby allowing high system capacities. To improve system performance, contra-rotating loop systems have been used which reduces maximum travel time for passengers. An example is the downtown people mover employed in Miami, Florida, USA. Multiple trains travel in a clockwise direction on one lane and in a counter-clockwise direction on the other. This system is currently being expanded to include north and south traveling routes accessed from the outer downtown loop. Also note that this system is fed by a north-south metro system thereby providing convenient connections for suburban passengers into the downtown area.

The final configuration commonly used with automated transit systems is the pinched loop or turnback system, as seen with most urban metro systems. The use of automation allows trains to follow each other at close headways for maximum system capacity. Turnbacks at the ends of the line typically employ multiple switches to allow for the short headways and to maximize the system reliability. Bypass switches and sidings are also employed along the guideway in the event of a train failure and to reduce system downtime.

Most automated loop and pinched-loop systems can be used bi-directionally for added flexibility. In off-peak hours, these systems can operate in alternate loop mode or in single tracking modes which allows for track and wayside maintenance to take place while still maintaining service.

4. MONORAIL CHARACTERISTICS

Monorail trains are of modular design using standard center and nose cars units to form a fully articulated train. Each car rests on its own bogie assembly rigidly attached to one end of the car. The other end receives a ball-joint coupler that is rigidly connected to the next car bogie.

The car body is a fully-enclosed, double-wall unit. It has a low-body frame design that permits a low floor to ground height of only 203 mm.

The all aluminum structure features generous use of glass and provides a pleasing appearance. Seating arrangements can be customized.

Lateral guidance is accomplished by four guidewheels per axle. Each guidewheel is preloaded against the side wall of the trackway beam by springs.

Train braking is accomplished by two independent braking systems, including an electrical regenerative system for normal service braking and a mechanical disk-type friction brake system for emergency braking.

The electrical propulsion control system is based on the use of thyrister control with continuously adjustable voltage regulating the DC traction motors to allow stepless acceleration and deceleration of the train.

Train propulsion power is obtained from a three phase rail at 500-600 volts.

The monorail guideway utilizes a standard composite steel box girder formed by welding standard rolled steel plates into the box beam configuration. All steel sections are shop fabricated, pre-cut and contoured requiring only one field weld near each column to form a continuous guidebeam structure. The beam is supported by steel columns that are securely fastened to baseplates anchored to concrete foundations. Typical spans are 24 to 28 meters with special spans up to 49 meters possible with the use of deeper box beams.

Guideway switches are used for route network designs. Switching is accomplished by automatically aligning a tangent or curved section of box beam at the switching area. Switches may be of the pivot-type or rotating-type.

Control of monorails is usually performed semi-automatically, with a cab attendant overseeing various functions, although monorails have been implemented as fully automatic as well as in full manual operation.

Monorail stations may vary a great deal depending on the requirements with respect to their location, architectural theme and passenger interface. Stations must be accessible to elderly and handicapped persons and typically include fare collection, TV monitoring, public address and other passenger information systems.

5. AUTOMATED PEOPLE MOVER CHARACTERISTICS

Automated People Mover systems are an innovative transportation technology adaptable to the requirements of any medium to high capacity activity center, including airports, industrial parks, recreation complexes, shopping/apartment or office complexes and business districts.

The automated people mover systems feature attractive, comfortable vehicles operating on a dedicated guideway under a totally automatic control system and are available in two standard vehicle configurations.

The CX100 vehicle is 12 meters long and accommodates 95 standing passengers (using 4 passengers per square meter). The C45 car design is smaller and lighter than the CX100, but utilizes virtually the same technology. The C45 is 8.5 meters long and accommodates 39 standing passengers.

For the purposes of this presentation, the CX100 model will be specifically discussed.

The people mover vehicles ride on rubber tires. Four pairs of horizontally-mounted guidewheels lock the vehicle to a steel guidebeam anchored to the concrete guideway to prevent derailment and to provide continuous guidance for each moving vehicle. The people mover system allows system designs to include turning radiuses as low as 22.9 meters. This combined with low weight and negligible vibrations allows systems to be placed in congested areas and even cantilevered from or through buildings.

The interior of the vehicle is designed for passenger comfort and easy maintenance. A wide open space is provided making the people mover extremely efficient in terms of its passenger carrying capacity as opposed to its empty car weight of about 14,787 kg. Since travel times are typically short, most internal configurations provide for a large standee area, although seats may be provided in varying configurations.

Side windows make up about 35% of the vehicle wall, and large windows at both ends of the car provide a wide open feel for the passengers. Two, wide door openings of over 2 meters each are provided on each side of the car. The wide openings allow for increased flow of passengers which in turn minimizes the time a train must be stopped in the station.

Cars may be operated in trains, typically up to 4 vehicles in length, and are designed to run in all weather conditions.

The CX100 car body consists of a steel underframe and floor pan, and aluminum body, with fiberglass end pieces. These pieces can be customized to suit a particular requirement.

The guideway for the CX100 system consists of twin concrete running pads on which the vehicle rides. The running pads are placed on top of a support structure which may be of steel or concrete construction. The steel guidebeam is set in-between the running pads.

Columns, typically of concrete, support the structure and are usually provided in span lengths of 27 to 30 meters, although varying span lengths can be used as required.

Propulsion and braking are performed in a similar manner as that on the monorail technology. Power is provided from guideway mounted power rails at 600V, 3 phase for the CX100 and 480V, 3 phase for the C45.

Switching is also accomplished in a similar manner as that for the monorail, with curved and tangent guidebeam section that pivot or rotate to complete the switching operation.

Control of people movers is completely automatic, utilizing a sophisticated, microprocessor-based train control system, with electronic equipment mounted on board vehicles, in station equipment rooms and in a central control facility. The automatic train protection system utilizes a fixed block approach for safe train separation, and headways as low as 70 seconds have been implemented.

In the central control facility, large screen, high resolution color monitors and touch-screen control consoles use real time display and control. Much of the implementation is by computer software, allowing for easier expansion capabilities.

The automated people movers are designed with service in mind. In fact, Adtranz provides operations and maintenance services on many of these systems after they are fully commissioned.

6. CONCLUSION

The Automated Transit Systems presented are designed to solve specific transportation problems and will complement higher capacity urban metro systems and light rail lines which may be implemented. You have seen how these systems can work in airport, downtown areas, amusement parks and fairgrounds, as well as in congested urban corridors. The systems approach to implementing modern transit systems is a powerful way to ensure that demanding requirements are met.

Delivering Customer Reliability

C514/051/96

Reliability – focusing the effort

P L DUNKERLEY BSc, MTech, MORS and **D M WALLEY** BSc, AFIMA
B R Research, Derby, UK

SYNOPSIS

Determining where effort should be directed in ensuring reliable railway operation, and thereby
delivering the advertised level of service performance to the customer, is a task made more simple
by a structured approach. This paper discusses the process of delivering reliability, illustratied with
a description of a high-level model to decide where effort should be focussed, and giving some
insights into the application of conventional engineering reliability analysis tools to examine the
performance of individual systems and compenents.

1. INTRODUCTION

For a railway business to invest confidently in new equipment and methods, its managers must be
able to see how the financial performance of the business will be improved, through operation of a
punctual train service which meets the promises of the advertised timetable.

The link between the reliability of individual components and sub-systems, and their effect on the
punctuality performance of the train service operating on a route, however, is an area in which a
gap in understanding was apparent.

There is a need for techniques to enable business planners to explore this inter-relationship, and to
examine how the train punctuality along the route is influenced by the periodic random failures
occurring to track, signalling and vehicles, and at a more detailed level, to enable engineers to
deliver equipment which keeps on working, and when it does fail, is simple to repair.

The paper discusses the scope of the problem, and then describes the present methods available for
gathering and monitoring railway reliability statistics in Britain, and the way in which the data
provided by these systems can be used as a basis for our analysis.

It shows how the aspects of route geography and operation, and modelling the detail of the
timetable, can be built into the model.

2. THE SCOPE OF THE RELIABILITY PROBLEM

Delivery of a reliable train service on a specific route can be represented as a five - stage process, as shown in Figure 1.

At the first stage, the business specification determines the basic reliability of the individual vehicle and infrastructure components of the system.

The procurement process which follows incorporates the design and development processes, and is strongly influenced by the environment in which the new investment is being made. To a large extent, it determines the value for money that the railway rolling stock or infrastructure operator obtains from its suppliers. It is at the first two stages of the process that the inherent reliability and maintainability of the equipment is set.

At the third stage, the maintenance activity determines how well the individual parts of the system are running, the frequency with which they break down, the detection of failures, and the consequent speed and effectiveness of repair.

By appropriate action, the probability of failure can be reduced, and the availability of equipment can be improved - with consequent effects on overall vehicle fleet size requirements and thus the initial capital investment.

At the fourth stage, the performance of all of the equipment together as a system is affected by operating and timetabling considerations. When failures do occur, flexibility in the timetable reduces the impact of failures by preventing one train failure from unduly delaying following trains.

The fifth stage is where performance is measured against the promises made in the timetable. The level of performance determines the revenue level, and in the case of poor performance, may also result in the need to compensate individual passengers.

At each of these stages, investment and improvements can be made. The extent to which they are worthwhile at each stage needs to be examined in detail.

As an extreme example, no action could be taken at the first four stages and then more and more compensatory payments made at stage five. By contrast, very much higher investment in vehicle development at stage one might result in excellent vehicles, in service too late to take advantage of market opportunities.

Each succeeding stage involves finding ways of living with the situation determined by the previous stages. Striking the right balance of emphasis of effort between the stages is vital, and it is the business manager's task to determine this balance.

There should thus be a feedback loop, from the end of the process to each of the intermediate stages, whereby action can be taken to affect the quality of the end product. Because of an inadequate understanding of the complete process, and all the interactions involved, this feedback is not always there.

To make the judgements necessary to get the balance right, the business must have a method of evaluating and predicting the effects of change. It must be able to examine the levels of reliability specified for components and sub-systems, the way they are maintained and repaired, and the way they influence the performance of trains operating to the timetable.

A model, called MERIT, has been developed jointly by BR Research and the Rail Operational Research Group (Rail OR) to enable these complex relationships and interactions to be explored, and is becoming an invaluable tool for the management of reliability within the railway businesses.

With the aid of MERIT, it is possible to answer a wide range of "what - if?" questions. For example, trade-off studies can be carried out to examine the balance between investment in better, more reliable (and possibly more costly) equipment, and the provision of better repair back-up during the lifetime of the operation.

3. PERFORMANCE AND RELIABILITY MONITORING

Much of the systematic recording of train punctuality statistics is based on the TOPS and TRUST systems, and some newer derivatives and similar substitutes. These use links to the train describers in the signalling system, comparing the times trains pass timing points against the planned timings specified in the timetable.

Variations against the timetable are flagged and recorded, and human operators monitoring the output from the system ascribe causes and responsibility for each incident of train delay. The data generated is analysed daily and over longer periods, using other computer - based systems, and this is then used as a planning aid by business staff.

Underpinning this analysis of actual train running are a number of engineering-based systems, recording failures and rectification work carried out to specific engineering systems. These include FRAME, which is use to monitor signal and telecommunications equipment, and RAVERS, used to monitor the performance and maintenance history of traction and rolling stock.

These systems-all yield management information and are in constant use today. They also provide a vital input for our new reliability model, in terms of basic data on the reliability of the individual components and sub-systems, and in terms of the overall punctuality of the service.

But to model the service in order to understand the interactions taking place, and study the effects of changes to the component reliability, their repair and the effects of changing the service intensity and timetable design, another level of complexity is needed.

The next section of the paper describes the way in which failures delay trains, in terms of what happens to the train first affected, and then to those trains following behind which may also be held up by the failure to the first train.

4. TYPES OF TRAIN DELAY

4.1 Primary and Secondary Delays

Two types of failure can interfere with train operation on a route:

     1. failure of a train itself

     2. failure of the infrastructure in front of a train

Calculation of the effects of failure on the first train affected in a sequence of trains following a timetable, is relatively straightforward. This is referred to as primary delay.

In the case of a failure to the train itself, it involves analysing the frequency of failures of all the different components and subsystems involved - time consuming but not difficult in principle - and for each type of failure, evaluating the delay to the train while the fault is repaired, or while another train can be used to assist the failed train.

Where the failure occurs to an item of infrastructure - a track feature such as a track circuit, or a broken rail - it is similarly relatively simple to calculate the effect. In this case, the frequency or probability of failure is again important, and the delay to the train must also be calculated.

This type of failure will often result in the train being halted at the signal behind (in the rear of) the faulty equipment, and told to proceed at a reduced speed until the faulty component is passed.

A repair team will be dispatched to attend to the fault, and some time later, the line will be cleared for normal operation. In some cases, the line may be blocked for a time, and depending on track configuration, it may be possible for a diversionary route to be set up.

Difficulties occur in both of these cases when trains are following at short headways behind each other. In this case, a train failing or stopping for a time and/or proceeding at a lower speed will inevitably eventually cause a bunching effect to the trains behind it. This is called this secondary delay. The total delay to trains on the route is the sum of all primary and all secondary delays.

The MERIT model has been designed specially to calculate the total delay, built up in this way.

4.2 Measurement of Primary Delays

The effects of delays to the first train can be summarised in three sets of parameters:

Probability or frequency of failure. This may be over a period, per component - for example the frequency of rail breaks per track mile per year, or of track circuit failures per circuit per year, or the frequency of vehicle failures per vehicle mile per year. In both cases, the expected number of incidents per year over the route segment is easily calculated given the route characteristics and the number of trains passing over a period.

Duration of delay. This is the length of time each train will be delayed in passing or clearing the incident. In the case of track failures, this allows for the time for the train crew to contact the signal box, and for the train to proceed at reduced speed past the site. For vehicle failures, this will be the expected time required either to repair the fault, or to "rescue" the train by sending a replacement locomotive, or by pushing the train to a place (such as a loop at the next station) where it can be repaired more easily.

Time to repair or recover. An infrastructure failure may be causing a delay relatively minor in duration, but which it may take several hours to repair. A repair team will need to be assigned and to travel to the site, obtain permission from the signalman so that it can have safe access to work on the fault, and then carry out the actual repair. In the case of locomotive failures, however, the time to repair will be the same as the duration of the primary delay, as far as train running is concerned.

5. STRUCTURE OF THE MERIT MODEL

The structure of MERIT is summarised in Figure 2. This shows four sets of input data on the left hand side, as follows:

Route Data - this describes the route or network being modelled, and includes geographical data such as the locations of all junctions and stations, and information on the number and types of all signals, track circuits and S&C.

Delay Data - this contains information on the failure rates, repair distributions and effects of failures to all infrastructure and rolling stock on the route or network.

Timetable Data - contains details of all train timings for movements being modelled.

Control Rules - this data enables the model to determine priority for allowing trains with conflicting movements at junctions to pass, if they are delayed. It is based initially on rules in force on the specific route being modelled.

The model is then run using the data inputs described. The model simulates trains according to the timetable presented:, and calculates delays due to equipment failures by statistical sampling. In turn, these primary delays generate secondary delays to following trains, and the model keeps a constant record of the way in which the total delay suffered by each train is building up.

Outputs are generated corresponding to the items listed on the right hand side of Figure 2. These are:

Delay causes -attribution of delays to individual trains, or groups of trains, caused by the different failure types input.

Service Punctuality - service performance, in terms of measures such as percentage of trains right time, and within ten minutes of time, can be plotted for individual trains or groups of trains, either at the destination station or at any point along the route.

Revenue Effect - given knowledge on the performance and charging regime in place, or the effect of delay on passenger revenue, the punctuality can be translated directly into a revenue effect for the route or train service operator.

By changing the model inputs, for example, to examine lower failure rates, or faster repair times, and re-running, the effect of the changes in terms of performance and revenue can be compared. Knowing the cost of such changes - in re-design of components or provision of more fault-teams for instance - the user can investigate the financial case for investment in reliability improvement. MERIT in effect provides the means of "closing the loop" in the process described in Section 2 and Figure 1.

The next section of this paper discusses the detailed analysis of reliability performance - methods of deciding how to improve the reliability and repair of specific equipments, having identified them as critical to service performance using the higher-level MERIT model.

6. TURNING ANALYSIS INTO ACTION - RELIABILITY ENGINEERING

MERIT enables the performance of the route or network to be examined, and enables the operator to:-

  a. Decide where action and/or investment will be most effective

  b. Calculate how effective it could be, in terms of revenue.

In this section, some aspects of the analysis of the engineering systems are briefly discussed. The techniques themselves are covered very well in the many textbooks on reliability engineering. Here some observations are made on the application of the techniques, drawn from experience. They will be illustrated in the Conference presentation with some examples.

Use all the data - there is a strong temptation to use the 'time to failure" data for a particular system, but this has the effect of ignoring probably more, extremely meaningful data on equipments which have not failed. By effectively throwing this data away, the analyst may be under-estimating the reliability of the system, and will almost certainly draw completely the wrong conclusions.

Look at the shapes of the distributions - merely considering parameters such as mean or median lifetime, mean time/distance between failures, loses much of the qualitative sense of the data. For example, whether the failure-time distribution is flat or peaked will be a key indicator of whether planned preventive maintenance will be worthwhile, or will merely cause cost and disruption to equipment working satisfactorily already.

Look for differences - many of the reasons underlying unreliability can stem from poor practice over only part of a fleet of vehicles or inventory of infrastructure equipment. Look for differences between builds and modifications, between manufacturers, and between depots and service teams. Otherwise these will be masked within the performance of the whole equipment population.

Talk to the people - data on a database alone, no matter how smartly analysed, may not give the full picture. Talk to the people who live with it, who drive and maintain it. They may have a different story to tell.

Feed the findings back into the design process - make sure the results of analysis made during the lifetime of one generation of equipment is used to develop the next generation. Avoid re-inventing the wheel.

7. CONCLUSIONS

This paper has discussed several aspects of reliability applied to getting the most from a railway system. It has introduced a simple model of the reliability process, going from initial specification of a system through several stages to its delivery and operation in the service on a route or network.

It briefly describes the MERIT model, linking reliability and operating parameters through a simulation model to calculate the effect on punctuality performance, and ultimately revenue, enabling cost and benefit assessment of the value of action to be carried out.

Finally, it has discussed the application of reliability engineering techniques - not in terms of the techniques themselves, but in drawing out a few lessons learned from experience of their application. In the Conference itself, examples based on recent studies will be used as illustrations.

Figure 1: Reliability - A Five-Stage Process

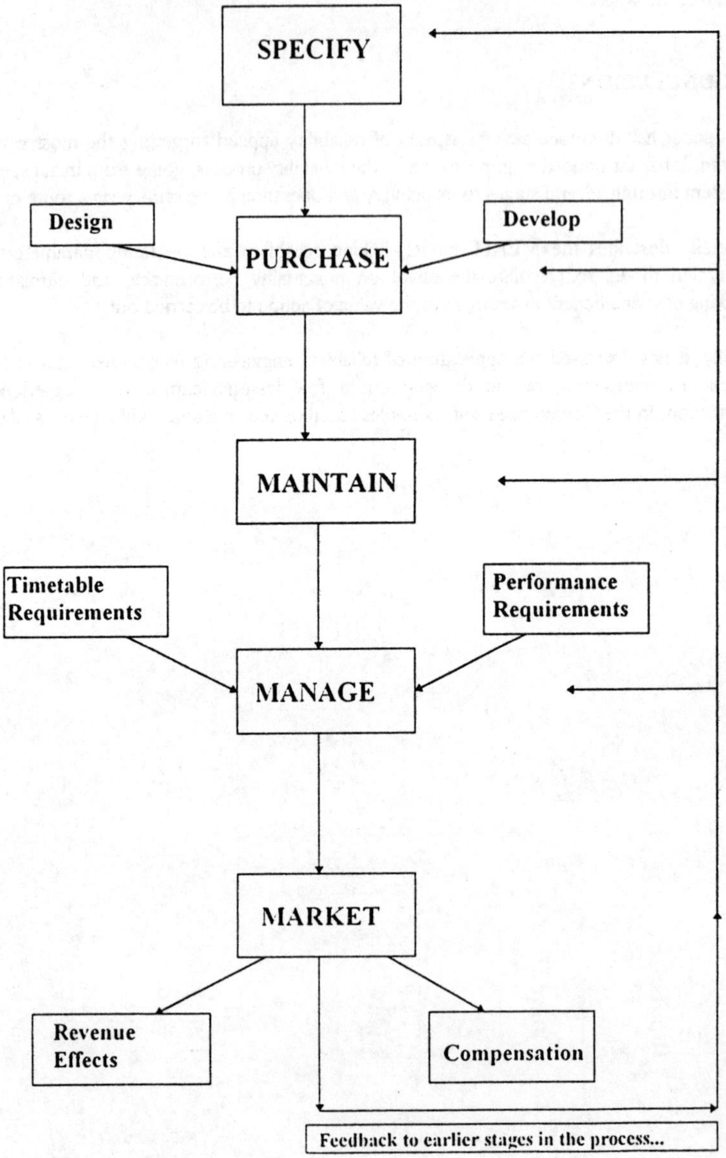

© IMechE 1996 C514/05

Figure 2: Schematic Diagram of Merit Process

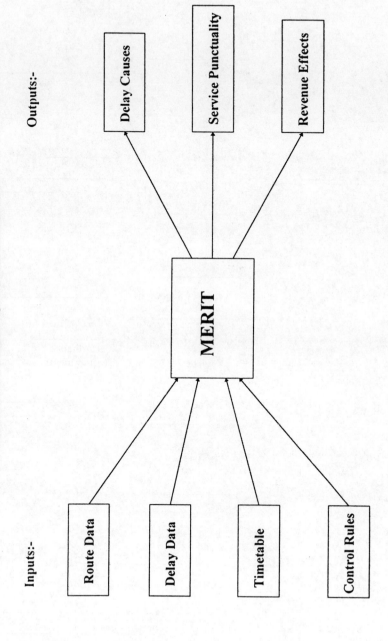

Inputs:-

Outputs:-

C514/034/96

TransiT – Putting the traveller into traveller information

J C BULLOCK MA, MSc, MBritCompSoc, and C A GOBLE BSc, MIEEE, ACM
Department of Computer Science, University of Manchester, UK
C M OSMAN
B R Research, Derby, UK

This paper describes a proposed user-oriented travel and tourism information system, which will help the traveller to choose their destination and plan their journey. By "packaging" information about places and transport services, it aims to promote the use of public transport for leisure pursuits.

Traditional travel information systems do not always allow appropriate information to be readily obtained, since their knowledge base is usually restricted to service-centred rather than passenger-centred information; they can thus answer queries about transport services but not about the places they serve. The traveller must collect the appropriate set of literature to select a suitable destination and plan their journey; journey planning becomes an exercise in cross-referencing and collating disparate information, often from different sources.

The proposed system will allow the user to pose queries, using touch-screen forms, which define the requirements of their destination and journey, and will then guide them through the journey planning process. Its knowledge base will contain both service-centred and passenger-centred material (details of places and their facilities), and the user may thus plan a journey starting from a request about services ("arriving in London by 9:30") or places ("somewhere with a beach, where I can sail").

An advanced knowledge representation will be used to to support flexible, intuitive querying facilities, and to structure the knowledge base to match users' perceptions. As well as allowing effective access by users, this may help the system to deliver information targeted to their need; this is beneficial to the user, who can locate relevant information, and to the information provider, whose information is delivered effectively.

1   INTRODUCTION

Travel and tourism information sources often do not adequately promote the use of public transport for leisure travel. They are frequently disparate and generic, leaving the traveller to collect exact details and plan their journey; many people may thus perceive public transport as inconvenient and inappropriate for their needs. In addition, information sources are rarely structured with respect to traveller requirements, making it awkward to locate services and facilities which satisfy individual needs.

If public transport is to thrive, it must present itself in a manner which will appeal to the private transport user; otherwise, improvements to services through better journey times and comfort will not be appreciated. The TransiT project is investigating the provision of user-oriented travel and tourism information, that is information which is structured to match the traveller's requirements and perceptions, and which integrates with and supports their journey planning activities. To achieve this, the project will use ethnographic observation techniques to study human travel planning activities, and an advanced knowledge representation to maintain a rich, flexible information model.

This paper introduces the concept of user-oriented information, and discusses the motivation for it by reference to traditional and contemporary sources of travel information. It then describes the TransiT project, specifically the ethnographic and knowledge representation techniques which will be used; finally the relation between user-oriented information provision and current research into information systems is discussed.

2   PROVIDING TRAVEL INFORMATION

This section describes the limitations of existing travel and tourism information sources, and how these may be alleviated by user-oriented information provision; it then discusses the benefits to public transport of user-oriented information.

2.1 The Problem with Existing Travel Information

Traditional travel and tourism information sources fulfil complementary roles, which do not adequately address the needs of the traveller:

- Brochures provide vivid descriptions of destinations, and usually include information on travel to them. However, the description is generally from the perspective of the destination promoter, detailing what they assume the reader will want to know, and the travel information relating to public transport is often static and generic, thus the reader must search elsewhere to obtain specific details.
- Destination marketing systems act as 'active brochures', allowing the traveller to browse and query a database of destination information; however public transport information is still static and generic (1).
- Travel systems provide public transport information which is dynamic, possibly real-time, and can be queried with respect to individual journey requirements; however, they frequently provide only unimodal travel information, and do not integrate destination information.

Although these sources provide extensive coverage, they provide diffuse information: the traveller must identify information relevant to their requirements, then collate this into a usable plan. Contemporary tourism information systems developed by recent projects have addressed this problem by combining the features of destination marketing systems and travel systems. They present an integrated, appealing, multimedia interface to tourism information and detailed, multimodal travel information; the traveller can now browse through descriptions of destinations, together with full details of how to reach them (2).

It is recognised that the modern traveller is becoming more sophisticated, having greater independence and a wider range of interests, and is less prepared to accept rigid options (3). In view of this, many tourism systems have adopted hypermedia technology, to allow the traveller to explore in one go a large range of destination, accommodation and travel information; a spate of animated articles predict that these systems will cause radical changes in the travel market, and will better serve the needs of the dynamic tourist (4, 5).

However, both traditional and contemporary information sources are structured around what the *information provider* offers, rather than what the *information consumer* wants to know. If the structure of a tourism information system reflects only the perspective of the information provider and their interpretation of traveller needs, its querying facilities may not adequately address actual traveller needs, and as systems expand beyond local coverage it may become increasingly difficult for the traveller to locate appropriate information. Moreover, experiences with the hypermedia systems indicate that hypertext and digitised media are not sufficient to support user needs, unless accompanied by adequate querying and collation facilities (6, 7); if a tourism system simply presents an increased diversity of information, this increases the burden of collating information and evaluating the constraints and compromises of different choices.

2.2 User-oriented Information - A Possible Way Forward

Promotional material is thus restrictive, since travellers have requirements when planning travel (8) but cannot query information sources in terms of these; however promotional material is necessary, since travellers may not exactly know their requirements and need suggestions. There is thus a mixture of needs, namely the ability to promote a destination based on what it offers, and the ability to choose a destination based on individual requirements.

A user-oriented tourism information system aims to address this by allowing the traveller to query its knowledge base in terms of their requirements, and to explore the available information in order to discover what they require. Since travel is usually the means to some other activity, such a system must also support travel planning in the specific context of the traveller's chosen activities, thus helping the traveller to balance their requirements with travel constraints. The TransiT project will investigate the provision of such an information system.

2.3 The Benefits to Public Transport from User-Oriented Information

Public transport is particularly disadvantaged by fragmented or incomplete information availability, since if information is unimodal or confined to a particular operator, the traveller will not perceive a coherent transport network (9). Previous sections have highlighted the problem of collating information and evaluating different choices; this problem particularly affects the patronage of public transport, since a potential traveller has often to plan their journey to a destination with reference to several mutually dependent constraints imposed by the service operator. For example, certain U.K. rail operators offer nine ticket types, whose availability and validity is dependent upon the day or time of travel, time of booking, availability of capacity and journey distance; the traveller must consult with the operator, and in effect 'ask permission' to travel to a destination at a particular time and for a particular cost. In comparison, a complete car itinerary can be planned independently with reference to an atlas.

Transport telematics projects (10, 11) have tended to focus on mode choice during the journey, thus public transport might only be considered when a private mode fails. A system which integrates travel and tourism information, and manages *all* the complexities of public transport in relation to tourism activities may favour public transport, since (9):

• Public transport can be considered from the outset, on an equal footing with private transport for providing an end-to-end package.
• The profile of public transport is raised in relation to tourism.

Otherwise, travellers may not see public transport as relevant to their leisure activities, and advances in the public transport infrastructure may not be appreciated.

2.4 User-Oriented Information and the Informed Traveller Programme

The Informed Traveller Programme, which is part of the U.K. Office of Science and Technology *Foresight* initiative (9), mainly concentrates on the provision of multimodal, real-time systems, but also sees tourism information as a component which integrates later; the TransiT project thus complements and depends upon advances made during this programme.

3 OVERVIEW OF THE TRANSIT PROJECT

The TransiT project aims to develop a prototype user-oriented information system, TransiT, and to investigate whether this can usefully employ an advanced knowledge representation scheme to encode its information model and support flexible querying and browsing facilities. This section identifies the information to be represented in the model, and describes how this will be collected and represented.

3.1 What Exactly is User-Oriented Information?

Previous sections have described in a fairly abstract way the requirements of a user-oriented information system: it should provide information which is structured to match the traveller's requirements and perceptions, and integrate with and support their journey planning activities. In practical terms, this requires the system to have knowledge both of the descriptive details of destinations and transport, namely their significant features and attributes, and also *functional* details, namely reasons for choosing them. Previous sections have also described how destination and service selection are only one aspect of travel planning, and the system must help to manage the complexities of the complete travel planning task; it thus also requires a model of this task.

3.2 Collecting Information

Building a user-oriented model requires large quantities of information to be collected; this information cannot come solely from the developer's intuitions, otherwise the system will be meeting assumed traveller needs. A range of sources are being employed to obtain the relevant information:

a) British Tourism Association/English Tourist Board research.
b) Discussions with railway operators and passenger transport executives.
c) Tourism research literature.
d) Ethnographic studies in railway travel centres and visitor information centres.

These sources are being used to provide a range of information: (a) to (c) for information on travellers' requirements and perceptions of destinations, and reasons for travel; (d) for information on reasons for travel and the structure of the travel planning task. Since the present project aims to develop only a prototype system, the scope of this research is restricted to determining the kinds of information involved, rather than conducting an exhaustive survey.

3.3 Ethnographic Studies

Ethnography in Software Engineering

It is increasingly recognised that 'social' factors of system design cannot be ignored by software engineering. Traditional methods of requirements capture, such as structured interviews and questionnaires, may miss or obscure these aspects (12), or may impose the researchers preconceptions onto their results (13).

Ethnographic study involves carrying out observation fieldwork in the environment where the proposed system is to operate; by observing events and interactions in their natural setting and recording them in detail, it aims to build up a detailed description of practice, whilst minimising the risk of recording preconceived results. The use of ethnography to inform systems design is becoming widespread, particularly in the computer-supported cooperative work (CSCW) arena (12, 13, 14). CSCW systems aim to integrate with and support work in an existing, social context, and it is increasingly recognised that 'not only *ought* systems to be resonant with the human world of work ... but they are more *effective* if they are designed to be so in the first place' (13). Whilst the TransiT project is not seeking to develop a multi-user

CSCW system, it is seeking to develop a system which integrates with and supports the *existing* task of planning leisure travel.

Whilst it can be argued that ethnography may produce too rich a description from which it is hard to present clear findings to guide systems design (12, 13), projects which have used ethnography to inspire their designs cite a range of benefits:

- By studying a natural scenario, ethnography can highlight the truly required system functionality, rather than that perceived by the system designers (13); it can also scope the functionality, by highlighting areas of the task which are not amenable to computerisation, and can also highlight differences between individuals' needs and preferences, avoiding the 'if I like it, then others also will' attitude (14).
- Ethnography describes an integral task, rather than prematurely decomposing it into a set of sub-tasks (e.g. task analysis); it thus retains the coherence of the task, and may also record subtleties of its performance e.g. the use of tacit knowledge by the actors involved (13, 14).

Planning leisure travel is a social task, often conducted by groups of people, and often with the support of a travel or tourism agent; it is also a highly personal task, since there are a vast range of reasons for, and circumstances of leisure travel. It could be argued that the requirements resulting from an ethnographic study will be too broad ever to be captured by a computer system; however it is felt that recording such diversity may guard against the assumption that there is *a* procedure for planning leisure travel, and a resulting system design which is too rigid and simplistic.

Ethnography in the TransiT Project

Fieldwork will be performed at three venues, the travel centre at Manchester Piccadilly railway station, Manchester Visitor Centre and Chester Tourist Information Centre. Interactions between customers and staff will be observed to provide an indication of the structure of travel and tourism planning dialogues.

A preliminary study has been carried out at Manchester Piccadilly, in the two weeks around Christmas 1995; at this time the travel centre was busy, and even this short study highlighted a number of important aspects of the journey planning task:

- Some customers had exact journey requirements, specifying a particular train and ticket type; others were less exact, for example knowing only that they wanted a return ticket to London. Moreover, the order in which requirements were presented was non-deterministic: it was not simply destination followed by date, time and fare.
- Customers often did not know what fare options are available. Customers can purchase normal 'Saver' return fares (off-peak tickets valid for one month), by specifying their departure date; if they also specify their return date and exact travel times, they become eligible for a lower priced APEX return fare. However in the study, most customers only supplied the additional information, and thus purchased a cheaper ticket, when prompted by the booking staff.
- Customers sometimes did not know the terminology used to describe tickets. One customer referred to a Saver return as a 'monthly return'; some customers did not know the significance of the tickets they were offered, until an explanation was given by the booking staff.
- Booking staff supported the journey planning process. Once a journey had been booked, the staff confirmed all the details to the customer. When the customer's requirements could not be met, staff explained this to them, searched for the closest alternatives, *and* explained what they were doing.
- The process had a social nature. Customers often told the booking staff why they were travelling, even though this provided no additional details of requirements. When the requirements of groups could not be met, they often held an independent discussion to decide

suitable options.

This study highlights that even the process of arranging a train journey is highly involved and varied; thus, a system which adopts a rigid task structure and cannot present information with respect to user requirements is unlikely to be adequate.

The Ethnography Methodology

The TransiT project will adopt a 'quick and dirty' (12) ethnographic approach, whereby preliminary, small-scale ethnographic surveys will be used to guide initial system development; this could act as the first iteration of a concurrent ethnographic cycle (12) if required, namely fieldwork $\Rightarrow$ debriefing $\Rightarrow$ prototype development $\Rightarrow$ fieldwork... .

3.4 Representing Information

The TransiT project will use GRAIL[1], a terminological logic based knowledge representation scheme to maintain its user-oriented information model; GRAIL was developed during the GALEN[2] project by the Medical Informatics Group at the University of Manchester. Terminological logics are increasingly being applied in the information retrieval arena. The TAMBIS project at the University of Manchester is applying GRAIL to the task of accessing biological information sources; its use of GRAIL to support indexing and querying facilities is analogous to that in the TransiT project, however it will go further in using GRAIL to mediate between, and allow a generic interface to heterogeneous databases; other terminological logic based schemes are also being applied to the task of database integration and document retrieval (15, 16).

One aim of the TransiT project is to assess the suitability of GRAIL for supporting a user-oriented information system. To date, GRAIL has been applied exclusively and successfully to the management of medical information; although medicine is a highly complex domain, it has been found that the classification facilities of GRAIL can be successfully applied to it. Specifically, GRAIL has been used to mediate between a variety of medical terminology coding schemes used throughout Europe (17). This function has direct applicability to the domains of travel and tourism, since a range of independent information systems already exist, based on varied knowledge representation schemes; an integrated system must be able to mediate between these heterogeneous systems, since it is unrealistic to assume that they will all adopt a 'standard' representation.

An Introduction to GRAIL

The GRAIL language allows primitive concepts and the relations between them to be defined; the GRAIL classifier maintains a subsumption hierarchy of concepts, supporting multiple inheritance. Complex concepts can be created from simpler concepts, using relations:

Concept which hasRelation **Value**

where Value is another, possibly complex concept; a relation-value pair is a *criterion*. Rather than manually define a subsumption hierarchy, it is more usual to allow the classifier to infer this: it can dynamically classify a new concept (primitive or complex) with respect to an existing concept model, updating the model to ensure that all subsumption relations pertaining to the new concept are in place.

GRAIL was originally developed to support a user-oriented clinical workstation, PEN&PAD (18); the functionality of this system exploits the facilities of the GRAIL classifier,

---

1 GALEN Representation And Integration Language
2 Generalised Architecture for Languages, Encyclopaedias and Nomenclatures; a European Union funded programme.

namely:

a) A concept can be built compositionally, and may potentially cut across multiple subsumption hierarchies i.e. multi-axis classification.
b) For any given concept, the classifier can indicate which criteria can be applied to it. This is used to guide the construction of predictive data-entry forms for describing conditions. Given the criteria which can be sensibly applied to a concept, a form can be dynamically constructed to allow these *and only these* criteria to be entered; it is thus impossible to specify a meaningless condition such as 'a fractured eye'.
c) Given a concept, the classifier can indicate which concepts it subsumes (is a more general form of); conversely, the classifier can indicate the subsuming parents of a concept.

Applying GRAIL to the TransiT Project

The TransiT project aims to exploit the facilities of GRAIL in a similar manner to PEN&PAD, with the items (a) to (c) above having the following practical significance for the project:

a) Classification axes will relate to the attributes, perceptions, or functions of a destination or service; a concept representing the traveller's requirements can thus be constructed in terms of these. For example, 'a destination which is relaxing, with sandy beaches and good restaurants'.
b) Once such a requirement concept has been constructed, the classifier can be used to determine what other requirements could sensibly be added, and the system will offer only these.
c) The traveller can pose requirements at various levels of detail, and members of a set of requirements need not all be at the same level; a vague (high-level concept) requirement will identify all more detailed (lower-level concept) requirements which it subsumes. For example, 'Sports' might subsume 'Sailing, Swimming, Horse Riding, Golf ...', but 'Water Sports' would subsume only 'Sailing, Swimming ...'.

In this way, the traveller is able to present their requirements at any level of detail, and in any order; requirements will be specified using mouse-driven forms, constructed from the model as in PEN&PAD, thus hiding the complexities of the GRAIL mechanism. The GRAIL concept model will also serve two additional functions:

- It will support the browsing of information. Since the model details the relations between concepts, it can be used to indicate hyper-links between information items.
- It will act as an index of available information. Each item in the system's information base will be associated with a particular GRAIL concept; once a requirement concept has been constructed, the system will identify from the model which concepts, and hence which items are suitable.

4   USER-CENTRED INFORMATION MEETS INFORMATION RETRIEVAL

This section discusses the knowledge-based approach to information provision adopted by the TransiT project, and relates it to other research into information retrieval and intelligent information systems.

4.1 The Price of Knowledge

Developing a user-oriented information model requires a large quantity of information to be collected and analysed. For a full-scale model this would be a costly task, however the cost of putting knowledge into the system should provide benefits when retrieving information. A rich, flexible representation scheme is able to capture a wider range of knowledge, and it should be possible to construct more elaborate models to reflect traveller needs, for example meta-models which provide a user-oriented perspective onto an underlying conceptual model.

However, the major benefit of a knowledge-rich system is that it has more knowledge of the information that it stores, which provides a sounder base on which to carry out additional inference; this becomes an important factor when the system is considered in relation to other research into information retrieval and intelligent information systems.

4.2 User-Centred Information meets Information Retrieval

On their own, user-centred querying and browsing are not sufficient to ensure a usable system. The system must also maintain an appropriate dialogue with the user, present its output in a coherent manner, and possibly adapt its behaviour to accommodate the user's individual requirements; this section presents an overview of other research into these issues.

Managing the Information-Seeking Dialogue

The travel centre ethnographic study identified a sophisticated, cooperative dialogue structure involved in planning travel, which must be adequately addressed by a travel planning system. By building on previous linguistic research into cooperative dialogues, various contemporary research projects are investigating the support of such dialogues: (19, 20) have developed a system to provide intuitive information retrieval facilities, with cooperative responses to user actions; (21, 22) are developing a system to act as an interactive, collaborative flight planning agent.

In addition, the structure of human information seeking activity has been extensively documented in (23), which presents an interesting view of 'intelligent support systems'; since fully modelling an intelligent agent is beyond the scope of present technology, an information system should seek to 'stimulate the user rather than simulate the user', for example by the use of interface metaphor (24).

Presenting Information

Research into document retrieval has investigated ranking output for relevance in relation to the user's query (26), and research into multimedia presentation has investigated 'intelligent' media selection and page layout (26). The user interface is particularly important for travel information systems (9); user-centred design, as carried out during the development of the PEN&PAD system (27), can produce sophisticated interfaces which are appropriate to user needs.

Adapting the Interaction

Much work has been done in computational linguistics (28, 29) into 'user modelling', namely maintaining a model of the user's knowledge and needs, and adapting the system's behaviour to account for this; here the system benefits from a rich information model, since it has more knowledge from which to draw inferences. Although current technology cannot support models broad enough to cover all aspects of individual travel and tourism needs, it may be possible to successfully model certain aspects, as long as the user can understand the reasoning behind any inferences drawn (30).

These functions are information intensive, depending upon knowledge of the task or of the user, and complement those provided by the TransiT project, whose knowledge-based strategy is better able to support integration with them.

5 CONCLUSIONS

This paper has identified problems with existing travel and tourism information sources, and proposed a solution using ethnographic studies and an advanced knowledge representation system. Realisation of this project presents significant challenges. However, these are challenges which must be addressed if significant advances in traveller information systems are to be made.

6 REFERENCES

(1) SHELDON, P.J. Destination Information Systems. Annals of Tourism Research, 1993, 20, 633-649.

(2) BLACKLEDGE, D., PICKUP, L. Public Transport Passenger Information Systems - The Potential of Advanced Transport Telematics, Proceedings of IVHS America Third Annual Meeting, Washington, 1993, 295-301.

(3) FOSTER, P. The Retail Travel Shop, Horwath Book of Tourism, 1990, (The MacMillan Press, London) 145-152.

(4) HOFFMAN, J.D. Emerging Technologies and Their Impact on Travel Distribution, Journal of Vacation Marketing, 1994, 1, 95-103.

(5) POLLOCK, A. The Impact of Information Technology on Destination Marketing, Travel and Tourism Analyst, June 1995, 66-83.

(6) NIELSEN, J. The Art of Navigating Through Hypertext. Communications of the ACM, 1990, 33, 296-310.

(7) THURING, M., HANNEMANN, J., HAKKE, J.M. Hypermedia and Cognition: Designing for Comprehension. Communications of the ACM, 1995, 38, 57-66.

(8) WITT, C.A., WRIGHT, P.L. Tourist Motivation: Life after Maslow, Choice and Demand in Tourism, 1992, (Mansell Publishing, London) 33-56.

(9) OFFICE OF SCIENCE AND TECHNOLOGY. Technology Foresight: Progress Through Partnership, Volume 5: Transport, 1995, (HMSO, London).

(10) SOMMERVILLE, F., TATE, A. The Testing of a Promise in Traveller Information, First World Congress on Applications of Transport Telematics and Intelligent Vehicle-Highway Systems, Paris, 1995, (Artech House, London) 2249-2256.

(11) ALLOUCHE, J-F. Pilot Experiment in Multimodal Information in the Ile-de-France Region. First World Congress on Applications of Transport Telematics and Intelligent Vehicle-Highway Systems, Paris, 1995, (Artech House, London) 2337-2345.

(12) HUGHES, J., KING, V., RODDEN, T., ANDERSEN, H. Moving Out from the Control Room: Ethnography in System Design, CSCW '94: Transcending Boundaries, Chapel Hill NC, 1994, (ACM Press) 429-439.

(13) HUGHES, J., RANDALL, D., SHAPIRO, D. From Ethnographic Record to System Design. Some Experience from the Field, Computer Supported Cooperative Work, 1993, 1, 123-141.

(14) ROGERS, Y. Exploring Obstacles: Integrating CSCW in Evolving Organisations, CSCW '94: Transcending Boundaries, Chapel Hill NC, 1994, (ACM Press) 67-77.

(15) CARENINI, G., PIANESI, F., PONZI, M., STOCK, O. Natural Language Generation and Hypertext Access, Applied Artificial Intelligence, 1993, 7, 135-164.

(16) AGOSTI, M., CRESTANI, F. A Methodology for the Automatic Construction of a Hypertext for Information Retrieval, ACM/SIGAPP Symposium on Applied Computing, Indianapolis, 1993, (ACM Press) 745-753.

(17) RECTOR, A.L., ZANSTRA, P., SOLOMON, D. GALEN: Terminology Services for Clinical Information Systems, Health in the New Communications Age, 1995, (IOS Press, Oxford) 90-100.

(18) KIRBY, J. A System for Engineering Detailed Clinical Data, To be presented at HealthCare Computing, Harrogate UK, 1996, (BJHC, Weybridge UK).

(19) BELKIN, N.J., COOL, C., STEIN, A., THIEL, U. Cases, Scripts and Information-Seeking Strategies: On the Design of Interactive Information Retrieval Systems, 1994, (Technical Report No. 875, GMD, Sankt Augustin, Germany).

(20) STEIN, A., MAIER, E. Structuring Collaborative Information-Seeking Dialogues, Knowledge-Based Systems, 1994, 8, 82-93.

(21) RICH, C. Window Sharing with Collaborative Interface Agents, SIGCHI Bulletin, January 1996.

(22) SIDNER, C.L. Building a Collaborative Interface Agent. To appear in: Fourth International Colloquium on Cognitive Science, (Kluwer).

(23) INGWERSEN, P. Information Retrieval Interaction, 1992 (Taylor Graham, London).

(24) PEJTERSEN, A.M. A Library System for Information Retrieval Based on a Cognitive Task Analysis and Supported by an Icon-Based Interface, SIGIR '89, Cambridge MA, 1989, (ACM Press) 40-47.

(25) HARMAN, D. Ranking Algorithms, Information Retrieval, Data Structures and Algorithms, 1992, (Prentice-Hall, Englewood Cliffs NJ) 362-392.

(26) KAMPS, T., REICHENBERGER, K. Automatic Layout as an Organisation Process, 1994, (Technical Report No. 825, GMD, Sankt Augustin Germany).

(27) NOWLAN, W.A. Clinical Workstations: Identifying Clinical Requirements and Understanding Clinical Information, International Journal of Bio-Medical Computing, 1994, 34, 85-94.

(28) CAWSEY, A. Explanation and Interaction, 1993, (MIT Press, London).

(29) CARENINI, G., MITTAL, V. O., MOORE, J. D. Generating Patient-Specific Interactive Natural-Language Explanations, Eighteenth Annual Symposium on Computer Applications in Medical Care, Washington, 1995, (Hanley and Belfus Inc., Philadelphia) 5-9.

(30) KARLGREN, J., HOOK, K., LANTZ, A., PALME, J., PARGMAN, D. The Glass Box User Model for Filtering, 1994, (Technical Report R94:14, Departments of Computer and Systems Sciences, Computational Linguistics, and Psychology, Stockholm University).

Suspension Developments

C514/055/96

Active tilting control of series E991 e.m.u. experimental train (development of third-generation active tilting control)

K SASAKI, H KATO, T KONOKAWA, Y SATO, and Y KAKEHI
East Japan Railway Company, Tokyo, Japan

Synopsis
The Series E991 train, nicknamed TRY-Z, of East Japan Railway is
manufactured for research, development, and field tests of component
technologies. From last September we started the field tests of the
newly developed three type tilting mechanisms.
      The curve negotiating velocity is aimed to be 120km/h at 400m
radius curvature, and then cant deficiency is 200/1067 rad. For
improving running safety the tilting center height is lowered and the
active control method is also installed.
      The first type of bogie has the hydro-pneumatic/coil suspension
and hydraulic tilting mechanism, the second one has pneumatic
suspension/tilting mechanism, and the last one has link type tilting
mechanism.

Notation
h_b: Gravity center height of the car body
h_t: Gravity center height of the bogie
μ : Ratio of the bogie mass (including unsprung mass) $2m_t$ to
      car body mass m_b
αu: Excessive centrifugal acceleration in curve
g: Gravitational acceleration $9.8m/s^2$
G: Track gauge 1067mm
y_b: Lateral displacement of car body's gravity center in
      consideration of the tilting mechanism
h_{tc}: Tilting center height
ϕt: Tilting angle [maximum 7 degrees in TRY-Z]
y_s: Lateral displacement of car body's gravity center due to
      the suspension of bogie
Q: Lateral force
P: Wheel load
P_o: Static wheel load
ΔP: change of wheel load in curve

1 PURPOSES OF TRAIN DEVELOPMENT

Manufacture of the Series E991 train [TRY-Z] has been continued as
a train for technological development in pursuit of the 'realization

of the 21st century's ideal railway system', a futuristic subject in conventional lines. This train is an AC/DC electric multiple unit (e.m.u.) train having a 3-car composition of Mc1, T and Mc2, and has been manufactured for developing the component technologies unallowable for a train to be used in commercial service, and for conducting a leading-edge development while adopting the latest technological trend and advance. The basic concepts of development are listed below (1).

(a) COST
Aiming at reduction of total cost for providing stable service in our low-growth society.

(b) INFORMATION
Aiming at materialization of a safe and comfortable railway system by positive introduction of the information technologies under rapid progress.

(c) ENVIRONMENT
Aiming at establishment of environment-friendly railway technologies through energy saving, noise reduction and creation of comfortable railcar space.

Enhancement of curve running performance is one of the principal development subjects of the Series E991 [TRY-Z]. Outlined below is the 'third-generation tilting mechanism' under development as a successor to the 'natural pendulum type vehicle, 'actively controlled natural pendulum vehicle' and other.

2 BOGIE AND TILTING CONTROL

The Mc1 car uses a bolster-less type bogie having the hydro-pneumatic/coil suspension mechanism developed as a next-generation suspension substitutive for pneumatic suspension. On the other hand, Mc2 and T cars use a bolster-less type bogie having the conventional pneumatic suspension mechanism (2).

Each car is provided with hydraulic active or semi-active oscillation control (3)(4)(5). In particular, the bogie of T car is an inside frame type intended to reduce the unsprung mass and its wheelbase has been shortened to expect a fall in lateral force.

The cross section of car body has the dimensions determined for the lower gravity center and tilt angle, 7 degrees in consideration of tilting control. Roof height is 3300 mm above the rail face in the coach room area excluding the operator's deck. While the floor is as low as 1000 mm, the ceiling of coach room has a height of 2100

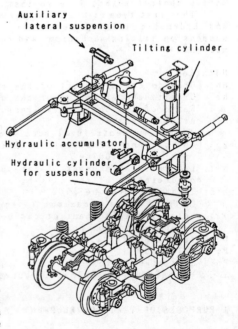

Auxiliary lateral suspension

Tilting cylinder

Hydraulic accumulator

Hydraulic cylinder for suspension

Fig 1 Mc1's DT955 Bogie

mm at the central area so as not to degrade amenity.

2.1 Hydraulic suspension type bogie [type DT955]

Figure 1 shows the Mc1 car's bolster-less type bogie with the hydro-pneumatic/coil suspension mechanism. For this bogie, the hydraulic cylinder right above the coil spring and the hydraulic accumulator provided with the tilting bolster are piped with a restrictor placed in between to configure a 'hydro-pneumatic' suspension. Configuration has been selected so as to obtain an appropriate spring constant in series with the coil spring. Also, the coil spring allows the bogie to turn and the spring constant is compensated for shortage with the auxiliary lateral suspension.

This mechanism is advantageous in that the area occupied by the suspension equipment on the horizontal plane can be reduced substantially differing from the pneumatic spring. This makes possible the bogie configuration in which the cylinder for tilting the car body is accommodated within the height of the floor.

For tilting, either left or right tilting cylinder extends. Therefore, the tilting center is located on the tilting bolster on the side where the tilting cylinder is not extended. In addition, the tilting bolster is connected to the car body by means of the bolster anchor facing the car end, so that the bogie turns to the inside of a curve at the time of tilting.

In the roller rig test carried out before completion of the train, an excellent running stability was confirmed up to 360 km/h though a temperature rise in the running gear had a restricting effect.

2.2 Inside frame type bogie [type TR913]

Figure 2 shows the T car's bolster-less bogie unique in the inside frame. The inside frame type refers to the configuration in which the primary suspension equipment is provided at the inside of wheel. Though this type has a drawback that the roll stiffness of axle spring is reduced at a high curve negotiating velocity, its advantage can be thought to be a weight reduction of the wheelset under the spring.

In this type of bogie configuration, the direction of axle bending due to the vertical load applied to the wheelset is reversed. When a lateral force is also applied during running under the above condition, the axle is bent further in usual cases. However, axle bending is moderated with the inside frame type. Therefore, it can be thought that the axle may be designed more finely.

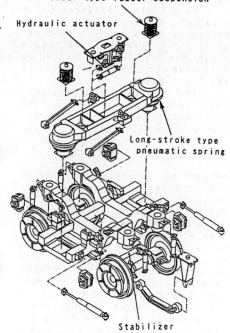

Shear type rubber suspension

Hydraulic actuator

Long-stroke type pneumatic spring

Stabilizer

Fig 2 T's TR913 Bogie

Each bogie of this type has a weight of only 3.5 t including the tilting equipment thanks to use of the inside frame mechanism, shortening of the wheelbase for reducing the lateral force, etc. And the primary suspension equipment has an axle beam type configuration for functioning also as a stabilizer to compensate for a shortage in the roll stiffness.

Because the long-stroke type pneumatic spring under the tilting bolster is expanded/compressed for tilting, the tilting center is located at the height of pneumatic spring and coincides with the center of car body. Bogie turning is dependent on shearing deformation of the laminated rubber provided on the tilting bolster.

With this bogie, an excellent running stability was also confirmed up to 360 km/h in a roller rig test. And for improving on vibration and riding comfort, active oscillation control is effected with the hydraulic actuator installed in parallel with the anti-swaying damper. In a running test, nearly the same damping effect as calculated has been confirmed (3).

2.3 Hydraulic link type pendulum bogie [type DT956]

Figure 3 shows the Mc2 car's hydraulic link pendulum bogie. This bogie is structured to suppress the change of wheel load on the inner rail side during curve negotiation by locating the tilting center 675 mm lower than the conventional pneumatically driven roller bearing type pendulum bogie (6).

The bogie is under active oscillation control with the lateral and vertical hydraulic actuators installed in parallel with the pneumatic spring.

As a bogie configuration, a greater importance is placed on practical use than technological advance. However, the bogie weight has been further reduced by reviewing the structural design of the bogie frame.

The result of a running test has verified that the active oscillation control has nearly the same damping effect as calculated.

2.4 Tilting control system

Attempt to run through a curve at a high speed has begun with the 'natural pendulum type train' in Japan, but the initial Ser. 381 e.m.u. trainset was not evaluated highly because of a delay in the movement of pendulum, its swing back, etc. As a solution to those problems, the 'actively controlled natural pendulum system' has been put into practical use, by which the pendulum is moved through prior detection of the curve entry and exit positions.

The problems to be resolved on the 'actively controlled natural

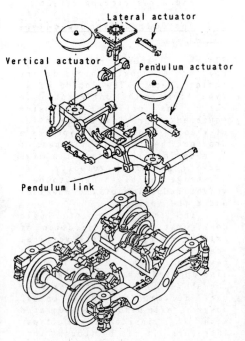

Fig 3 Mc2's DT956 Bogie

162

© IMechE 1996 C514/055

pendulum system' include the change of wheel load on the inner rail
side in a curve because the rotation center of pendulum is far higher
than the gravity center.

In the Series E991 [TRY-Z] project, the development of a new
tilting control system is under way targeting the following.
(a) Minimization of such a change in wheel load as above
(b) Taking a measure for damping the car body rolling oscillation
rather than tilting control only for reducing the stationary lateral
acceleration
(c) Reduction of lateral force during curve negotiation
Three types of tilting systems are shown in Fig. 4.

2.5 Tilting center height and change of wheel load

For the conventional pendulum
railcar, it has been considered
proper from the viewpoint of
riding comfort that the tilting
center height is located at the
head level of passengers sitting
on seats. Further, it has been
necessary to locate the tilting
center height above the gravity
center of car body due to a
precondition of the natural
pendulum system.

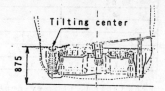

Mcl's DT955 bogie

However, excessive
centrifugal acceleration becomes
1.9 times for increasing the
running velocity of the present
pendulum railcar from 100 km/h to
120 km/h in a curve having radius
400 m, for example. Therefore,
further consideration is required
for factors related to running
safety such as change of wheel
load.

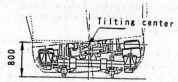

T's TR913 bogie

The quasi-static change rate
of wheel load can be expressed as
follows.

Mc2's DT956 bogie

Fig 4 Tilting Systems

$$\Delta P/Po = 2\{(hb + \mu\ ht)\ \alpha\ u/g + yb\ \}/(1+\mu)G \qquad (1)$$

where,

$$yb = (htc - hb)\ \phi\ t + ys \qquad (2)$$

In these formulas, it is proved that the effect of the gravity
center height of the car body hb and the lateral displacement of car
body's gravity center in consideration of the tilting mechanism yb
is significant for the quasi-static change of wheel load. Although
yb includes the lateral displacement due to the suspension of bogie,
the difference in the tilting center height htc becomes significant
for that at running in a curve.

Table 1 lists the result of calculation assuming that only the
tilting center height differs among railcars while other factors such
as mass, gravity center height, suspension characteristics are

completely the same.

Table 1 Tilting center height and quasi-static
change rate of wheel load (calculated)

Velocity	120			km/h
Curve	Radius:400 Cant:0.105			m
α u	0.185			g
Axle load	7.25			t
μ	0.45			
h b	1.300			m
h t	0.400			m
φ t	7			deg.
y s	0.048			m
h t c	2.275	1.600	0.800	m
Δ P/Po	0.57	0.46	0.34	
Equi.Velocity	90	105	120	km/h

From Table 1, it is evident that the tilting center height   exerts
influence over the suppression of wheel load change.   It can be
understood that there is a significant difference of 23% in the
quasi-static change rate of wheel load between the conventional
railcar higher in the tilting center height and the one lower in that
height.
When calculating the velocity condition where the change rate
of wheel load becomes the same, the result is obtainable as shown in
the row of equivalent velocity in Table 1.   Thus, it can be said that
a difference in the tilting center height   brings about a velocity
difference of 30 km/h at maximum.
Actually, the dynamic component of wheel load change is added
due to the car body rolling oscillation at running in a transient curve
and the local track irregularity.   Therefore, in case a condition
undesirable for running safety can be presumed, the gravity center
height of the car body and the tilting center height need to be lowered.

Table 2 Change rate of wheel load
(TRY-Z, quasi-static, calculated)

	Mc1	T	Mc2	
μ	0.45	0.32	0.3	
- - hb	1.35	1.37	1.36	m
htc	0.875	0.8	1.6	m
Δ P/Po	0.35	0.36	0.5	

From such a viewpoint as above, 3 types of tilting mechanism
different in the tilting center height have been developed for TRY-Z,
and their advantages and disadvantages have been compared.
Table 2 lists the result of predicting the quasi-static change
rate of wheel load in TRY-Z when running conditions are the same as
in Table 1.   The height of Mc1 car body's gravity center in Table 2
is a value given in consideration of an increase in the gravity center
height of the car body at running in a curve because the tilting center
is not located at that of the car body.
In Mc1 [DT955] and T [TR913] systems, the tilting center is lower
than the gravity center of car body as shown in the above table.   This
means that tilting will not occur naturally.   Therefore, it is the
matter of course that the tilting control system must have a high

reliability and fail-safe provision. Yet, because the center of tilting is lower than the floor surface, advantages such as a reduction of the change of wheel load on the inner rail side in a curve can be expected.

In contrast, the mechanism of the Mc2 [DT956] system is an extension of the natural pendulum system. So the reliability required for the tilting control equipment may be comparatively low. However, for instance, the working force necessary for tilting is larger than the conventional pendulum railcar's because of incorporation of the link mechanism.

3 RUNNING TEST RESULTS

To verify the performance of tilting systems of TRY-Z, the running tests were held on the Chuo main line, between from the Saruhashi station to the Sasago one until last December. In this section of Chuo main line, there are 25 left or right curves, which have a radius of 400 meters, and there are many upward inclines of 25/1000 continuously. The distance between these stations is about 14km.

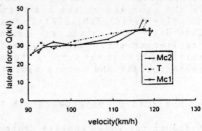
Fig 5 Lateral Force Q varying with Running Velocity

The wheel load P and the lateral force Q, which are concerned to the running safety, are measured by special equipment at the leading axle of each car. Figure 5 shows their maximum values of lateral force Q, varying with the running velocity, in the No.97 curve, typically has a radius of 400 meters. Also figure 6 shows the maximum values of the derailment coefficient Q/P in the same manner. And Figure 7 shows the relation between the maximum values of wheel load's change rate $\Delta P/Po$, the average one and the running velocity in the same curve.

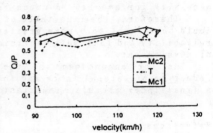
Fig 6 Derailment Coefficient Q/P varying with Running Velocity

The lateral force Q and the derailment coefficient Q/P, measured at each Mc1, T and Mc2 car during the running tests, were within the tolerable value. Q should be within 5 or 6 ton, which is determined by the fatigue strength of track fastener, and Q/P should be within 0.95.

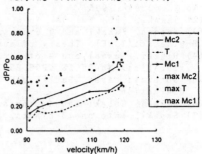
Fig 7 Change Rate of Wheel Load varying with Running Velocity

Only the wheel load's change rate $\Delta P/Po$ of Mc1 car [DT955] could be within the tolerable value in all curves of the test section. $\Delta P/Po$ should be within dynamically 0.8 and quasi-statically 0.6. T and Mc2 car exceeded that dynamic limit in some curves, as a result of that their tilting movement were not successful.

The quasi-static values of $\Delta P/Po$, shown in Figure 7, were proved to coincide with the estimated ones, calculated in table 2. And it would be necessary that the maximum value of $\Delta P/Po$ should be taken account of about 40% increase over the quasi-static one.

4 CONCLUSION

The authors describe that the experimental train Series E991 [TRY-Z] of East Japan Railway has 3 types of tilting mechanisms differing in the tilting center height. It is also described that the lower tilting center height is an important factor for the running safety in a curve, in the same way as the lower gravity center of car body.
     The Series E991 [TRY-Z] had been under the running test on the Chuo main line until last December for confirmation of the curve negotiating performance. The test results show that the running safety of Mc1 car was superior to the others, which has the lower tilting center height and secure hydraulic tilting mechanism.

5 SUBJECTS OF DEVELOPMENT IN FUTURE

Although the lower gravity center of car body is an important element for enhancing the curve running performance of a train with the tilting mechanism, the gravity center lowering range is restricted due to the necessity for amenity in coach rooms and weight reduction.
     Therefore, the importance of lowering the tilting center height could be thought to become greater in future. In this case, higher reliability and fail-safe provision of equipment become the subjects of development.
     As for enhancement of the curve running performance, the reduction of lateral force and the improvement on riding comfort as a final index are also important subjects.

References

(1) Hirose, Japan Railway Engineers Asociation (JREA) Vol.38, No.2, p22998,1995
(2) Sasaki, JREA Vol.38, No.5, p23153,1995
(3) Negoro, Sasaki and 3 others, Japan Society of Mechanical Engineers (JSME), Theory of mechanism, No.95-1, 1995, p157
(4) Higaki, Sasaki and 4 others, JSME, Control of motion and vibration, No.95-28,1995,P317
(5) Konokawa, Arai, Sasaki and 3 others, JSME, lecture at 73rd Convention
(6) Higaki, Sasaki and 4 others, JSME, lecture at 73rd Convention
(7) Sasaki, Kato and 3 others, J-Rail'95, p185

C514/017/96

The economic case for tilting trains

J CARTMELL MA, FIMechE, MIM, FHKIE
Adtranz Total Rail Systems Limited, Derby, UK

SYNOPSIS

Tilting Trains go round existing curves faster, leading to better journey times, better business. Everyone has an economic value of their time - as well as enjoying absolute benefits from the quicker journey such as not having to travel the day before. Rail can provide better journey times in competing with road and air - but usually only at large cost and project lead time. The saga of the Channel Tunnel Rail Link is well known.

By investing in tilting trains, together with some relatively modest improvements to track and alignment quality, limiting speeds round curves can, on twisty routes, significantly improve journey times. The technology is available: does the economic case stack up?

1.   WHY TILT?

Transport competition is all about journey time, cost, comfort, reliability and personal safety. This conference, and this paper, focuses on the first of these factors, journey time. There is a strong relationship with the second, cost, and there are tangible benefits in comfort through the use of tilting trains.

As every schoolboy cyclist knows, it is the speed up the hill that rules journey time, not the speed down the other side. This same mathematics is used to good effect in the design of frequently stopping urban transport systems where top speed is seen to influence journey time much less than acceleration and braking rates and dwell times. To reduce the top speed by coasting shows significant energy savings for small journey time penalty. To reduce dwell times gives worthwhile improvements in commercial line speed.

So it is that if we can reduce or eliminate speed restrictions due to curves on main line or cross country routes, significant journey time improvements can be made. One approach to this problem, and that adopted by the French in their programme of construction of "Lignes a Grande Vitesse" is to build new, straight alignments. While there may in fact be some relaxation of gradient standards (depending principally on whether freight trains are also to use the new route), horizontal radii are severely restricted and civil engineering costs are correspondingly large - in addition, that is, to land acquisition costs. So any new line has to be financially and politically extremely well supported or it will never happen. An alternative is to tilt.

Trains on existing tracks are limited in speed round curves by passenger comfort, not safety. On most standard gauge (1435 mm) railways cant is limited to 150 mm lest a train stops on a curve or has to negotiate it very slowly. Passenger comfort is then a problem in the other direction and flange climbing a danger. There are further restrictions in high cant application where slow freight trains use the line.

The acceleration towards the centre of a curve, as our schoolboy cyclist knows, is V^2/R (figure 1) and by tilting the train we allow gravity to provide some, or most, of the force needed to produce that acceleration towards the centre. The passenger now experiences a more nearly balanced turn as in an aeroplane or on a bicycle.

Figure 1: Rounding a curve

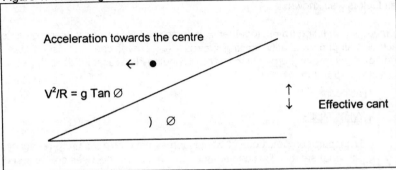

The train bogies and the track both experience an increase in the curving force to the tune of the difference , with and without tilt, in the squares of the speeds. The limit of train operating speed is now safety and the economics of long term track and bogie maintenance but a significant increase in curving speeds can nonetheless be obtained.

2.   EXTRA SPEED AVAILABLE

So much for the mechanism. Let us now look at the sort of journey time reductions that can be achieved.

The West Coast Main Line is a route that has been well studied for improvement., as have cross-Pennine routes. In Australia, the Sydney to Canberra line has been shown to have potential for time savings through the use of tilt. In each case, the cost of curve straightening has been found to be prohibitive and tilting trains have been shown to offer much more cost effective improvements.

A maximum cant of 150 mm is already applied on main line curves and passenger comfort limits cant deficiency at the curving speed to a further 110 mm. By comparison, a tilt of only 8 degrees from the normal to the plane of the track gives a further 200 mm of effective cant for balanced curving. Some cant deficiency is beneficial on top of that to maintain positive passenger orientation and comfort, but safety and maintenance considerations now begin to predominate. The structure gauge on most UK routes means, however, that tilt has to be restricted to 6° if excessive narrowing at the top of the bodyshell is to be avoided. 6° of tilt gives an extra 150mm of effective cant. The total effective cant, for 70 mm cant deficiency as perceived by the passenger, is then 370 mm as illustrated in Table 1. Table 2 shows the extra curving speed available at different curve radii for each of the values of total effective cant show in Table 1. At 370 mm total effective cant, or 220 mm cant deficiency as seen by the track, lateral forces are double those when operating conventionally at 110 mm of cant deficiency. This issue is discussed later.

	Conventional	Tilting 6°	Tilting 8°
Track cant	150	150	150
Train tilt	-	150	200
Perceived cant deficiency	110	70	70
TOTAL EFFECTIVE CANT	260	370	420

Table 1: Cant and cant deficiency (mm)

Total effective cant (mm)	260	370	420
Curve radius m	Train km/h	Speed km/h	(max) km/h
400	96	114	122
800	136	162	173
1200	166	198	211
2400	235	281	299

Table 2: Increases in curving speed

Stech '96 Tilting Trains
GBTRS/J Cartmell

The speed increase obtained benefits not only in journey time but also in a reduction in energy costs otherwise incurred to slow down and re-accelerate the train. While the speed increase available in curves is in the range 19%-27%, the overall reduction in end to end journey time will always be somewhat less than that. It depends on the ratio of the length and number of speed restricted curves to the length of unrestricted track as well as on any increase in power available on the tilting train for improved acceleration out of curves and station stops, improved gradient climbing ability, etc. On the West Coast Main Line, end to end journey time savings of the order of 15% are predicted if 225 km/h tilting trains are used. That this is achievable was demonstrated by the late lamented APT which covered the journey in 3 hours 53 minutes, a saving of over 20% against current time-table. On the Sydney to Canberra route, a journey time of 3½ hours could be time-tabled, a similar time saving of 20% without any route investment. So these time savings are real.

3.   COMPETITION AND THE ECONOMIC VALUE OF TIME

The existing minimum timetabled journey time from Euston to Glasgow
Central, northbound and southbound, is 5 hours. A 15% saving in that time
represents 45 minutes, but how can that be priced for the assessment of
improved revenues to be compared with the capital and maintenance
investment required?

The Demand Forecasting Handbook uses the concept of generalised
journey time for journeys of comparable quality - and tilting trains may well
enjoy a "sparks effect" of improved perceived and actual quality - which
have an elasticity of 0.9. A 10% reduction in journey time, for instance, can
be expected to result in a 9% increase in patronage. In Sweden, a 25%
reduction in journey time between Stockholm and Gothenburg led to a 20%
increase in patronage to give load factors of 80%. Here, a 15% time
reduction would thus lead to about a 13.5% increase in passengers. An
average train load of 300 could thus be expected to increase to 340.

If, for the single journey envisaged, the average fare is £55 then an extra
revenue of £2200 per train journey could accrue to the rail operator before
making any allowance for a premium fare which the "sparks effect" of tilt may
permit. Alternatively, people are reckoned to be willing to pay between £10
and £20 per hour for time saved, depending on the purpose of their journey,
so our 45 minutes would be worth an extra £7.50 to £15 on the price of the
ticket. At an average of £10, that equates to £3000 extra per train journey
for the existing number of passengers. So something around £2200 to
£3000 extra appears to be available per train journey. Such an increase
looks particularly attractive when set only against variable costs.

On this route, the competition is from motorway coaches for the leisure
traveller (a cheaper but much longer and less comfortable journey) and from
the air for the business traveller. Air travel suffers, as ever, from transfer
time to and from, and waiting time at, airports together with the uncertainty of
air traffic and weather delays. At 4 hours 15 minutes city centre to city
centre, the preference for many will be the train.

But benefits are not restricted to high speed main lines. Tilting trains on
regional railways, which frequently run through twisting, hilly, routes, can
also show benefit to passenger and operator. Since May 1992 the VT610
tilting DMU has been operating a number of routes out of Nuremberg. In
increasing the permitted line speed from 120 km/h to 160 km/h, which is
maintained through all but the sharpest curves, a 30% reduction in journey
time has been achieved.

4.   CAPITAL INVESTMENT

The investment to upgrade an existing route to be suitable for the operation of tilting trains is clearly very much less (by more than an order of magnitude) than that for reconstructing the line, or building a new one, for the operation of high speed conventional trains, as illustrated by Table 3. While some curve straightening may be possible at reasonable cost, there are usually almost insurmountable problems of land acquisition, property demolition and topographical constraint. The very object of tilt technology is to avoid those costs and to spend relatively modest sums only on the improvement of track quality.

New alignments

- French LGV $10M/km
- German ICE $18M/km
- Spanish AVE $9M/km

Upgrade for high speed tilt

- Sweden $0.5M/km
- WCML estimate $0.8M/km

Table 3: Capital Cost Comparison

Much has been made of the cost of reinforcing the electrical system on the West Coast Main Line for use by higher speed trains. While higher speed does indeed demand more power, the use of tilt will in fact reduce the local power consumption by avoiding the need to re-accelerate after braking for a sharp curve. Trials in USA between an X2000 tilting train and the AEM-7 locomotive hauled Metroliner claimed as high as 40% energy saving due to this effect. The use of modern pantographs can in any event often provide reliable current collection from existing systems at higher speeds up to, say, 250 km/h.

On other routes, diesel traction is readily applied to tilting multiple units or locomotive hauled trains. German Railways' experience with the VT610 DMU operating out of Nuremberg led to investment in its VT611/612 successor, this time with an electric tilting mechanism. The new trains enter service from September 1996: 200, 2 car sets have been ordered of which the first 50 are currently confirmed.

© IMechE 1996 C514/0

For either electric or diesel traction, the extra cost on the rail vehicle of the tilting mechanism is as low as 5 or 6%. However, other more luxurious features may be considered appropriate in order to complete an altogether more attractive total package for the passenger. Increased power and a higher performance/lower wear bogie will also probably be installed. It is therefore difficult to compare train prices on a strictly like for like basis, but the planned costs of the vehicle package must clearly be taken into account in calculating any financial or economic rates of return. Of course, if the shorter journey time releases a trainset from the diagram, the economics begin to look altogether more attractive again.

It is interesting to note that the French, too, have commissioned a tilting train prototype "as the standard TGV is very expensive". GEC - Alsthom will deliver at the end of 1997 a train that is capable of 350 km/h on LGV's and 220 km/h on other routes.

5.   MAINTENANCE COSTS

The tilt mechanism on the train requires maintenance which can be programmed to coincide with existing vehicle maintenance routines. The reliability of hydraulic tilt mechanisms has been shown to be of a high order and even lower maintenance electric versions are now being introduced. Control is solid state and virtually maintenance free. This is illustrated by the reliability now being achieved by the X2000 - 1.7 failures per million kilometres.

Bogie running gear maintenance, including wheel turning, can in fact be reduced - thanks to parallel improvements in bogie design. Modern radial steering bogies impose significantly reduced track forces over the whole route - while there will be an increase in lateral forces on curves at higher speeds. In spite of that, tests in Sweden, supported by computer simulation, demonstrated that track forces from an X2000 train running at 230-245mm cant deficiency were no higher than from a conventional electric locomotive running at 100 mm cant deficiency. This was explained as being due to a combination of lower static axle load, lower centre of gravity and lighter unsprung mass on the X2000. Increased track maintenance is nonetheless likely to be required, chiefly to maintain the higher minimum track quality standards demanded by higher speed running. Again, however, experience with existing tilting train operation in Sweden and Germany has been good and revised track maintenance regimes have been accommodated in existing procedures.

6.   <u>CONCLUSION</u>

The demand for travel by both business and leisure users has been seen, particularly in the last twenty years, to be linked directly to growth of the economy. While the use of EMail, the Internet and video conferencing may be improving communications, they have yet to make any real inroads into our need to meet face to face. Air travel is generally expensive and, in this country, is applicable only to longer journeys. The private car, of course, has much to offer in flexibility and convenience, particularly over shorter distances and its use has mushroomed in recent years. The resulting pressure on our road network is experienced by us all most days of the week and must be arrested. We have to find alternative and more economic means of transport to relieve that pressure wherever we can.

The railways have long been shown to be effective at intermediate distances of between 200 and 500 km. That general distance is extended if journey times can be reduced by the use of modern tilting train technology which takes advantage of the laws of physics when applied to existing, twisty, cross country routes. Compared with other investment alternatives, tilting trains allow shorter journey times and improved passenger comfort without the need for costly and time consuming infrastructure upgrading. The benefits can start to flow immediately.

TECHNICAL DETAILS OF TYPICAL TILTING TRAINS

	Locomotive, Hauled, Electric	Diesel Multiple Unit
Maximum service speed	225 km/h	160 km/h
Number of seats	1st class car 51 seats 2nd class car 76 seats	148 seats per 2 car unit
Train configuration	1 power car + 2-5 intermediate cars + driving trailer	2 car DMV
Car length	Power car 17,6m Interm. Car 25,0m Driving trailer 22,6m	25.4 m/each
Car width	3,08m	2.85m
Car height	3,8m	3.97m
Tilt acuators	hydraulic	electrical
Power cars tilt	no	yes
Curving compensation	appr. 0,65	app 0,65
Maximum tilt	8 degrees	8 degrees
Traction	15 or 25 kV ac	diesel hydraulic
Distributed traction	no (4 motors in power car)	1 diesel drive unit/car
No pantographs	1 of 2 raised (on power car)	None
Running gear	2 bogies per car 2 axles per bogie	2 bogies per car 2 axles per bogie
Self steering wheelset	yes	yes
Vehicle weight	Power car 73 tonne Interm. Car 47-49 tonne Driving trailer 55 tonne	100 tonne (2 car unit)
Power	4,0 MW max 3,3 MW cont.	2 MTU diesels @ 540 kW Giving 890 kW per 2 car unit
Max axle load	18,5 tonne	16,5 tonne
Acceleration rate	0,4 m/s²	0,8 m/s²
Braking rate	1,1 m/s² (service)	0,9 m/s²

C514/026/96

Development of unsymmetric self-steering truck using soft–hard axle springs

A HASHIMOTO BMechEng, **K YAMADA**, and **T METOKI**
Rolling Stock Department, Central Japan Railway Company, Nagoya, Japan
Y SUDA MJSME, MASME
Institute of Industrial Science, University of Tokyo, Japan

SYNOPSIS

In order to reduce traveling time on Japan's conventional railways, which contain many curves, it is often more effective to increase the speed on curves than to attain a higher maximum speed on straights.

To decrease lateral force, we have developed and implemented an unsymmetric selfsteering truck using soft–hard axle springs. This new design has been proven in field tests using actual trains to reduce lateral force in curves and to maintain running stability in straights. Subsequently, a new type has been developed for bolsterless trucks that does not require soft–hard switching based on the direction of travel.

1. INTRODUCTION

Japan has relatively little flat land and many mountains. Consequently, the conventional railway lines in Japan contain many curved sections. For this reason, increasing the speed at which trains can negotiate curved sections of track is an important issue in attempting to reduce travel times. In order to control the increase in the load force applied to the tracks (lateral force), which is one of the factors limiting faster running speeds on curves, JR Central began development work in 1989, based on a new concept, on a self–steering truck design for 2–axle bolsterless trucks used in passenger trains. The new trucks were success- fully implemented on the new series 383 pendulum trains.

The series 383 EMU trains were developed as a successor to the old series 381 trains. These series 381 EMU trains, which were used for the old "Shinano" conventional limited express trains, were introduced in 1973 as Japan's first pendulum trains. Of the 250 km of this train's run from Nagoya to Nagano, 120 km (or roughly half) is subject to speed limitations due to curves. The series 381 is able to maintain a commercial speed of approxi- mately 90 km/h through mountainous areas containing many 300 m or 400 m small–radius curves. With the aim of attaining a running speed on curves even faster than that of the series 381 in the series 383, JR Central began development work on a steerable truck.

There are several practical examples of steerable trucks developed by railways overseas. In Japan, however, there were as yet no examples of steerable trucks that had reached the stage of practical implementation. The reason for this was the barrier problems such as structural complexity, increased unsprung mass, additional maintenance work, and higher initial cost presented to implementation. To overcome these problems, the new concept (the Suda theory) of making the truck's suspension constantly unsymmetric was incorporated into the design to make the truck self-steering. In this way, excessive structural complexity was avoided and a simple self-steering truck using soft-hard axle springs was achieved.

Initially, a "soft-hard switching system" (jointly developed by JR Central and Sumitomo Metal Industries), which switched the axle springs' suspension unsymmetrically soft or hard depending on the running direction was used. It consisted of a axle spring soft-hard switching mechanism and a switch monitoring system.

Subsequently, as a result of tests conducted on the series 383 prototype (6-car train, 3M3T), which lasted from 1994 through 1995, a soft-hard fixed system that did not require soft-hard switching was discovered. This design (which was jointly developed by JR Central and Associate Professor Suda) required neither the axle spring soft-hard switching mechanism nor the switch monitoring system, allowing the overall structure to be further simplified.

This paper describes the process leading to the practical implementation of the unsymmetric self-steering truck using soft-hard axle springs and the results of that implementation.

2. BASIC DEVELOPMENT CONCEPT FOR THE NEW SERIES 383 EMU TRAIN

The goal of the series 381 train, which was commercially developed in 1973, was the ability to negotiate curves at higher speeds, and it therefore incorporated features such as a lightweight aluminum body and a passive pendulum system. It has a maximum running speed of 120 km/h and a speed on curves (R400 m) of 20 km/h above the basic speed. The series 381 has already demonstrated a high level of running performance on sections of track running through mountainous areas.

More than 20 years had passed since the series 381 was first introduced, the accommodations it offered were beginning to look the worse for wear, and maintenance was also becoming a problem. Therefore, the decision was made to develop the series 383 as a successor to the series 381.

The basic concepts guiding the development effort were "faster," "more comfortable," and "easier to use."

To make the train "faster," targets of 35 km/h above the basic speed on curves (R 600 m) and a maximum running speed of 130 km/h were set. The question of how to overcome the limitations associated with raising the speed on curves therefore became an important technical development factor.

The main factors limiting increased speed on curves, and the specific measures taken to deal with them, are as follows:

- Increase in load force applied to tracks (lateral force)
  -> Development of lighter car and steerable truck
- Decreased riding comfort
  -> Use of controllable pendulum
- Decreased stability against side winds
  -> Car with lower center of gravity

Of these, the development of a steerable truck (something that had not yet reached the practical implementation stage in Japan) was a particularly important issue.

3. PROCESS BY WHICH THE SELF–STEERING TRUCK WAS INTRODUCED

When the running speed on curved sections of track increases, the increase in excess centrifugal force brings with it an increase in the lateral force applied to the tracks. The speed is therefore limited by the track standards. Also, noise increases due to friction between the wheels and rails, even if the speed limitation is not reached. Since freight trains also run on the conventional lines in Japan, it is not possible to use a large amount of cant. In most places, the cant is already the maximum value (105 mm).

For these reasons, the main measures that can be taken are the following two:
- Decreasing the wheel load by reducing the overall weight of the cars
- Making steering easier on curves by using steerable trucks

The new series 383 employs design elements such as a lightweight stainless steel construction, a VVVF inverter device, and bolsterless trucks to achieve a weight (average weight per car) of 35.6 tons. This is a reduction of two tons from the 37.6 ton weight of the old series 381, which used an aluminum body.

The use of steerable trucks makes it possible to reduce the attack angle between the rails and wheels by actually turning (steering) the wheels on curves. This improves turning performance on curves. In this way the load force applied to the tracks (lateral force) is reduced, making it possible to run at a higher speed on curves.

However, from a logical viewpoint, it is difficult to design a truck that provides both running stability on straight sections of track and excellent steering performance when negotiating curves. The key to a good design is therefore determining how well these opposing goals can be achieved at the same time. Two well–known practical examples are the Scheffel truck from South Africa, which uses a link mechanism to connect the leading and trailing axles, and the guided steering truck from Canada, which uses steering links to connect the body, truck frames, and axles.

Most previous practical implementations of steerable trucks have used mechanical link devices, or the like. They therefore have the demerits of increased unsprung mass and additional maintenance requirements. Also, such designs are difficult to adapt for the power truck. Faced with these difficulties, we set out searching for a steering method that would satisfy the requirements listed below while reducing lateral force.
- Fail–safe and having good running stability
- Simple structure and easy to install on power trucks
- Easy maintenance

– Limited increase in unsprung mass

In order to satisfy the above conditions, we began development work on an unsymmetric self–steering truck using soft–hard axle springs based on a theory proposed by Associate Professor Suda.

Study of the proposal began in 1989 and, following the development of prototypes and fixed tests, running tests in actual trains were conducted in 1990 to check lateral force performance. In 1994 a prototype of the new series 383 was completed. Running tests on the series 383 were carried out in 1994 and 1995, and in 1995 the new train went into commercial use. Subsequently, a soft–hard fixed type that does not require soft–hard switching for the axle springs has been developed.

The main aims in incorporating steerable trucks into the series 383 were the following two points:
– To achieve an even greater speed on curves (35 km/h above basic speed on 600 m
   curves radius) than that of the old series 381 and recent pendulum cars developed by
   the various JR companies
– To not increase above–ground maintenance costs compared with the previous series 381
   even while increasing speed on curves

4. THE CONCEPT OF THE UNSYMMETRIC STEERING TRUCK

In the past, truck design has emphasized running stability on straight sections of track. For this reason, designs have kept wheel equivalent conicity small and made the hardness of the leading and trailing axle springs supporting the wheels rather hard. This meant that no steering of the wheels took place and the large angle at which the wheels contacted the rails (the attack angle) was maintained on curves.

Then self–steering performance was improved by softening the primary longitudinal stiffness of the axle springs for the leading wheelsets, thereby lessening the restraining force on the leading wheelsets. As a reflection of the importance given to running stability on straight sections of track, the primary longitudinal stiffness of the axle springs for the trailing wheelsets was left hard, as before. The main points of the Suda theory are introduced below.

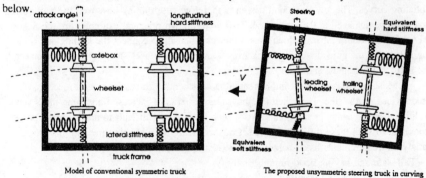

Fig.1 Model of Conventional symmetric truck and Unsymmetric truck

The primary longitudinal stiffness ratio of the leading and trailing wheelsets is the basis of this steering theory. If the leading wheelset stiffness is expressed as k_{x1} and the trailing wheelset stiffness as k_{x2}, the unsymmetric index a_s indicating the axle spring unsymmetricalness can be expressed in an equation (in which a is half of the wheelbase) as follows:

$$a_s = -a \frac{k_{x1} - k_{x2}}{k_{x1} + k_{x2}}$$

If a_s satisfies all of the following equation's conditions as defined by the truck data (wheel radius (r), wheel equivalent conicity (λ), equivalent bending stiffness (k_b)), the longitudinal creep coefficient (κ_{11}) (which expresses the contact force characteristics between the wheels and rails), and the lateral distance between wheel contacts ($2b$), ideal steering performance should be obtained.

$$a_s = \frac{k_b r}{2 \kappa_{11} \lambda b} .$$

Since this equation is not affected by factors such as speed or curve radius, there is no need to adjust the primary longitudinal stiffness while the train is running. This means that it is possible to construct a steerable truck using a fixed primary longitudinal stiffness ratio. Based on this steering theory, it was determined that the primary longitudinal stiffness ratio of the axle springs for the new series 383 should be approximately 1:4.

5. UNSYMMETRIC SELF–STEERING TRUCK WITH SOFT–HARD SWITCHING DEVICE

5.1 Structure and system

Since, the leading and trailing wheelsets can change depending on the direction in which the train is traveling, it is necessary to be able to switch the axle springs between soft and hard in order always to provide soft support for the leading wheelset and hard support for the trailing wheelset. This type of steering device consists of an axle spring soft–hard switching mechanism and a switch monitoring system.

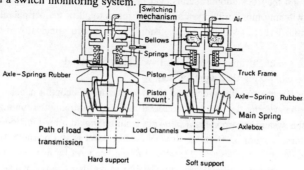

Fig.2 Structure of steering device

The axle spring soft–hard switching mechanism engages (or disengages) longitudinally soft axle spring rubber in order to change the axle spring support stiffness.
Switching of the primary longitudinal stiffness is performed all at once for the entire train when the train reverses direction at stations, etc., using the driver's normal operation controls.
The command "Soft" causes compressed air to be supplied to the bellows in the steering device of the leading wheelset. This causes the piston to descend and disengage from the piston mount, releasing the restraint on the axle spring rubber. When the truck is in this status, a series link is formed with the main spring as shown in the primary longitudinal path of load transmission indicated in Fig. 2, resulting in a soft supporting state. Conversely, air is released from the trailing wheelset side, causing the piston to set itself in the piston mount and restraining the axle spring rubber. In this case support is provided by the main spring alone, resulting in a hard supporting state.

The leading wheelset is provided soft support because there is a series link between the axle spring rubber and the main spring. The trailing wheelset is provided hard support because there is no linkage. Fig. 3 is a model of this steerable truck structure.

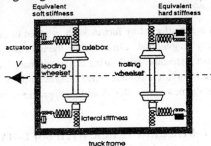

Fig.3 "Soft–Hard Switching" Self steering truck

During normal operation the wheelsets are arranged in soft–hard sequence in a truck based on the direction of travel. We also discovered, based on the results of simulations and tests, that the train could still run if the sequence was hard–hard or hard–soft. A fail–safe structure, in which hard support will result should the pressurized air system fail was used. However, a switching monitoring system was added as a backup due to the possibility of instability should the truck support status ever become soft–soft.

The switching monitoring system actually refers to all of the peripheral devices in the axle spring soft–hard switching mechanism. It is composed principally of the following devices.
– Device for sending primary longitudinal switching commands
– Device for monitoring the switching status
– Device for transmitting the monitored status to the driver's control panel

Since the steering device plays a crucial role in determining the running performance of the train, the system incorporates a fail–safe feature. Should a malfunction occur in the electrical system, monitoring from the driver's control panel is still possible. In addition, all wheelsets are automatically switched from soft to hard support by releasing the pressurized air. The electrical system's reliability and resistance to weather conditions and noise has

© IMechE 1996 C514/026

been confirmed, and in the more than one year since the introduction of the series 383 no malfunctions have occurred.

5.2 Effectiveness

Fig. 4 illustrates an example of the results of our simulations. The wheelset steering angles on a typical curve radius of 420 m (transition curve: 75 m, cant: 60 mm) are plotted. In this case, a linear model was used for truck suspension characteristics and the Kalker's non-linear theory was used for the contact geometry between the wheels and rails. The results show a greater amount of steering of the leading wheelset (a reduced attack angle) for the unsymmetric self-steering truck than for the conventional truck. This was confirmed to be approximately the ideal steering status relative to the rail (see Fig. 1, above) when the truck's insufficient bogie angle is taken into account.

Also, the results of evaluations of running performance using test trucks mounted onto the old series 381 cars shows that running stability on straight sections is maintained, while ideal steering was achieved on curves. This indicated that it was possible to reduce lateral force compared with conventional trucks. Subsequent long-term tests also confirmed that, sharpening of the flange was also reduced.

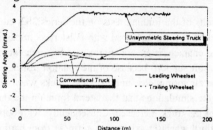

Fig.4 Steering angles (Result of curving simulations)

The actual performance testing and performance improvement testing on series 383 cars fitted with the steerable trucks began in September 1994 and lasted for about six months. The results showed that the initial targets had been met, both for running stability and also for lateral force performance.

With regard to running stability on straight sections, no abnormal vibrations occurred even in the 200 km/h bench tests and above. Running stability on straight sections at the maximum speed of 130 km/h was confirmed for the series 383 as well.

Steering effectiveness was evaluated using the maximum lateral force and the aver-age lateral force. According to the results of all tests, maximum lateral force was reduced significantly compared with the series 381, though these measurements were made only at the rail joints. Since the track standards were met, running at the target speed was possible. As for average lateral force, comparison of the results obtained using hard-hard (no steering) and soft-hard (steering activated) settings indicated that the biggest effects were achieved on small or medium radius curves(R500m or less). By adding the steering function, the average lateral force was reduced by approximately 30 %.(See Fig.5 for an examples)This confirmed that the new trucks could be expected to help reduce the amount of track maintenance work.

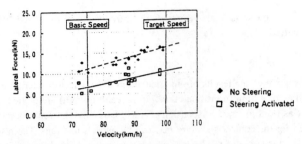

Fig.5 Average lateral force of the leading wheelset of first car
(on 400 m curve radius and 90 mm cant)

6. SOFT-HARD FIXED SELF-STEERING TRUCK

6.1 Creation of the soft-hard fixed configuration

With regard to the lateral force characteristics of individual cars in a train equipped with two-axle trucks, it has been known from experience for some time that the lateral force is greater for the first wheelset in each truck (the first and third wheelsets, proceeding in the direction of travel).

However, the results of the running tests on the series 383, etc., show that the lateral force on the third wheelset of a car equipped with bolsterless trucks is generally lower. This is because ideal steering characteristics are approached to some extent, even if the third wheelset does not have soft support, because of the rotation resistance between the body and the trucks caused by the air spring, etc., in bolsterless trucks when negotiating turns. We also confirmed in separate simulations that the lateral force on the third wheelset dropped as the rotation resistance of the truck increased. Consequently, with bolsterless trucks there is no need to change the third wheelset axle spring stiffness from hard to soft in order to re-duce the lateral force on the third wheelset. This led to the idea of a new configuration in which the support characteristics for each car would be "soft-hard-hard-soft," instead of switching the axle spring stiffness to "soft-hard-soft-hard" based on the direction of travel.

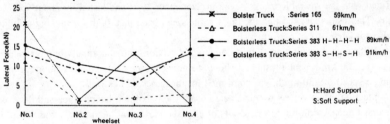

Fig.6 Comparison of Lateral Forces (on 300 m curve radius)

On the other hand, in conventional bolster-equipped trucks the rotation resistance is applied to the side friction block as changes in the bogie angle. While traveling through a curve, the changes in the rotation angle are small so there is little rotation resistance. It is thought that for this reason the trailing truck does not pull very much on the body and the steering performance is equivalent to that of the leading truck. Therefore, a "soft–hard–soft–hard" arrangement in which the first and third wheelsets have soft support is recommended for bolster–equipped trucks.

The new "soft–hard–hard–soft" sequence for bolsterless trucks results in an un–symmetric relationship between the leading and trailing wheelsets in each truck. For each car, however, there is a symmetric relationship between the leading and trailing trucks with the soft–hard fixed configuration, so no soft–hard switching is necessary. Since the soft–hard fixed configuration does not require a soft–hard switching mechanism or switch moni–toring system, it results in a much simplified truck structure and makes the cars far easier to maintain. Fig. 7 illustrates the "soft–hard–hard–soft" axle spring support pattern for a single car, including the trailing truck.

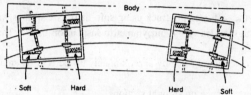

Fig.7 "Soft–Hard Stable" Self–Steering Truck

6.2 Verification

It was necessary to verify the following points with regard to the soft–hard fixed steering method which creates a "soft–hard–hard–soft" pattern for each car:
- Verification of the running stability of the trailing truck on straight sections of track due to its "hard–soft" (third wheelset: hard support, fourth wheelset: soft) configuration
- Confirmation of performance on curves with hard support for the third wheelset

To test running stability, a running simulation with worn wheels(150,000 km run) was performed. The results of an evaluation of vibration convergence characteristics for the yawing (swaying) mode of the (soft support) first and fourth wheelsets and the lateral vibra–tion mode of the body (which can result from the swaying) showed that the unstable (diver–gence) range for the fourth wheelset was 173 km/h and above. This was almost equivalent to the results for the "soft–hard–soft–hard" steering method. No problems were encountered in bench tests of running stability at 200 km/h and above, in running tests using actual trains, and in measurements of truck frame lateral vibration, etc., at the maximum speed of 130km/h.

The results of running tests using the series 383 indicated comparable performance negotiating curves to that of the soft–hard switching steering method. As for the average la–teral force, third wheelset lateral force was comparable to an arrangement with no steering and hard support, however, as shown in Fig. 8, was not higher than first wheelset lateral force. Our measurements indicate that lateral force was evenly distributed between the various wheelsets.

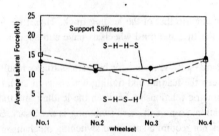

Fig.8 Average of Lateral Forces(at 90 km/h speed on 300 m curve radius and 105mm cant)

7. CONCLUSION

As described above, the self–steering truck using soft–hard axle springs was developed and implemented based on the idea of the "unsymmetric suspension truck." This design provides the following advantages:

- Reduction of average lateral force by approximately 30 percent on small or medium radius curves
- Stable running performance on straight sections
- No links or axle spring switching devices required
- Can be applied to drive and non–drive wheelsets

By using this type of steerable truck, travel times on conventional railway lines can be reduced by increasing the speed on curves. Series 383 EMU trains are supposed to provide service to persons attending the Nagano Winter Olympics scheduled for 1998.

REFERENCES

(1) Suda,Y., "Improvement of High Speed Stability and Curving Performance by Paramater Control of Trucks for Rail Vehicles Considering Independently Rotating Wheelsets and Unsymmetric Structure. " JSME Int.J., 1990, 33–2 Ser.III, pp.176–182.

(2) Suda,Y., Anderson, R.J., "Dynamics of Unsymmetric Suspension Trucks with semiactive control. " Vehicle System Dynamics, 1993, pp. 480–496.

(3) Suda,Y., "High Speed Stability and Curving Performance of Longitudinally Unsymmetric Trucks with Semi–active Control. " Vehicle System Dynamics, 1994, 23, pp.29–52.

(4) Suda,Y., Anderson,R.J., "Dynamic Characteristics of Unsymmetric, Light–Weight, Trucks for High–Speed Trains. " JSME STech'93, 1993, Vol 2, pp.389–394.

(5) Yamada,K., Hino,K., "Radial–Steering Bogie. " JSME STech'93, 1993, Vol 2, pp.407–412.

(6) Yoshikawa,K., Yamada,K., Katumata,Y., Metoki,T., Nakata,M., Hino,K., "Control system of Steering Trucks. " ASME, 1994, TER–94–31. (In Japanese)

(7) Suda,Y., Yamada,K., Hino,K., Siiba,T., "Improvement of Steering Ability by Unsymmetric Radial Trucks. " JSME, 1993, NO.930–81. (In Japanese)

(8) Suda,Y., Yoshikawa,K., Yamada,K., Metoki,T., "Development of a new Steering Truck System. " J–RAIL'95, 1995. (In Japanese)

(9) Suda,Y., Yoshikawa,K., Yamada,K., Metoki,T., Nakata,M., "Running Performance of Self Steering Truck. " J–RAIL'95, 1995. (In Japanese)

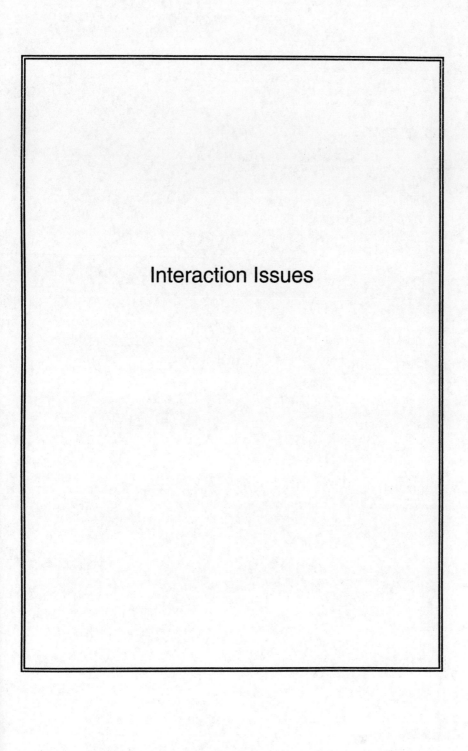

Interaction Issues

C514/030/96

Track maintenance for tilt train operation on sharp curves

M UCHIDA MJSCE, **H TAKAI** MJSCE, **T ISHIKAWA, E YAZAWA,** and **M MIWA**
Railway Technical Research Institute, Tokyo, Japan

Japan is a very mountainous country and raising the curving speed is the most effective way to diminish train schedule time. Recently, Japan Railway(JR) companies are pushing speedup projects on sharp curves with tilt trains which compensate lateral stationary acceleration and improve passengers' riding comfort on curves. Then, the following two research objects in the track maintenance become important. One is confirmation of track maintenance standards to sustain vehicle running safety and passengers' riding comfort. The other is estimation of train load on tracks, track irregularity growth and countermeasures for those problems. We studied the relation between vehicle running characteristics and track or vehicle conditions on sharp curves in actual running tests. In this report, we show an estimation method of vehicle lateral force and lateral vibration, track maintenance targets and track lateral deformation characteristics.

1 INTRODUCTION

On JR(Japan Rail) narrow-gauge lines, several projects of increasing the train curving speed by adopting a tilting car are aggressively underway using existing track facilities. Generally speaking, the following two points are important from the viewpoint of track maintenance.
    (1) Track maintenance standard which ensures the train running safety and good riding quality.
    (2) Estimation of increasing train load and track irregularity growth, and study of countermeasures like strengthening of track structures.
    On curves, static forces like unbalanced centrifugal force or wheel lateral force act depending on main features of curves(curve radius, track cant) and train speed. At the same time, lateral acceleration and changing lateral force occur by track alignment irregularities on outer rail because trains are pushed outward by centrifugal force(Fig 1). Especially in the case of tilting cars, these phenomena are more prominent because tilting cars run faster than non-tilting cars, and they generate higher centrifugal forces. Then some problems emerge like growth of track alignment irregularity, wear and defect of track components, and riding discomfort. Improving the track maintenance method on sharp curves in tilting car operation is the main theme of this report. We study the track maintenance standard considering the riding comfort; estimation of lateral force and growth of track alignment irregularity; cause of corrugation wear on outer rail, based on the results of actual running tests.

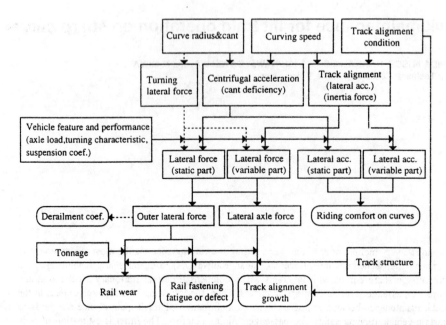

Fig 1 Vehicle running characteristics and influence on track on curves

2 ESTIMATION OF LATERAL VIBRATION AND STUDY OF RIDING COMFORT ON CURVES

2.1 Estimation Method of Lateral Vibration Using Standard Deviation of Track Alignment

To confirm the relation between track irregularity and vehicle vibration quantitatively, we used to employ maximum value of track irregularity or P-value and maximum value of vehicle vibration or riding comfort level. Maximum value of vehicle vibration is not sufficient to show general characteristics of a special section because it depends on special track irregularity. P-value and riding comfort level are also items of estimation, and they restrict the application area. We try to get more general characteristics by using standard deviation(STD) of track irregularity and vehicle vibration.

From actual running test results shown in Fig 2, we assume that curving speed V(km/h) is in proportion to STD of lateral vibration σ (m/s^2). The proportion coefficient k_v shown in Fig 3, is also in proportion to STD of track alignment. Then, the STD of lateral vibration σ_{aH} is calculated by the following formula.

$$\sigma_{aH} = k_z \cdot \sigma_z \cdot V \tag{1}$$

From several running tests, the value of kz is obtained as follows.
- High-performance electric train(tilt) $\cdots$ 0.0006~0.0010
- High-performance electric train(non-tilt) $\cdots$ 0.0008~0.0012

Here, maximum value of track irregularity or vehicle vibration can be estimated as three times the STD according to general statistical characteristics.

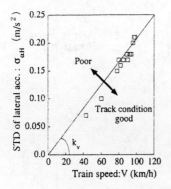

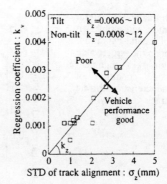

Fig 2   Train speed and lateral vibration(STD)

Fig 3   Track alignment(STD) and vehicle vibration coefficient

2.2   Actual Condition of Riding Comfort of Tilting Trains on Curves

Tilting cars decrease the lateral acceleration parallel to car floor shown in Fig 4 thanks to their carbody tilting device. That value is roughly calculated by the following formula using cant deficiency Cd, tilt angle Φ and suspension coefficient s.

$$\overline{\alpha_H} = (C_d/G - \Phi) \times (1 + s)g$$

$$C_d = GV^2 / gR - C \quad\text{--}\quad (2)$$

And when the tilting device is set under secondary suspension, lateral vibration acceleration decreases dramatically as shock against lateral movement stopper decreases. This is the reason for difference of k_z in section 2.1.

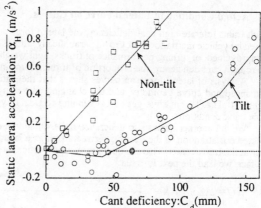

Fig 4   Cant deficiency and static lateral acceleration

2.3   Track Alignment Maintenance Target Considering Riding Quality

Riding quality of the passengers on a high speed curving train is estimated correctly with the combination of static lateral acceleration and vibration lateral acceleration. We set the following target value(Fig 5).

(1) When static lateral acceleration is below 0.6m/s², vibration lateral acceleration (peak-to-peak) should be 2.0 m/s².

(2) When static lateral acceleration exceeds 0.6m/s², the sum of static and vibration lateral accelerations should be 1.6m/s². This means:

$$\alpha_H = 3.2 - 2\,\overline{\alpha_H} \quad\text{--}\quad (3)$$

With this idea of lateral riding comfort and estimation method of lateral vibration, we can obtain the target value of STD. of track alignment under actual condition of vehicle characteristics or train speed or track geometry.

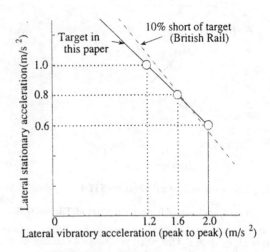

Fig 5   Target value of lateral riding comfort on curves

3   ESTIMATION OF LATERAL FORCE AND TRACK STRENGTH

3.1   Actual Condition of Lateral Force on Curves

As tilt trains tolerate high cant deficiency, the lateral force mainly consists of axle lateral force caused by vehicle inertia force. It grows parabolically with its speed. At the same time, by virtue of improvement of turning performance of trucks, light weight and decrease of inner wheel load due to high cant deficiency, the proportion of turning force in total lateral load becomes small.

Analysis of lateral force on curves shows that the static part of lateral force depends on train speed and curve geometry, and the changing part of it depends on train speed and track alignment. That is, once we get actual running test data, we can estimate lateral force under every track condition.

We did another analysis to generalize the estimation method using field data. At first, we directed our attention to the relation between front axle lateral force and lateral vibration on the floor at the truck center. There is obvious by a proportional relation between them(Fig 6). From this fact, we lead the next formula.

$$\Delta Q = 2M_0 \cdot \alpha_H \cdot K_H \quad \text{---} \quad (4)$$

Here the correction coefficient K_H indicates the load ratio of front axle of vehicle to inertia load of one truck. This coefficient is influenced by cant deficiency and curvature. Fig 7 shows the ratio of inner lateral load, and wheel load which means trucks curving performance. It is obvious that a newly designed truck works effectively, which improves self-turning function by lowering the spring constant in front-rear direction.

Fig 8 shows test data of impact lateral load caused at rail joints, in relation to train speed and curve radius.

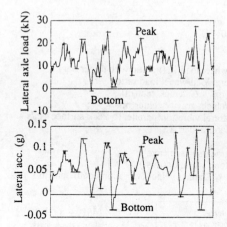

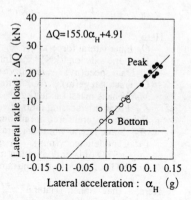

Fig 6   Lateral vibration(0-peak) and axle lateral force

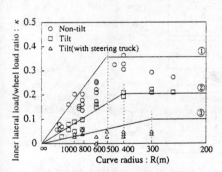

Fig 7   Curve radius and inner lateral
force/wheel load ratio

Fig 8   Impact lateral load at rail joint

3.2   Estimation Method of Lateral Force

We propose an estimation method of front lateral force and axle force, which is important when
we consider the track strength on curved track. This method consists in summing up the static
part which depends on curve geometry and the variable part which depends on track irregularity.
The most important parameter is train speed.

Static part consists of an uncompensated centrifugal force by cant deficiency and a
curving lateral force. Variable part consists of a vehicle inertia force by lateral vibration and a
shocking lateral force at rail joint (Fig 9)

$$Q_o = \overline{Q_i} + \Delta Q \text{ --} (5)$$

$$\overline{Q_i} = \kappa \cdot \overline{P_i}$$

$$\overline{P_i} = \frac{W_0}{2} \left\{ \left(1 + \frac{V^2}{gR} \cdot \frac{C}{G} \right) - \frac{2H_G^*}{G} \left(\frac{V^2}{gR} - \frac{C}{G} \right) \right\}$$

$$\Delta Q = 2 M_0 \cdot \alpha_H \cdot K_H + S \qquad \alpha_H = \overline{\alpha_H} + \Delta \alpha_H$$
$$\overline{\alpha_H} = C_d / G \qquad \Delta \alpha'_H = 3 \sigma_{\alpha H}$$
$$\sigma_{\alpha H} = k_z \cdot \sigma_z \cdot V \qquad K_H = 0.6 + 80 / R$$

Here,

Q_i :Inner lateral force(kN), $\overline{P_i}$:Inner wheel load(kN)(static with upper bar)
$\overline{W_0}$:Static axle load(kN), κ :Inner lateral load/wheel load ratio,
V :Train speed(m/s), R :Curve radius(m), C :Track cant(m),
G :Track gauge(m), HG^* :Effective vehicle gravity center(m),
M_0 :Vehicle mass(1 axle)(kg), α_H :Lateral vibration(parallel to track surface)(m/s²)
K_H :Front axle bearing ratio of vehicle lateral inertia
S :Impact lateral force at rail joint(kN),
$\overline{\alpha_H}$:Static lateral vibration(parallel to track surface)(m/s²),
Cd :Cant deficiency(mm), $\Delta \alpha_H$:Lateral vibration(m/s²),
$\sigma_{\alpha H}$:STD of lateral vibration(m/s²), k_z :Lateral vehicle vibration coefficient(m/s²)
σ_z :STD of track alignment(mm)

Here, the inner lateral force/wheel load ratio K is given in Fig 7. Front axle bearing ratio K_H is calculated by the formula considering curve radius. Shocking lateral forces at rail joints are given in Fig 8. These values decrease to a half at welding points of long welded rail(LWR) track.

Fig 10 shows an example of comparison between estimation results and actual data. Estimation data are a little larger than field data. This means the estimation method outputs a safety-side value, which is appropriate to apply for actual works.

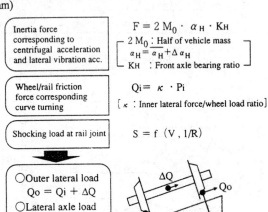

Fig 9   Estimation of lateral force on curves

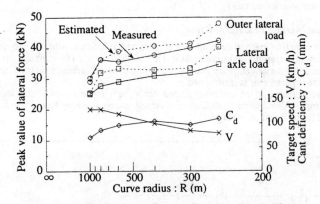

Fig 10   Comparison of estimated and measured lateral forces

3.3 Setting Method for Target Value of Track Alignment

Here, we propose a setting method for targets value of track alignment considering lateral track strength against maximum lateral force. We check two items. One is axle lateral load for lateral track shift, the other is outer lateral load for limit of rail fastening devices.

First, for lateral shift, we substitute estimated value of dynamic axle load into formula (6), and obtain lateral axle load delta Q that makes safety margin against lateral force zero. Then we calculate STD of alignment irregularity by the method mentioned in section 3.2.

$$p = 0.85(a+bW) - \Delta Q \quad\text{---}\quad (6)$$

In the formula, parameters a and b are decided considering the difference in track structure (Fig 11). Dynamic axle load is estimated as 90% of static axle load in running test.

Second, lateral force limit is also calculated by the method in section 3.2.

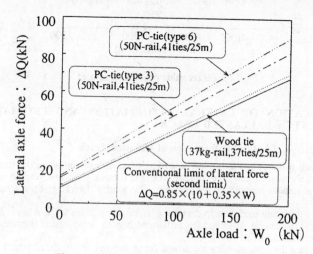

Fig 11   Limit of lateral force for track shifting

4   TRIAL CALCULATION OF TARGET VALUE FOR TRACK ALIGNMENT

Objects of our investigation are tilt and non-tilt super express trains. Calculation conditions are shown in Fig 12. Here, train speeds are set considering the cant deficiency 70mm for non-tilt train, 110mm for tilt train.

(1)   As train speed is higher on wider curve, target value from the view point of riding comfort is smaller for wider curve. For tilt train it is 0.5 to 1.0 mm smaller than for non-tilt train.

(2)   In case of tilt train, target value for lateral track shift is nearly constant regardless of curve radius because track strength on curves is high. But it becomes more severe than for riding comfort. In case of non-tilt train, it is more loose because train speed is low and cant deficiency is small.

(3)   Target value for limit of lateral force to rail fastening device is almost the same regardless of curve radius because track structure on sharp curves is strong. The limit for tilt train is a little smaller than for non-tilt train on account of curving performance.

(4)   For lateral maximum force, target value on tangent track is fairly smaller than on curved track. But for riding comfort, it is almost the same as on large curve.

(5)   Generally speaking, target value for track alignment depends on riding comfort on large curves, maximum lateral force on sharp curves.

514/030 © IMechE 1996

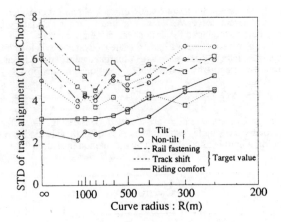

Fig 12 Target value of track alignment

5. QUANTIFICATION OF LATERAL DEFORMATION AND ESTIMATION OF ALIGNMENT OF BALLASTED TRACK

5.1 Quantification of Lateral Deformation of Ballasted Track

To study the track strength against an increase of axle lateral load following the train speed up on curves, not only direct track deterioration like lateral track shift, but gradual deformation like growth of alignment irregularity is important. This kind of lateral gradual deterioration is not clarified.

We made a full-scale track test to determine lateral force under vertical load. The result is shown in Fig 13. In general, after sudden settlement due to compression, increase of lateral deformation becomes constant.

Fig 14 shows the relation between lateral force and maximum deformation after initial deformation. The figure shows that alpha depends on lateral pressure from tie, and does not depend on vertical load. It shows that initial displacement α depends on lateral load(lateral pressure to tie), and does not depend on vertical load(rail pressure).

Fig 15 shows the relation between lateral force and increase of lateral deformation beta. The figures show that large lateral load and small vertical load contribute to lateral displacement growth. Also they show that if lateral load is smaller than a specific value, displacement never occurs.

And we confirm that lateral spring constant of ballast is influenced by load strength or initial deformation. According to the above mentioned results, we propose a formula to calculate initial deformation and deformation growth coefficient β.

$$\alpha = 1.24 \times 10^{-3} \cdot Q_t \quad (7)$$

$$\beta = a \cdot Q_t / K_b' - (b \cdot P_{r12} - c) \quad (8)$$

$$K_b' = \frac{d}{e \cdot Q_t^{1.8} + f} + g \cdot P_{r12} + h$$

$$P_{r12} = P_{r1} + P_{r2}$$

Here,

Q_t :Lateral pressure to tie(kN)
K_b :Coefficient of ballast lateral spring(MN/m)
P_{r1}, P_{r2} :Right and left rail pressures(kN)

In these formulas, the coefficients are as follows.
$a=3.90\times10^{-5}$, $b=8.38\times10^{-8}$, $c=1.67\times10^{-6}$, $d=21.5$,
$e=1.24\times10^{-3}$, $f=4.27\times10^{-2}$, $g=1.30$, h=-26.4

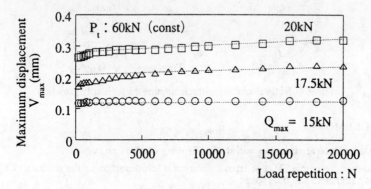

Fig 13   Lateral displacement under load repetition

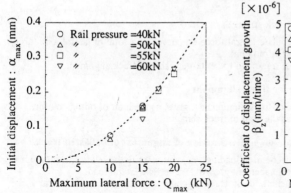

Fig 14   Lateral force and initial displacement

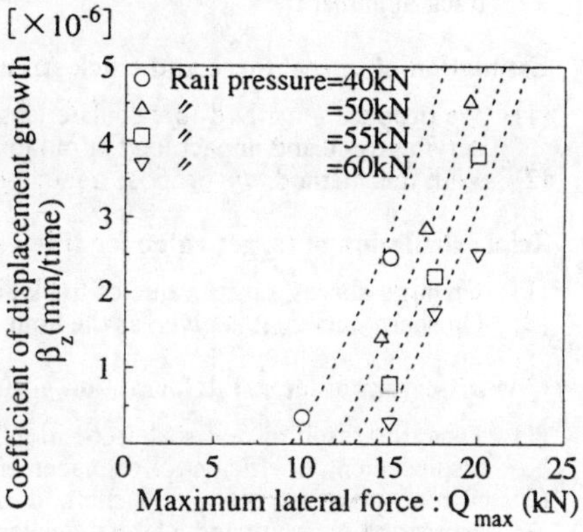

Fig 15   Lateral force and coefficient of displacement growth

5.2   Estimation Method of Alignment Irregularity Growth

We also propose a method to calculate alignment irregularity growth. First, we estimate maximum lateral force by the way shown in the chapter 3, and substitute that value into formulas (7), (8), then we obtain lateral deformation. Next we calculate lateral alignment growth by summing up the deformation of each axle. Fig 16 shows the calculation results. Alignment irregularity growth becomes larger according to train speed.

This method of estimation of alignment irregularity growth enables us to consider track structure and to decide optimum target value of track irregularity considering train speed, traffic volume, vehicle/track condition.

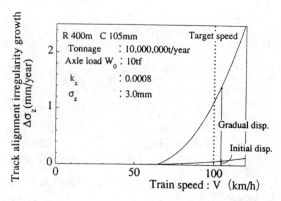

Fig 16   Trial calculation of track alignment irregularity growth

6. CONCLUSIONS

Estimation of lateral vibration and study of riding comfort on curves

(1) Standard deviation(STD) of lateral vibration is in proportion to the product of STD of track alignment and train speed.
(2) Tilt train realizes fairly good riding quality in comparison with non-tilt train.
(3) We set target value of riding comfort on curves, and proposed a corresponding target track alignment.

Estimation of lateral force and track strength

(1) We propose a method to calculate lateral force on curves as a total of lateral axle force, curving force and impact load at rail joint.
(2) With this method, we propose a method to set target value of track alignment.

Trial calculation of target value for track alignment

(1) On large curves, target value of track alignment is given as the limit of riding comfort.
(2) On sharp curves, it is given as the limit of track shift.

Quantification of lateral deformation and estimation of alignment of ballasted track

(1) From the result of real-scale repetition loading test, we quantify the relation among initial displacement, coefficient of displacement growth and vertical/lateral load.
(2) We propose an estimation method of track alignment growth using a lateral force estimation equation and a lateral displacement calculation equation.

REFERENCES

(1) Uchida, Nagato, Takai, Ishikawa, Track Maintenance on Sharp Curves for Tilting Car Operation(in Japanese), J-Rail '94, Tokyo, 1994, pp.85-90
(2) Uchida, Takai, Yazawa, Miwa, Setting Method of Track Alignment Target Value Considering Riding Comfort and Lateral Force on Curves(in Japanese), RTRI Report, 1995, Vol.9, No.12, pp.31-36
(3) Ishikawa, Uchida, Lateral Cyclic Deformation Characteristics of Railroad Ballast(in Japanese), RTRI Report, 1995, Vol.9, No.4, pp.31-36
(4) Uchida, Ishikawa, Miwa, A Study on the Effect of Speedup at Curves on Track Irregularity Growth(in Japanese), J-Rail '95, Tokyo, 1995, pp.121-124

C514/028/96

Track dynamic behaviour caused by high frequency rail/wheel contact forces

S MIURA MS, MJSCE and M ISHIDA BE, MJSCE
Railway Technical Research Institute, Tokyo, Japan

SYNOPSIS

Large dynamic forces act between wheel and rail due to rail surface irregularities at rail welded parts and/or wheel flats. With regard to such dynamic forces and track responses caused thereby, various models have been so far proposed, with which a lot of findings have been obtained. Considering the current status of an increasing train speed, however, it is necessary for the design of track and vehicle components to make prediction by those models more precise than ever. In fact, it is very important to understand dynamic track and vehicle behaviour due to such dynamic forces for the purpose of estimating various components of any kind of track structures. The authors proposed an analytical track dynamic model in which rail is considered to have a distributed mass and to be supported discretely at sleepers above ballast divided into three layers, which can cover the dynamic track and vehicle behaviour in the high frequency range. This paper also describes the dynamic response of track and vehicle caused by trains running on track irregularities whose waves were about 0.03 to 2.00m long using the analytical track model .

1. INTRODUCTION

Dynamic forces interacting between rail and wheel have an essential influence on the increasing track irregularities, wear and/or deterioration of materials of track components and environmental noise and vibration associated with track structures. Such dynamic interaction has been so far investigated and a lot of achievements has been obtained from their analytical results using various dynamic models (1)-(6). However, it has recently come to be very important to predict the characteristics of rail/wheel interaction, the vibration and stresses of track components including the effect of the measures for reducing or preventing them from the view point of the measures for environmental issues and maintenance of track due to an increasing train speed.

In this study, an analytical track model to be able to solve such problems has been proposed and its adequacy verified on the basis of measurements to some extent. In supporting the increased train speed, it is very important to develop the maintenance technology of track for the sake of controlling the deterioration and fracture of track materials and their influence on environmental issues such as rolling noise, vibration and others, resulting from suppressing the increase in dynamic wheel loads. Also, the estimated results of dynamic wheel loads caused by short or long pitch irregularities on rail surface whose amplitudes are 1/1000 of their wavelength are assumed are reported using the track model of a continuous beam discretely supported at sleepers.

2. BACKGROUND ON ANALYTICAL MODELS

The authors have so far analysed dynamic forces between rail and wheel, track vibration and others caused by trains running on track irregularities whose wavelengths are relatively short like rail surface irregularities occurring at rail welded parts with track subsiding caused by those irregularities, using a simple model which consists mainly of the unsprung mass of vehicle and the effective mass of rail. With regard to adequacy of such a simple model, it verified that the masses of carbody and truck and the stiffness of primary suspension don't very much influence the dynamic interaction between rail and wheel, and the dynamic behaviour of one wheel in a wheelset doesn't very much influence that of the other wheel, in comparison with a model whose degree of freedom is much larger than that of the simple model. The achievements so far obtained indicate that the simple model which consists of the unsprung mass of one wheel and the effective mass of rail has enough precision to predict the phenomenon of track dynamic behaviour at least under about 100Hz of excitation frequency(7).

Such a simple lumped mass model is, however, known to be unavailable for more than several hundred Hz of excitation frequency. Because the effective mass of rail in the simple model must be changed in response to excitation frequency. Thus the analysis using the continuous beam track model is considered indispensable to revealing the dynamic behaviour of rail and wheel at frequencies of over several hundred Hz. The model takes into account rail distributed mass and flexural rigidity in its longitudinal direction.

3. TRACK DYNAMIC MODEL

From the above discussion, the track model was established as a continuous beam of Bernoulli-Euler or Timoshenko as shown in Fig.1. In this figure, rail is assumed to be a continuous beam supported discretely at the fastening points where sleeper, ballast and substrate are elastically joined. Considered as mass of grains, the ballast is divided into three layers(8). Also, the Hertzian contact spring is assumed to act between rail and wheel in the model.

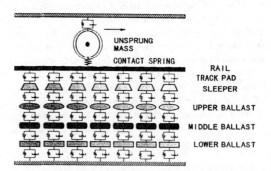

Fig.1 Continuous beam model

This Timoshenko beam model is expressed as follows:

$$EI\frac{\partial^2\varphi(x,t)}{\partial x^2} + \frac{GA}{k}\left[\frac{\partial y(x,t)}{\partial x} - \varphi(x,t)\right] - \rho I\frac{\partial^2\varphi(x,t)}{\partial t^2} = 0 \qquad \cdots\cdots (1)$$

$$\rho A\frac{\partial^2 y(x,t)}{\partial t^2} - \frac{\partial}{\partial x}\left\{\frac{GA}{k}\left[\frac{\partial y(x,t)}{\partial x} - \varphi(x,t)\right]\right\} = -\sum_{j=1}^{M} R_j(t)\delta(x-x_j) + \sum_{i=1}^{N} P_i(t)\delta(x-x_i) \qquad \cdots\cdots (2)$$

where,

$y(x,t)$: Rail deflection
EI : Flexural rigidity
I : Second moment of rail area

$\psi(x,t)$: Rotational angle
ρ : Density of rail
ρA : Mass per unit length of rail

$R_j(t)$: Rail/sleeper interacting force (rail pressure) x_j : Position of the jth sleeper
$P_i(t)$: Rail/wheel interacting force (dynamic wheel load) x_i : The position of the ith wheel
x_i : The position of the ith wheel $\delta(x)$: Delta function of Dirac
GA/k : Shear rigidity (where $1/k$ is corresponding to Tiomshenko shear coefficient)

Assuming mode summation transformations as the solution of the above equations, the rail deflection $y(x,t)$ can be expressed with the normalized vibration modal function $\Phi_n(x)$ and modal time coefficient $Y_n(t)$ and the rotational angle $\psi(x,t)$ can be expressed with $Z_n(x)$ and $\Psi_n(t)$ as follows:

$$y(x,t) = \sum_{n=1}^{\infty} Y_n(t)\Phi_n(x) \quad \cdots\cdots (3)$$

$$\varphi(x,t) = \sum_{n=1}^{\infty} \Psi_n(t)Z_n(x) \quad \cdots\cdots (4)$$

where, $\Phi_n(x)$, $Z_n(x)$: Modal shape function, $Y_n(t)$, $\Psi_n(t)$: Modal time function
Substituting eq. (3) and (4) to eq. (1) and (2) and integrate eq. (1) and (2) with respect to time, the displacements of track components and others will be gained.

Also, the equation of Bernoulli-Euler beam is simpler than that of Timoshenko beam since Bernoulli-Euler theory is omitted shearing deformation and rotational inertia of beam which are taken into account in Timoshenko theory. The equation of Bernoulli-Euler beam is expressed as follows :

$$EI\frac{\partial^4 y(x,t)}{\partial x^4} - \rho A\frac{\partial^2 y(x,t)}{\partial t^2} = \sum_{j=1}^{M} R_j(t)\delta(x-x_j) + \sum_{i=1}^{N} P_i(t)\delta(x-x_i) \quad \cdots\cdots (5)$$

where notation follows as the above shown equations of Timoshenko beam.
Substituting eq. (3) to (5) and integrate (5) with respect to time, the displacements of track components and others will be gained.

4. ANALYTICAL RESULTS AND VERIFICATION ON THE MODEL

Dynamic wheel loads measured with strain gauges on a revenue line and analytical results using Bernoulli-Euler beam model on the basis of the same rail surface irregularity as that of revenue lines are compared in Fig.2. Since the dynamic wheel load is derived from rail shearing strains in the measurement on track side, shearing strains corresponding to the output of the strain gauges set in rails are calculated from a rail deflection curve in the analytical model. As shown in Fig.2, the results coincide with each other not perfectly but quite well, which testifies to the adequacy of the Bernoulli-Euler beam model.

In addition, the dynamic forces interacting between rail and wheel obtained from the model are shown in Fig.2. The both waveform of dynamic forces doesn't coincide with that of shearing strains. This means,asmentioned before, a rail deflection doesn't

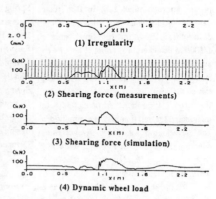

(1) Irregularity

(2) Shearing force (measurements)

(3) Shearing force (simulation)

(4) Dynamic wheel load

Fig.2 Verification by measured data

respond enough to a dynamic wheel load at high frequencies and it agrees with the findings(2) so far obtained that the vertical direct receptance of track for high frequency excitation comes down.

5. RELATION BETWEEN DYNAMIC WHEEL LOADS AND IRREGULARITIES

Bernoulli-Euler beam model was adopted in this analysis to understand the influence of some parameters on dynamic wheel loads. The technical details of vehicle and track components shown in Table 1 and short or long pitch irregularities on rail surface whose amplitudes are 1/1000 of their wavelength and waveforms are sinusoidal wave shown in Fig.3 used for the calculation. The reason why such irregularities are assumed is that power spectrum of the amplitude of irregularity is in inverse proportion to the third power of spatial frequency which is defined as one over its wavelength(9), which is considered to apply to the case of more than several meters of irregularity wavelength but is doubtful whether it can do to less than about a quarter meter of wavelength or not. The authors are of opinion that mechanism of more than several meters wavelength irregularity being formed is different from that of less than a quarter meter irregularity; the former depends mainly on ballast subsidence and the latter on wear or plastic deformation of rail, in fact, their damage mechanisms are different from each other. Then, for the time being , 1/1000 of wavelength is adopted here as an amplitude of the irregularity only for this study. On the other hand ,very few study has so far been carried out. However, from the above discussion, further study to understand the relation between amplitudes and wavelength of track irregularities covering the broad range of high frequency will be devoted to developing the track maintenance technology.

Train speed of 150km/h was set as a standard. The maximum values of dynamic wheel loads caused by rail head surface irregularities , hereinafter refereed to as irregularities, taking the values of train

Table.1 Technical details for analysis

Static wheel load	45.0	(kN)
Unsprung mass of vehicle (M_w)	830	(kg)
Contact spring (new profiles of rail and wheel)		
Curvature of wheel		
longitudinal direction of track	1/(860/2)	(1/mm)
lateral direction of track	0.0	(1/mm)
Curvature of rail		
longitudinal direction of track	0.0	(1/mm)
lateral direction of track	1/300	(1/mm)
Contact spring (worn profiles of rail and wheel)		
Curvature of wheel		
longitudinal direction of track	1/(860/2)	(1/mm)
lateral direction of track	-1/1700	(1/mm)
Curvature of rail		
longitudinal direction of track	0.0	(1/mm)
lateral direction of track	1/1500	(1/mm)
Rail mass (Type 50N)	50.4	(kg/m)
Rail flexural rigidity (Type 50N)	4.05×10^3	(MPa·m²)
Rail mass (Type 60kg)	60.8	(kg/m)
Rail flexural rigidity (Type 60kg)	6.34×10^3	(MPa·m²)
Track pad stiffness (K_r)	110000	(kN/m)
Track pad damping coefficient	98	(kN·sec/m)
Sleeper mass	100	(kg)
Spring coef. of supporting sleeper	1.78×10^6	(kN/m)
Damping coef. of supporting sleeper	980	(kN·sec/m)
Upper ballast mass	52.5	(kg)
Spring coef. of upper ballast	0.89×10^6	(kN/m)
Damping coef. of upper ballast	980	(kN·sec/m)
Middle ballast mass	52.5	(kg)
Spring coef. of middle ballast	0.81×10^6	(kN/m)
Damping coef. of middle ballast	980	(kN·sec/m)
Lower ballast mass	90.3	(kg)
Spring coef. of lower ballast	2.50×10^6	(kN/m)
Damping coef. of lower ballast	980	(kN·sec/m)

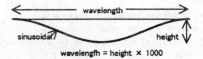

wavelength = height × 1000

Fig.3 Irregularity for analysis

speed, unsprung mass, track stiffness, rail/wheel contact spring coefficient, hereinafter refereed to as contact spring coefficient, and rail flexural rigidity as parameters are shown in Figures 4 to 8. In every case, there are two peaks of dynamic wheel loads around 0.03 to 0.07m and 0.6 to 0.8m of wavelength. The former peak corresponds to the natural vibration of effective mass of rail in contact with wheel with a Hertzian contact spring. The latter does to the natural vibration of unsprung mass coupled with effective mass of rail supported by track stiffness. The frequencies of those two peaks are about 900Hz and about 70Hz. Since the former vibration in the higher frequency range causes the high frequency variation of rail/wheel contact force to increase, it is not clearly enough but to a certain extent considered to exert an influence mainly on rolling noise and rolling contact fatigue(RCF) failures. The latter vibration is considered to have an influence on track deterioration and fracture of track materials, infrastructure noise and ground vibration.

5.1 Influence of train speed

Fig.4 shows the influence of train speeds on dynamic wheel loads. As shown in the figure, dynamic wheel loads at two resonance wavelengths in the case of higher speed become larger than those in the case of lower speed. The positions of the two peaks shift to longer wavelength in the case of higher train speed. It is particularly important to pay more attention to longer wavelength than so far done in both peaks of dynamic wheel loads. Also it is necessary to reduce the amplitude of irregularities and make track stiffness small.

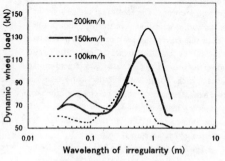

Fig.4 Influence of vehicle velosity

5.2 Influence of unsprung mass

Fig.5 shows the influence of unsprung mass on dynamic wheel loads. In this figure, since the peak of shorter wavelength is influenced by the natural vibration of rail effective mass, it is not influenced by unsprung mass. However, since the other peak of longer wavelength is influenced by the natural vibration of unsprung mass coupled with effective rail mass, the position of the peak shifts toward longer wavelength and the peak value gets largerdue to the increase in unsprung mass. Since the deterioration of track and track components are considered to be caused by dynamic wheel loads around the peak of longer wavelength, it will be veryeffective to reduce the unsprung mass in reducing such dynamic wheel loads.

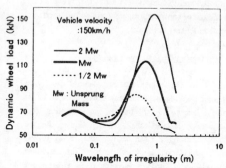

Fig.5 Influence of unsprung mass

5.3 Influence of track stiffness

Fig.6 shows the influence of track pad stiffness on dynamic wheel loads. In this study, only the spring coefficient of track pad is changed to investigated of track stiffness. In this figure, the peak of longer wavelength shifts toward shorter wavelength and its value gets larger with the increase in track stiffness, which means the track spring gets stiffer. The other peak value of shorter wavelength slightly decreases with the increase in track stiffness. This is because the peak of shorter wavelength is influenced by the effective mass of rail and the effective mass decreases due to the increase in track stiffness and the peak value of longer wavelength gets larger due to the increase in track stiffness.

5.4 Influence of contact spring

Fig .7 shows the influence of contact spring on dynamic wheel loads. In this study, the legend. "new profile" means that both profiles of rail and wheel are new and "worn profile" does
that both are worn. The coefficient of "new profile" contact spring is smaller than that of "worn profile" one. In this figure, the peak of shorter wavelength influenced by the effective mass of rail shifts toward shorter wavelength and its value gets larger with an increase in the coefficient of contact spring. On the other hand, the peak of longer wavelength influenced by unsprung mass coupled with effective mass of rail is not influenced at all by the increase in the coefficient of contact spring.

This discussion indicates it is important for the sake of reducing dynamic wheel loads in the range of high frequencies to design and control the interface of rail and wheel in contact with each other because the coefficient of contact spring depends on the interface of rail and wheel in contact.

5.5 Influence of rail flexural rigidity

Fig.8 shows the influence of rail flexural rigidity on dynamic wheel loads. Both peak values get a bit larger withthe increase in rail effective mass resulting from the increase in rail flexural rigidity. Considering the effect of enlarging a rail section on reducing rolling noise, an enlarged rail surely reduces the amplitude and velocity of vibration in comparison with the one before being enlarged if the same magnitude of wheel loads act on theabove-mentioned two rails. However, as shown in Fig.8,the larger rail gives rise to slightly larger dynamic wheel loads than the other rail on the

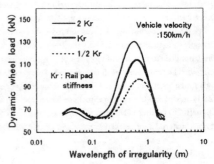

Fig.6 Influence of rail pad stiffness

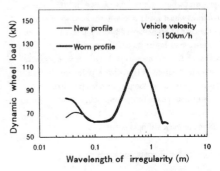

Fig.7 Influence of contact spring

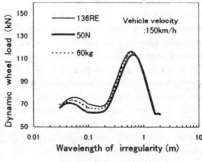

Fig.8 Influence of rail type

same irregularity of rail tread surface in the high frequency range of excitation which has great influence on the noise. The velocity of 60kg rail vibration is, hence, nearly equal to that of 50N one. Since the rail vibration velocity is pointed out to play an important role on rolling   noise(10), enlarging the rail section would not necessarily be effective to reducing noise ecause a largerail has a wider noise-emitting area than a small one has.

204

6. DIFFERENCE BETWEEN BERNOULLI-EULER AND TIMOSHENKO

The influence of some characteristics of track components on track dynamic response was so far investigated using Bernoulli-Euler beam model in this paper. The difference between Bernoulli-Euler beam and Timoshenko beam is hereafter described.

6.1 Rail vibration velocity

Regarding rail behaviour, the centre of irregularity was set in the middle of sleeper span to study the deference between Bernoulli-Euler beam and Timoshenko beam because the rail deflection in the case the centre of irregularity is set in the middle of sleeper span becomes larger than that in the case the centre of irregularity is set above a sleeper. Fig.9 shows the variation of rail vibration velocity of both Bernoulli-Euler beam model and Timoshenko beam model. In this figure, the both tendencies of Bernoulli-Euler beam and Timoshenko beam are almost the same as each other in the region near the peak of larger wavelength , though values of Timoshenko beam are slightly larger than those of Bernoulli-Euler beam.

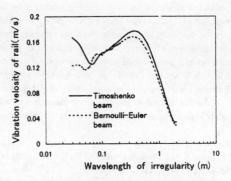

Fig.9 Difference of rail vibration velocity

On the other hand, the tendency of Timoshenko beam is quite deferent from that of Bernoulli-Euler beam near the region of the peak of shorter wavelength and the values of Timoshenko beam are much larger than those of Bernoulli-Euler beam, which is very interesting from the aspect of the influence of rail vibration velocity on rolling noise(10).

6.2 Ballast displacement

Regarding ballast displacement, the centre of irregularity was set above a sleeper to study the deference between Bernoulli-Euler beam and Timoshenko beam from the same reason as above mentioned. Fig.10 shows the variation of ballast displacement of upper layer. In this figure, the values of Timoshenko beam are as a whole slightly larger than those of Bernoulli-Euler beam. With regard to the deference between the two beam models, the deference in the side of shorter wavelength is larger than that in the side of larger wavelength. And almost no deference is observed in the region of larger wavelength.

From the aspect of track deterioration and/or ballast subsiding, there seems to be not so large difference between Timoshenko beam and Bernoulli-Euler beam. Further study for verifying it experimentally will be needed.

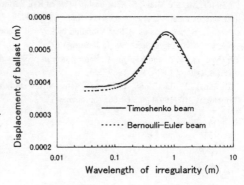

Fig.10 Difference of upper ballast displacement

7. CONCLUSIONS

Continuous beam track model which consists of Bernoulli-Euler beam or Timoshenko beam as a rail and is supported discretely at sleepers joined with ballast divide into three layers has been proposed to predict the dynamic forces interacting between rail and wheel, and the response of track excited by the forces. In comparison with the measured results in the actual track, it is made clear that the continuous beam model has enough precision to predict the dynamic phenomena of rail track components. Accordingly, this study proves the possibility of the continuous beam model to predict the phenomena discussed here.

In this study, it was made clear that computer simulation model which was able to predict the real phenomena precisely enough was very much useful for estimating the influence of large dynamic wheel loads on track structure and its components from the aspect of track maintenance toward the increase in train speed. Also, it was confirmed that Timoshenko beam would be needed to study rolling noise from the difference between Bernoulli-Euler beam and Timoshenko beam in high frequency excitation.

Further study about specifying some parameters of ballast and considering non-linearity of track materials will be expected to make use of such a computer simulation model more effectively

References
(1) Kunieda, M., Theoretical study on the dynamic effect of unsprung mass of railway vehicle, RTRI Research Materials, 1954, pp.11-24 (in Japanese)
(2) Grassie,S.L.,Gregory,R.W. Harrison,D.and Johnson,K.L.,The dynamic response of railway track to high frequency vertical excitation, J. Mech. Engrng. Sc. 24(1982), pp.103-111
(3) Knothe, KL. and Grassie, S.L., Modelling of Railway Track and Vehicle/Track Interaction at High Frequencies, Vehicle System Dynamcs, 22(1993),pp.209-262
(4) Nielsen, J., Train/Track Interaction Coupling of Moving and Stationary Dynamic Systems, Ph.D dissertation, Chalmerse University of Technology, 1993
(5) Miura, S., Abe, N., and Ishida, M., Dynamic Load on Rail Welding Joint, Proc. of S'Tech '93, Nov. 1993
(6) Cai, Z. and Raymond, G.P., Modelling of the dynamic response of railway track to wheel/rail impact loading, Structural Engineering and Mechanics, Vol.2,No.1(1994), pp.95-112
(7) Ishida, M., and Miura, S.,Relationship between Rail Surface Irregularity and Dynamic Wheel Load, Proc. of 10th International Wheelset Conference, Sidney, Sep 1992
(8) Miura, S., Analysis on Ballast Vibration by Layered Model, Proc. of the 46th Annual Conference of Japan Society of Civil Engrs, Sep 1991 (in Japanese)
(9) Evenson, D., A. and Kaplan, A., Some problems of Wheel/Rail Interaction Associated with High Speed Trains, Bull. of IRCA, Sep. 1969
(10) Sunaga,Y.and Ushida,M., Current Status on the study and Reducing Measure on Wheel/Rail Noise , Tribologist , 10 (1993) ,pp.860-865 (in Japanese)

C514/053/96

Extending the limits of pantograph/overhead performance

G A SCOTT MA, MIMechE and M COOK
BR Research, Derby, UK

SYNOPSIS

With the rising level of international high speed rail traffic in Europe and the resulting need for locomotives to be able to run from one railway system onto another, compatibility between the pantograph and the overhead line is becoming a significant problem. Overhead line equipment is a complex arrangement of wires and cables which are disturbed into complex dynamical motions by the passage of a pantograph. This paper gives examples of the factors which influence the dynamic behaviour of the pantograph/overhead system and describes techniques which can be used to take full advantage of the system's performance capability.

INTRODUCTION

At the present time, the level of international rail traffic in Europe is steadily rising and current plans envisage a high speed rail network connecting the major cities of the European Union. If trains are to run at speed from one railway to another, a significant problem which needs to be addressed is the compatibility of the pantograph and overhead line. In general, each country in Europe has developed its own overhead line system and has optimised pantographs for use on it. Historically therefore, international trains have stopped to change locomotives when crossing from one system to another, with consequent delays which would be otherwise unnecessary. In view of increasing business demands to minimise journey times, these time penalties are no longer considered to be acceptable. However, the problem of designing a pantograph which is capable of operating satisfactorily on a range of different overhead designs is one which first needs to be solved.

Overhead line equipment is a complex arrangement of wires and cables which are disturbed by the passage of a pantograph into a complex series of wave motions and, since the speed of transmission of these waves is greater than the train speed, the pantograph is continually running into its own wake. The dynamic interaction of pantographs and the overhead line is therefore an important aspect of the system design and plays a major part in determining the maximum speeds which are achievable.[1]

The railway businesses are always keen to increase line speeds whenever it becomes feasible. From the current collection point of view, the ideal solution to higher speed operation will always be a new higher specification overhead system with a considerable safety margin. However, the high capital costs of installing new overhead line equipment can often be difficult to justify financially, and therefore, there is a considerable incentive to obtain the optimum performance from existing equipment. Even when new high speed equipment has been installed, there is still the challenge of maintaining speeds when running off the new system onto conventional routes.

SYSTEM CONSTRAINTS

In Britain, electrification has used a relatively lightweight, low tension design of overhead line equipment with considerable development going into a sophisticated pantograph in order to maintain current collection quality at speed. In France, good quality current collection on TGV lines has been achieved by the use of a high tension overhead system. In Germany, the tension used by DB for its ICE lines is midway between that used by BR and SNCF, and a combination of a stitched overhead design and a lightweight pantograph has been used to maintain performance. Historically, there has been a tendency for pantograph and overhead designs to evolve simultaneously, such that each pantograph tends to become optimised for operation on its own national overhead system. It cannot automatically be assumed that a high speed pantograph which performs well on its own overhead will perform acceptably at line speed on another overhead design; this is particularly the case when overhead systems are optimised for different levels of uplift force.

A particular source of incompatibility between systems is whether they run with ac or dc power. With high voltage ac systems, the amount of current required to power the train is relatively low, and such systems are usually based upon lightweight overhead conductor wires and light pantograph designs. The need to avoid excessive movements of the light and often relatively low tensioned wires means that uplift forces are generally kept as low as possible on ac systems. In contrast, with low voltage dc systems, much greater levels of current are required, and correspondingly heavier conductor wires and pantograph contact strips are necessary. In such circumstances, continuity of contact can be better guaranteed with higher levels of uplift force. There is also a widely held belief that dc current flow can become interrupted, even without actual contact loss, unless reasonable levels of contact force are maintained. Accordingly, dc systems are usually operated with much greater levels of uplift force, and dc overhead equipment tends to be designed such that it gives its optimum performance under these conditions.

Thus, dc pantographs are in general dynamically inferior to ac pans, on account of their greater dynamic mass. But, more crucially, if operated on ac equipment with their normal uplift force characteristics, they are liable to cause dangerous levels of wire displacements. The operation of an ac pantograph on a dc overhead system is a more realistic proposition, but it may not necessarily be able to perform satisfactorily at line speed, without increasing uplift force level beyond its normal specifications.

ASSESSMENT OF CURRENT COLLECTION QUALITY

In order to make clear decisions regarding the compatibility of any particular pantograph/overhead combination, it is necessary to be able to define current collection quality in unambiguous numerical terms.

In general, the dynamic performance of the system will be considered acceptable if the current flow to the locomotive is not interrupted, if the forces or electric arcing generated between pantograph and overhead line are not so large as to cause damage or high rates of wear, and if the physical displacements induced in the system do not cause mechanical or electrical clearances to be exceeded.[2] Expressing these requirements in a quantitative way can be a problem.

Historically, BR technical specifications have stated that loss of contact between the pantograph and overhead line should be less than 1% at normal operating speeds. However, with today's strict requirements concerning electromagnetic compatibility, the interference caused by stray currents associated with loss of contact needs to be kept to a minimum, and for many modern applications, a figure of 1% loss of contact would be considered highly unsatisfactory.

Large contact forces can cause problems in a number of ways; possible physical damage to the overhead or pantograph, the resulting large deflections in the wires increasing the risk of infringing clearances, and increased wear of components.

The physical clearances and limits of deflection of the overhead are relatively easy to determine. However, these values cannot be used directly in assessing performance, as some safety factor needs to be incorporated to take account of the inevitable deterioration in performance which will occur under severe operating conditions.

Recent discussion among European railways has resulted in statistical analysis of the contact force being adopted as opposed to a loss of contact measurement. A major disadvantage of using loss of contact is that a measurement of zero gives no clear indication of whether the system is performing well within its maximum speed capabilities, or whether only a small further speed increment would cause the contact performance to break down.

The statistics of contact force can easily be obtained from measured or calculated force values. In statistical terms, if the force is assumed to have a Gaussian distribution with mean value M and standard deviation σ, the dynamic range of contact force can be considered as extending between an upper limit of $M+3\sigma$ and a lower limit of $M-3\sigma$. Loss of contact is then indicated if $M-3\sigma < 0$. Strictly, the range between $M+3\sigma$ and $M-3\sigma$ will contain 99.73% of all contact force values. Thus if $M-3\sigma = 0$, contact loss will occur 0.135% of the time and the force will exceed 2M for 0.135% of the time.

In fact, experiments and simulations have each shown that the contact force variation cannot truly be described as having a Gaussian distribution. Nevertheless, comparisons between experimental measurements of loss of contact and statistical analysis of simulated results have shown a remarkably good agreement, provided that the frequency bandwidth of the simulation is sufficient to include all the major forcing frequencies of the system (Figure 1).

In recent years, BR Research have devised a graphical representation of current collection performance, as shown in Figure 2. Thus, in a single illustration, it is possible to concisely display the following key performance parameters for a train with two raised pantographs:

Mean Contact Force
Dynamic Range of Contact Force: $M+3\sigma$, $M-3\sigma$
Statistical Occurrence of Loss of Contact
Statistical Occurrence of Low Contact Forces
(below a specified safety margin, eg 30 N or 50 N)
Average Wire Displacement at Registration Arms

SENSITIVITY TO KEY PARAMETERS

Many of the key system parameters vary depending on the specific conditions present at a particular time. During its lifetime, the system will be subjected to a whole range of these parameter variations and it is important to consider each of these when assessing the system performance limits. The sensitivity of a typical overhead/pantograph system performance to a range of key parameters is illustrated in Figures 3 to 7.

It is generally accepted that the single most critical parameter affecting overhead performance is the contact wire tension. However, when specifying the wire tension necessary to achieve a desired level of performance, it needs to be appreciated that auto–tensioning devices are far from perfect; the actual operating tension may vary considerably, particularly during periods of rapid temperature fluctuation. Figure 3 demonstrates the great sensitivity of the system to contact wire tension, and in particular, the extent to which performance can deteriorate if loss of tension occurs.

In contrast to contact wire tension, Figure 4 shows that performance is much less sensitive to tension variations in the catenary wire. However, these results show only the direct effects of changes to catenary wire tension; an important secondary effect will be the influence of catenary wire tension upon contact wire sag.

The sensitivity of the system to contact wire sag is shown in Figure 5. The basic principle of designing a system with a sagged contact wire is to compensate for the variation in overhead flexibility between the end of span and mid–span positions, such that the pantograph runs as near as possible with a level trajectory. However, the optimum value of sag may depend upon operating speed and may differ for each design of pantograph which operates on the system, particularly where the pantograph uplift force characteristic varies significantly.

Figure 6 shows the sensitivity of the system to errors in setting the contact wire height. Simulations of overhead response normally assume that the heights of the dropper bottoms are set to fit a smooth parabolic profile. To examine the importance of installation and maintenance tolerances, profile errors have been introduced to the model, such that successive droppers have positive and negative height errors of the values indicated. It can be seen that whilst errors of ± 1 or 2 mm are fairly insignificant, errors of ± 10 mm can have a very adverse effect upon system performance.

One of the most important parameters affecting pantograph/overhead system compatibility is the static uplift force applied to the pantograph. Provided that the overhead design is fairly robust, an increase in uplift force will usually result in a better performance, in terms of loss of contact. In general, operating at a higher mean force can be expected to improve the safety margin of M−3σ above zero, as long as the system performance is reasonably stable at the speeds being considered. Once the speed is reached where M−3σ becomes negative, it is unlikely that an increase in uplift force will offer any significant improvement.

Pantograph uplift force is also a major factor influencing overhead system displacements. As force levels rise, greater movements of the overhead lines will be experienced, leading to a greater possibility of mechanical and electrical clearances being infringed. The optimum level of uplift force is therefore often a compromise between on the one hand, a high enough force to minimise the risk of contact loss, and on the other hand, a low enough force to avoid any risk of excessive wire movements.

However, any optimisation of uplift force on the pantograph also needs to take into account aerodynamic forces. The air flow acting on the pantograph on any particular day, in any particular location and at any operating height is a factor which can be extremely inconsistent. Aerodynamic forces acting on the pantograph can be controlled to a large extent provided that the wind flow is smooth and the direction is constant. In practice, wind conditions are extremely variable and unpredictable. Particular problems can arise when the pantograph is subject to high levels of crosswind and upward air flow. Figure 7 shows the effect that crosswinds acting on a pantograph may have upon the wire uplift at registration arm positions. These simulated results confirm practical experience that wire uplifts in windy conditions can be in excess of twice those which occur during calm weather.

OPTIMISATION OF SYSTEM PERFORMANCE

This kind of sensitivity study will identify the most critical parameters affecting the performance of any particular overhead/pantograph system. Information concerning the likely variations of these parameters can be obtained from various sources: overhead line maintenance teams may routinely acquire measurements of contact wire tension, height and stagger; wind tunnel tests can be carried out to assess aerodynamic forces likely to act on the pantograph under a range of conditions; on−site measurements of wind speed and wire uplift can be made at specific locations.

A knowledge of the practical tolerances on these key parameters can then be used to give a more realistic assessment of the system performance, as compared to the idealised performance which will be indicated by simulations of perfectly maintained design case equipment. A better understanding of the tolerances on performance which are achievable can lead to design criteria being more precisely defined. A limit on wire uplift can be specified for nominal conditions such that available clearances will not be exceeded in worst case conditions. Similarly, a target value of M−3σ can be set for design case simulations such that M−3σ will not fall below zero in practice.

Having a more precise understanding of the achievable performance of an overhead system can lead to substantial cost savings. Faced with a need to upgrade an overhead system to operate at greater speeds, there may be a tendency to overdesign to a much higher performance standard than is actually required. Alternatively, when only a modest increase in operating speed is required, it may be more cost effective to focus attention on optimising the key performance parameters, or reducing the practical tolerances which exist.

Having confidence in the overall capabilities of the basic system design, it is then possible to identify the weak points in the system and examine the possibilities of improving performance in certain specific areas. Usually, the maximum speed capability of the system is limited by performance at discrete features, such as neutral sections, wire gradients, level crossings, overbridges, etc.

Figure 8 shows pantograph response over a typical overbridge section on a high speed route. It can be seen that pantograph performance on the initial level wire section is quite acceptable. As the pantograph negotiates the down gradient, mean force levels rise, and this will inevitably result in increased wire uplifts. Performance at the bridge itself is particularly poor. Then, as the pantograph rises on the exit gradient, there is a noticeable lowering of force levels, which will increase the likelihood of loss of contact occurring.

There is much scope for using simulation techniques to improve performance at the overbridge: optimising dropper spacings, the grading of the contact wire height, the bridge arm support positions and the dynamic properties of the resilient arms. Performance on the gradiented sections is often a compromise between providing sufficient uplift force on the pantograph to maintain contact on the up gradient, whilst limiting uplift force on the down gradient such that contact wire movements are not excessive. Where this compromise cannot be achieved, there is the possibility of considering a partial system upgrade (to a heavier, higher tensioned overhead, for instance) at critical locations.

CONCLUSIONS

Dynamic interaction of pantographs and overhead line equipment plays a fundamental part in the safe operation and reliability of the system.

Good system performance can be obtained by building new high speed lines with a high quality overhead design. However, this is very expensive, and if existing lines can be used, savings can be made which might enable financial justification where otherwise there is none.

New expensive systems can be designed to perform well with large factors of safety over nominal performance. Confidence in the performance of the existing system requires more detailed knowledge of its behaviour in a wider range of conditions.

A combination of experimental information and computer simulation enables this information to be obtained, and therefore, full use can be made of the capabilities of the system.

REFERENCES

1.   Gostling, R. J. and Hobbs, A. E. W. "The Interaction of Pantograph and Overhead Equipment: Practical Applications of a New Theoretical Technique". Proc. I. Mech. E. Vol. 197 No. 13, Jan 1983.

2.   Betts, A. I., Holmes, R. and Hall, J. H. "Defining and Measuring the Quality of Current Collection on Overhead Electrified Railways". I. E. E. Conference "Electric Railways for a New Century", Sept 1987.

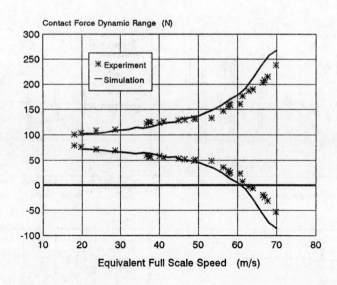

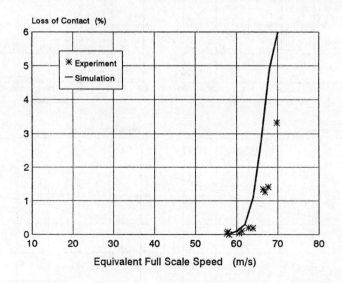

FIGURE 1   RELATIONSHIP BETWEEN CONTACT FORCE
DYNAMIC RANGE AND LOSS OF CONTACT

COMPARISON OF SIMULATED AND
EXPERIMENTAL RESULTS

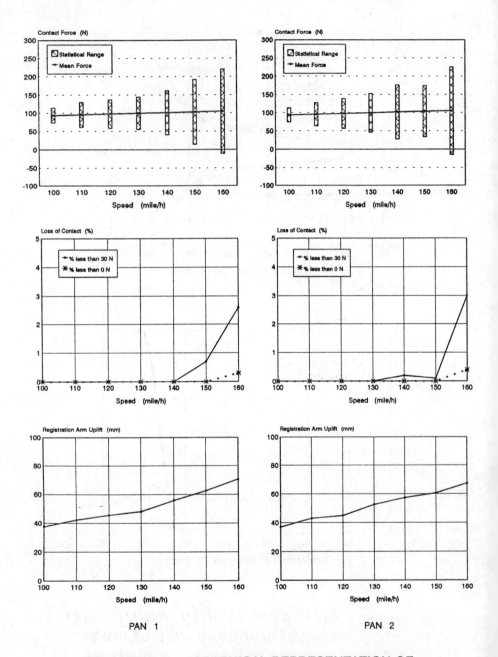

PAN 1   PAN 2

FIGURE 2   GRAPHICAL REPRESENTATION OF
                              KEY PERFORMANCE PARAMETERS

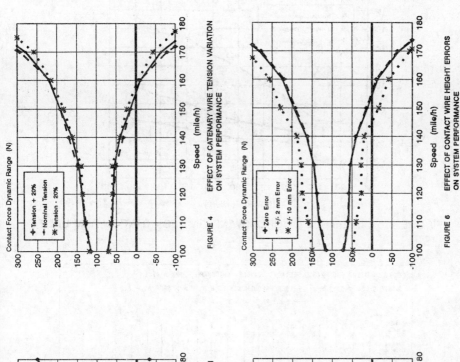

FIGURE 3 EFFECT OF CONTACT WIRE TENSION VARIATION ON SYSTEM PERFORMANCE

FIGURE 4 EFFECT OF CATENARY WIRE TENSION VARIATION ON SYSTEM PERFORMANCE

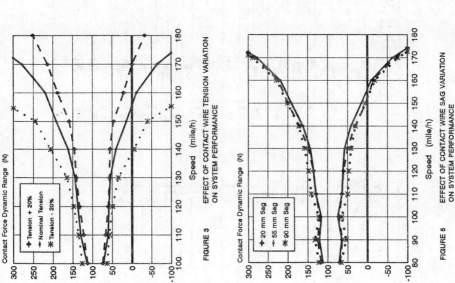

FIGURE 5 EFFECT OF CONTACT WIRE SAG VARIATION ON SYSTEM PERFORMANCE

FIGURE 6 EFFECT OF CONTACT WIRE HEIGHT ERRORS ON SYSTEM PERFORMANCE

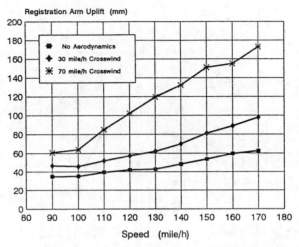

FIGURE 7   EFFECT OF CROSSWIND ON REGISTRATION ARM UPLIFT

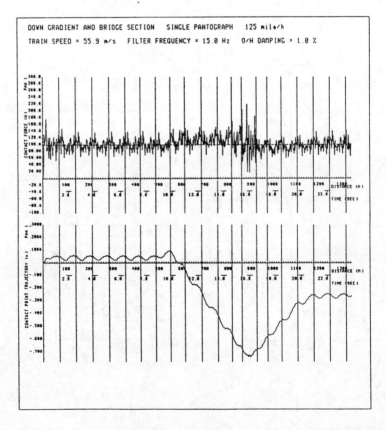

FIGURE 8   PANTOGRAPH RESPONSE ON HIGH SPEED BRIDGE SECTION

Developments in MAGLEV and Braking

C514/018/96

The role of brake technologies and optimizing brake systems for higher speeds on Shinkansen and conventional vehicles

N KUMAGAI MJSME and **I HASEGAWA** MJSME
Railway Technical Research Institute, Tokyo, Japan

SYNOPSIS

We would like to explain in this paper the role of brake technologies, optimisation of several brake control systems and developing technologies of Shinkansen and conventional vehicles in Japan.   In order to improve the brake performance, there are two aspects to be addressed; adhesion and braking force.   For example, in the aspect of adhesion, braking force pattern control, train set force control, abrasive block, injection device of adhesion-enhancing material, sophisticated wheel slide control method; in the aspect of braking power, hybrid rail brake, hydraulic brake, C/C disk brake.   Railway Technical Research Institute and JR railway groups have been using some of the above methods and the others are being developed now.

1.   INTRODUCTION

The primary purpose of brake technologies is to shorten braking distance safely and reliably. Improved brakes are necessary for trains to operate at higher maximum speeds. In urban areas higher deceleration is necessary for efficient commuter trains. In addition, brakes must be able to protect trains from sudden collapse of elevated railway bridge and track structures, as occurred in the Hanshin Earthquake in January 1995.

The important issue in increasing deceleration in the high-speed range is to ensure braking power in wet weather, when the adhesion between wheel and rail is low.   There are two methods for this. The first method is to increase the adhesion and make efficient use of it when wheels are sliding. The second method is to improve the present system's braking power as much as possible.

In order to increase the maximum speed of rolling stock, the Railway Technical Research Institute(RTRI) takes the following methods:
In the aspect of adhesion;
   1) optimising the braking force
   2) Increased adhesion method to raise the rail/wheel adhesion force
      For example, an abrasive block, an injection device for adhesion enhancing material
   3) Sophisticated wheel slide control method to avail the adhesion force as far as possible

In the aspect of braking power;
  1) Non-adhesion brake to supplement the shortage of a brake force
     For example, a rail brake
  2) Refinement of the basic brake device.
     For example, a hydraulic brake, a C/C disk brake
     With progress in technical developments, not only the improvement in function and performance, but also the reduction of the cost of introducing these developments should be kept in mind.

2.   THE EFFICIENT USE OF ADHESION

2.1   Optimising the Braking Force

(1) Braking force pattern control[1]
     In designing the performance of vehicle or planning the operation of Shinkansen trains, the following equation has been used since 1964,

$$\mu=13.6/(v+85)$$

where μ=adhesion coefficient, v=train speed(km/h).   This equation was assessed from the experimental results on the roller rig. The planned value determines the lower limit of adhesion coefficients applicable to the case where rails are under wet conditions. The values of deceleration for various series of Shinkansen trains have been set according to the above equation shown in fig.1.
     This is a type of speed/adhesion pattern control method in which the braking force is controlled as a function of speed. This is a method to decrease the brake cylinder pressure in the high-speed range and to increase it in the low-speed range. This method is applicable to the 130 km/h rolling stock.

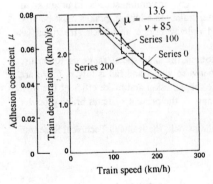

Fig.1   Designed values of
        Shinkansen trains

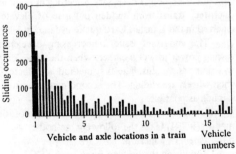

Fig.2   Wheel sliding occurrences
        (1988-1989, a Shinkansen train)

(2)Train set force control[1]

This concept is to utilize the adhesion force of vehicles at the middle and rear of a train. This approach is justified by the fact that,when a train runs on the rail surface moistened with rainwater, it is the wheels of leading vehicles that are most seriously affected by water. Meanwhile the further rearward a vehicle is located in a train, the less the vehicle's wheels are affected by moisture. The observation of wheel sliding occurrences at various vehicle locations in a train clearly shows that the further reaward a vehicle is located in a train, the less it suffers sliding, as shown in fig.2. Recently, in the light of such knowledge, field tests have demonstrated that full utilization of the higher levels of adhesion towards the rear of the train can improve the overall performance by 30%.

2.2 Adhesion Force Enhancing Method

On the Shinkansen, the distance necessary for the emergency brake to stop a train running at 270 km/h is about 3900 m (the average deceleration is 2.8 (km/h)/s).   If the braking power of the train were intensified in the high speed range, wheel slides would occur frequently in wet weather since adhesion between wheel and rail is low at high-speeds.   In that case the desired effect of shortening the braking distance could be lost.

(1) Abrasive Block[1]

The design value determines the lower limit of adhesion coefficients applicable to the case where rails are under wet conditions. The values of deceleration for various series of Shinkansen trains have been set according to the equation shown in fig.1.   However, in reality, sliding or spinning could occur even under normal braking or driving if surface conditions are particularly severe.

In the case of track devoted to high speed railway such as the Shinkansen, it is the presence of water on the contact surface that makes the speed effect on adhesion particularly sinificant.   In this condition, it is noted that surface roughness significantly affects the adhesion coefficient.   Consequently, wheel surface treatment can improve the adhesion properties during highspeed operation.

An example of the related measures is an abrasive block for increased adhesion(fig.3). The purpose of this device is to form many asperities on the wheel tread during braking, which leads to an increase of the adhesion force when water exists between wheel tread and rail surface. This was first introduced on the Tokaido Shinkansen electric multiple unit(emu). The comparison of wheel flat occurrence rate before 1972, when the block had not yet been adopted, with that after 1974 reveals a clear difference between them. The adhesion improvement effect was measured with a test train and verified the fact that the abrasive block made adhesion 20 to 30% higher than that of the conventional cast iron block.   At present the block is mounted on all the Shinkansen trains.

Fig.3   Abrasive block

(2) Adhesion enhancing material injection system[2]

As a means of increasing the wheel/rail adhesion coefficient, a method to spray ceramic particles on the rail surface is being developed(Fig.4). The spraying, particularly during high-speed operation, requires the following additional conditions: (a) Precise injection of the increased adhesion materials between wheel and rail; (b) Quick injection of an increased volume of adhesion materials; (c) No damage to the rail surface by the accumulation of injected materials.

Fig. 4   Increased adhesion material injection system

The increased adhesion material injection system for high-speed operation uses particles such as ceramics with a particle diameter of 10 to 200 μ m, and injects them into the wheel/rail contact area during high-speed operation at about 200 km/h. It is effective to continuously inject during emergency braking or to supply only at the time of wheel slide.

The performance laboratory tests confirm that it is a highly effective adhesion measure with an increase of adhesion coefficient from less than 0.1 to around 0.3 under wet weather conditions.

2.3   A High Performance Slip Rate Control System[3],[4]

The wheel slide lengthens the braking distance. So for the effective use of the wheel/rail force and an increased braking force, a sophisticated high response wheel slide control system is needed.

In order to suppress the wheel slide, a brake control system has been introduced for excellent high-speed rolling stock.   However, in the conventional control system, when the slide control was frequently applied, the braking distance tended to be longer. In order to avoid this, a high performance slide control that makes efficient use of adhesion between wheel and rail becomes necessary.

Recently, it has become possible to realise the change of adhesion force when a wheel of a vehicle slides.   Consequently, it has been confirmed that the slide control within the range of smaller wheel slide ratios, that is, less than approximately 5% is desirable as shown in fig.5. Therefore, 'a slip rate control wheel slide control system' based on the above principle has been developed.   This is a wheel slide control system that generates brake force while keeping the wheel slip rate under approximately five percent.

In the slip rate control, the number of pulses from the pulse sensor mounted on each axle and the pulse width are counted by a micro computer at a clock frequency of 100 MHz or more. This makes it possible to shorten the computing time for the wheel velocity and deceleration by one fourth and one fifth of the conventional computing time, respectively, while maintaining the accuracy of the detection.

In the slip rate control method, the level of air pressure reduction is changed according to the slide deceleration. The brake cylinder pressure is decreased stepwise in several stages with the extent of wheel slide constantly observed.

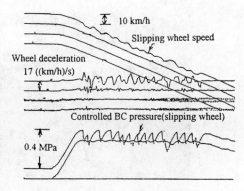

Fig. 5   State of emergency braking
with the slip rate control

That is to say that the goal of this new method is to keep the slide percentage (the differences in velocity) low and keep the level of air pressure reduction to a minimum(Fig.5). In the brake test using a train of Series-E501 EMU (East Japan Railway Company) with the slide percentage control, the velocity of the sliding wheel was controlled to keep the slide within a narrow range.   Moreover, when compared to the braking distance in dry conditions, its braking distance showed little difference.   From this result, it can be said that effective use has been made of the adhesion between the wheel and the rail.

3.   BRAKING POWER

3.1   Hybrid Rail Brake[5]

There are two kinds of rail brake, the eddy current rail brake which uses the magnetic drag of eddy current generated on a rail and the electro-magnetic rail brake which uses the friction generated when it adsorbs a rail. As a result of tests, the findings are that the eddy current rail brake requires a very large exciting current at a significantly high rail temperature. Therefore in cooperation with the East Japan Railway Company, the Railway Technical Research Institute has developed a hybrid rail brake(Fig.6) i.e. 'adsorption eddy current rail brake' which makes the best use of the advantages of both the eddy current rail brake and the electromagnetic rail brake. The following are the characteristics of an adsorptive eddy current rail brake.

- It requires only one tenth of the exciting current required by an eddy current rail brake to obtain an equal level of braking power.
- Eddy current and friction have little effect on rails and on factors related to signals.

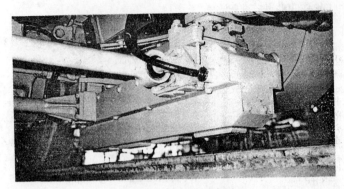

Fig. 6   Hybrid rail brake

The frictional materials of the adsorption eddy current rail brake are iron-base sintered alloys. This brake uses a coil exciting current of 30 A (DC 100 V).

A rail brake is to be used together with an adhesion brake.   Given that the goal is to stop a train running at 160 km/h within 600 metres by the emergency brake and the estimated adhesion brake deceleration is 4.5-5.0 (km/h)/s, the brake deceleration required for a rail brake will be 2.1-2.6(km/h)/s. If the weight of a car is 35 tons, 6000 N brake force is necessary for each rail brake to achieve this level of deceleration.

The deceleration of a train with adhesive brakes and hybrid rail brake (experiment using a train of JR Shikoku) was 6.7 (km/h)/s at 148 km/h initial braking velocity on the wet rail. Based on this deceleration, the estimated braking distance at 160 km/h is about 580 m.

The rise in rail temperature due to the hybrid rail brake is an important problem to be solved. For this reason, based on the test data, a simulation was conducted on an increment of the axial force in the rail in case of the rail brakes acting on an 11-car set. The axial force increment in case of the 11-car train comes to 4.8°C in terms of the mean rail temperature rise. Even if the temperature is added to the allowable rail temperature rise of 35°C, the total value will be lower than the temperature equivalent to the minimum rail buckling strength.   This fact indicates that the adsorption eddy current rail brake does not bring about a risk of rail buckling under the braking force conditions mentioned above.

3.2   Hydraulic Brake[6]

On the Shinkansen, a hydraulic diskbrake cylinder is employed to downsize the disk brake device. Air pressure is converted to oil pressure through an intensifying cylinder. As a result, the disk brake system becomes heavier and hinders the reduction of train weight. For this reason, a hydraulic control brake device was developed that directly controls the oil pressure in the Shinkansen's disk brake cylinder.

The hydraulic brake device is composed of a brake control device and a brake module. The brake module consists of an oil pressure generating part, a pressure accumulating part, a brake control part, and an emergency control part. A hydraulic brake module controls the disk brake pressure for two axles. The pressure generating part generates pressure by a hydraulic pump integrated with a motor. The pressure accumulating part has two accumulators for the service brake and the emergency brake. The pressure is controlled digitally by two valves of on-off control action, release valve, lockout valve. The emergency control part applies the second accumulator's oil pressure when the control system detects an unusual condition.

When a hydraulic brake system is mounted on a bogie(fig.7), the brake response improves because the hydraulic brake line is shorter. However, in this case it must be more vibration-resistant than the system mounted on a carbody. Jerk control is also possible if the control method is improved.

The primary gain from the introduction of a hydraulic brake is that it makes the train lighter. The mass is estimated to be one third of that of the conventional pneumatic brake.

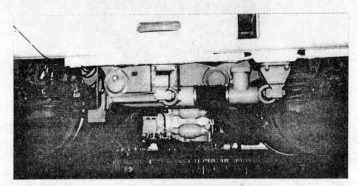

Fig.7   Hydraulic brake

3.3   C/C Multi-disk Brake[7]

With the mechanical brake system of Shinkansen vehicles, the heat resistance of the disk brake is one of the critical factors for speedup.   In order to consider a running speed of 350 km/h in the near future, a new type of disk device composed of new non-metal materials is being developed.   As a new material, Carbon/Carbon (C/C) composites which one used in the aircraft industry because of their light weight properties are being introduced into the rail industry.   C/C materials are crystals of graphite with specific gravity 1.8 which have been treated in temperature of 1500~1800°C.

There are a few differences with the use of C/C disk on railways compared with those used in aircraft.
- initial braking speed is high, braking energy is large but braking force is small according to adhesion force between rail and wheel
- bearing and disk are rotating at high speed, so the temperature of the bearing is an important factor
- repeated braking affects the life and temperature of the disk

Considering the above factors, a trial type multi-disk brake system was produced. This disk brake system is directly connected to the shaft of the main motor(Fig.8) and the whole brake system is mounted on the bogie for reducing the unsprung mass.   This system was tested on the brake test rig to consider its adoption for rail vehicles.

The instantaneous friction coefficient with C/C disk varies with the initial braking speed and the brake cylinder pressure. So the C/C disk must be used with brake force feedback control.

Fig.8   C/C multi-disk brake

4.   Future Brake System ---Closed Loop Brake System---

An air brake system(= a friction brake system) traditionally uses open loop control.   As a result, the deceleration through braking tends to vary because of friction material characteristics, initial braking temperature and adhesion between rail and wheel.   The ATC system, therefore needs enough braking distance to accommodate fluctuations in brake performance.   In order to achieve a constant brake effort, a brake torque feedback system is being developed.   In this system, it is important to detect the brake torque.   Several types of torque detecting devices with a unit brake or brake calipers have been tested using a high speed friction test rig, however, it is very difficult to sense the brake torque on the unit brake or brake calipers.   Because of the practical use of the torque feedback system, the torque detecting device is reasonably low cost, efficient and accurate.   In the brake control system a modern control or fuzzy control method is intended to be used.

5.   Conclusion

To promote railway services, faster train speeds are indispensable. In Japan, technological developments have been under way to achieve a maximum speed of 160 km/h on narrow gauge conventional lines and a maximum speed of 300-350 km/h on standard gauge Shinkansen. Since the brake performance is also related to the signaling system and train operating system, through the brake technologies mentioned in this paper, the quality and safety of railway transportation will be stepped up in the near future.

References

(1) OHYAMA,T. Adhesion Characteristics of Wheel/Rail System and its Control at High Speeds, Quarterly Report of RTRI,1992, Vol.33, No.1, 19.

(2) OHNO,K., BAN,T., OBARA,T., KAWAGUCHI, K. Adhesion Improvement with Jetting Ceramics Particles in High Speed Running, Quarterly Report of RTRI, 1994, Vol.35, No.4, 218.

(3) KUMAGAI,N.,OBARA,T.,HASEGAWA,I. A New Brake System for Speed-up Using Fine Anti-skid Brake Devices and Track Brakes, RTRI REPORT, 1994, Vol.8, No.3, 13

(4) KUMAGAI,N. Development of a Brake System for the High-Speed Rolling Stock on the Narrow-Gauge Lines, Japanese Railway Engineering, 1993, No.122, 16.

(5) OBARA,T.,TAKIGUCHI,T.,KUMAGAI,N. Development of Hybrid Rail Brake. RAILTECH'94, Birmingham, 1995, IMechE.

(6) UCHIDA,S., OBARA,T. The Brake System for Speedup of Shinkansen Vehicles, Railway Research Review, 1992, Vol.47, No.11, 15

(7) YASUDA,H., KUMAGAI, N. Development of C/C Composites Multi-plates Disk Brake of High Speed Shinkansen, The 4rd Transportation and Logistics Conferece, Tokyo, 1995, No.3206, JSME.

C514/033/96

Comparison of the suspension design philosophies of low-speed MAGLEV systems

J E PADDISON BEng, PhD, MIMechE, MAMIEE, H OHSAKI, and E MASADA
Department of Electrical Engineering, University of Tokyo, Japan

The suspensions of the two leading low speed Maglevs, the Japanese High Speed Surface Transport (HSST) Maglev and the GEC-Alsthom Birmingham Maglev are compared. The former employs a magnetic primary and an airspring secondary suspension enabling higher speed operation with satisfactory ride quality and canting. The latter uses a single stage magnetic suspension which provides the necessary ride quality without moving parts but its speed is limited to about 140km/h by the small airgap. Even so this airgap is twice as large as HSST's and requires four times the energy per tonne, nevertheless after 10 years of successful operation the energy costs still seem remarkably low.

NOTATION

B_o	Nominal flux density	L_a	Leakage reactance
G_o	Nominal airgap	N	Magnet turns
I_o	Nominal current	A, A_d, B, C	HSST Control variables
F_o	Nominal lift force	M	Magnet and supported mass
z_t	Change in track position	R	Magnet resistance
g	Small change in airgap	A_p	Pole face area
z	Small change in absolute position	b	Small change in flux density
i	Small change in current		

1. INTRODUCTION

Transport systems using Magnetic Levitation have been under development for the last 30 years. To date a great deal of money has been spent with little apparent impact on transport infrastructure. There are of course very impressive demonstration systems in Japan and Germany which are developed for the high speed transport market (>450km/h) but in this paper we will concentrate on the low speed transport range (<200km/h) where less spectacular but significant progress has been made. The latter systems use an electromagnet, fitted to the vehicle attracted up towards a steel track to provide suspension. Such a suspension is unstable because the reluctance of the magnetic field depends on the airgap between the magnet and track which gives a negative stiffness such that a change from the nominal operating point creates a force which augments the movement. Stable suspension is provided by a control feedback of some of the systems measurable parameters to provide a positive stiffness [1].

The GEC-Alsthom Transportation Projects Birmingham and High Speed Surface Transport HSST100 Maglev vehicles have both been designed for the low speed (<100km/h) transportation market. In addition to these two systems, which will be considered in this paper, the Magnet-Bahn Maglev system using permanent and a mechanical airgap control system operated in Berlin from 1988 until 1991 [2]. It is now under further development at the University of Braunschweig, Germany. The major advantage of low speed Maglev technology is that it offers a non-contacting and thus friction free operation which reduces the number of moving parts and improves the maintenance and reliability of the system. Also there is no wheel-rail interaction, reducing the track/wheel wear and noise which is particularly significant on transport systems requiring tight cornering where the guideway must pass around and close to existing structures. These systems have been designed for applications such as transport between airport terminals or between the airport and a communications terminus (as in the case at Birmingham), and for novel applications where quiet, environmentally friendly operation is required.

When compared with contemporary systems designed for low speed people mover applications ranging from moving walk ways to rubber tyred vehicles the Maglev technology has considerable advantages with respect to maintenance and reliability. More experience of Maglev is required in these areas before Maglev technology can be seriously compared with conventional steel rail/steel wheel systems.

Maglev has not as yet been successful on a larger scale because of its comparative newness and the availability of proven well understood alternatives which can be bought-off the shelf. With Maglev there has always been the need for some development costs during the first application of the technology.

At the time of writing the Birmingham Maglev is out of service but this should not detract from over 10 years of reliable service in which the system proved the validity of its low maintenance and high availability design. The inability to replace ageing components such as power transistors seems to have been the main reason for the shutdown. Hence there is a danger of the knowledge and experience gained on the Birmingham Maglev being lost or at least under-reported. The HSST Maglev system after considerable development is entering a phase where the company is preparing for its first commercial application on a 5.3km line between a commuter station and a theme park south of Tokyo when it must prove itself against conventional rubber and steel wheeled vehicles [3]. The Maglev technology is clearly at a significant turning point in the UK and in Japan.

2.   SUSPENSION REQUIREMENTS

ISO 2631 defines the ride quality requirements for transport systems and these are quantified in terms of the r.m.s acceleration experienced by passengers weighted to account for human susceptibility. For low speed systems of the type discussed here and used for relatively short journeys the acceptable level of r.m.s acceleration is taken as 4.5%g [4]. The airgap requirements are discussed for each specific case but in general terms the combined airgap changes due to random inputs and deterministic inputs, such as ramps, should not be large enough to endanger the magnet circuit integrity. In addition the suspension should be robust to changes in load of up to 30-40% of the unloaded vehicle.

3.   BIRMINGHAM MAGLEV DESIGN

3.1   System Description

The Birmingham Maglev could be regarded as a people mover operating horizontally over a distance of 623m between Birmingham Airport and railway station. The complete system was built for US$4.8 million in 1983. It operated for 20 hours a day with a journey time of 90 seconds and in the financial year April 1988 / April 1989 total energy costs were US$19.2 thousand and the cost of spare parts US$45.6 thousand. Over a typical year more than 1 million passenger journeys were made. Availability which is a comparison between the hours of operation needed for one vehicle compared with the actual hours performed was recorded as 99%. Maintenance was provided by two diagnostic technicians who also maintained the other airport electrical systems [5].

The key parameters of the Birmingham and HSST Maglevs are summarised in Table 1 which concentrates on those parameters relevant to the suspension. Each Birmingham vehicle consists of a single 6m long reinforced fibre glass cabin carrying about 40 people and their luggage. The only moving mechanical part of the vehicle is the automatic door in the cabin side. The cabin is mounted on a light weight welded aluminium chassis with a fundamental bending frequency of 40Hz which represents a very stiff chassis in comparison with a conventional railway vehicle which generally have a fundamental bending frequency of 8 to 15Hz.

The stiffness of the chassis is a result of the need to reduce its interaction with the high bandwidth action of the magnets and the vehicle body. This is particularly significant for a single stage suspension using magnets attached directly to the chassis in such a way that the vehicle is virtually wrapped around the guideway so that derailment is not possible as shown in Figure 1. Through the use of a stiff but lightweight chassis this problem was 'designed out' of the Birmingham suspension but at comparatively high construction cost. The vehicle itself sits above the track with its magnets slung beneath

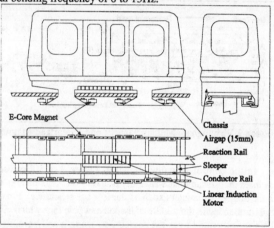

Figure 1 Birmingham Maglev (note direct attachment of the staggered magnets to the vehicle body).

the rails so that when energised the magnets are attracted towards the rails thus lifting the vehicle.

The vertical suspension requirements are provided by the 15mm airgap which is allowed to vary to accommodate the track roughness and varying loads as dictated by the suspension controller. This single stage suspension configuration limits the speed to a theoretical maximum of 140km/h due to the ride quality and canting requirements which are limited by the restricted movement in the airgap of 15mm [6].

Lateral suspension is provided by the inherent restoring forces due to the shearing of the magnetic field but this provides negligible damping. Higher damping is achieved by exciting pairs of magnets which are staggered or offset as shown in Figure 1 by ±10mm [7]. Such an arrangement requires about 13% more current to provide the same lift.

3.2   Birmingham Maglev Suspension Controller

The Birmingham Maglev suspension controller structure was formulated intuitively and tuned empirically to embody all the following suspension characteristics:

1.   to support the changing load.
2.   to follow the low frequency variations in the track - guidance characteristic
3.   to isolate the passenger from the high frequency component of the track such as the rail irregularities.

Before developing a controller, a linearized magnet model was derived for the magnet configuration shown in Figure 2 [8]. The non-linearities are relatively soft and the linearized model may be readily created. The model clearly shows how changes in airgap g produces an instability in the second loop and also produce a voltage feedback in the first loop which actually acts to oppose changes in flux density, which is a direct consequence of Faraday's law.

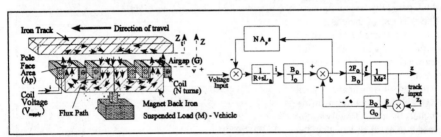

Figure 2 Birmingham Maglev Magnet and Linearized Magnet Model.

In the control structure shown in Figure 3, a 50Hz fast acting flux density control loop encloses the fundamental instability of the system and ensures the controller is robust to changes in the magnet parameters such as the leakage reactance. Flux density is measured using search coils embedded in the middle of the magnet pole face which gives a voltage proportional to flux. This is integrated over the search coil using a self-zeroing integrator to give a value for the change in the flux density. The flux can be controlled precisely enabling the suspension force to be adjusted as necessary. For low frequency motion such as the lift off condition when the change in flux is initially very small the flux signal is blended with the magnet current signal enabling a smooth take off. The main loop is a feedback loop composed of a pair of ride control filters fed from the magnet's absolute position, z derived from the acceleration measurement and a measurement of the airgap, g. The main loop's phase advance compensator is required for stability and its gain determines the suspensions reaction to load changes and is made stiff to ensure a deflection of 1mm in response to a 30-40% load change. The ride control filters are a complementary pair of high and low pass Butterworth filters. For low frequency track inputs the airgap dominates the feedback ensuring the vehicle follows the track whereas for high frequency track inputs the absolute position feedback dominates providing isolation. This control structure enables the response to track inputs to be kept separate from the response to the disturbance forces such as load changes.

The absolute position measurement is derived by double integrating the acceleration measurement but because of measurement drift the process must be self-zeroing at low frequencies down to 0.1Hz. This is not a problem for the controller as the high pass filter on the absolute position signal means that the low frequency component of the absolute position is not required.

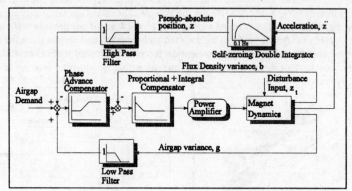

Figure 3 Birmingham Maglev one-degree-of-freedom suspension controller

The above control structure is easily converted from being a 1 degree of freedom magnet controller to being modal controller in which the rigid body bounce, pitch and roll modes are controlled independently through generalised position coordinates derived from the acceleration and airgap measurements. Each magnet retains its individual flux density feedback loop.

The advantage of modal control is that it prevents individual dynamically coupled magnet controllers from 'fighting' each other in order to preserve the airgap under their command. Also the ride quality, deflection and stability maybe considered for each of modes enabling them to be optimised without fear of conflict. The disadvantage of the modal control approach is that it relies on either rigid coupling between the magnets or a sophisticated control model which accommodates the flexible bending modes. The chassis despite being relatively stiff has a fundamental bending frequency of 40Hz. This bending appears in the bounce mode as unwanted acceleration and is ameliorated by a notch filter tuned to 40Hz inserted in the bounce control loop. Additional resonances in other body modes are dealt with in a similar manner.

4. HSST MAGLEV

4.1 HSST System Description
HSST have at present two vehicles; the 8m long HSST 100 S (short) and the 14.1m long 100 L (long). Both are designed to be used as part of a 4 vehicle train set with articulated connections. Both HSST vehicles are quite a lot bigger than the Birmingham Maglev carrying 75 and 121 standing and sitting people respectively. The basic suspension design is the same for both HSST vehicles. The HSST design uses a magnetic primary suspension with an 8mm

magnet airgap with a mechanical clearance of 6mm and an airspring secondary suspension to provide ride quality and guidance requirements. Four magnets are mounted in extruded cast aluminium and welded 2.5 m long module structures, two of which make a bogie. The modules are connected by anti-roll bars to provide a certain degree of decoupling. On the latest design two airsprings connect each module to the vehicle chassis via intermodule linear bearing sliding tables and thrust rods.

The vertical suspension then consists of two distinct parts; the magnetic primary providing track following to within ±1mm r.m.s and an airspring secondary reducing the effects of the bogie acceleration of 30%g to about 4.0%g in the vehicle. The use of the secondary suspension will enable the HSST Maglevs to travel at speeds of over 140km/h and provide sufficient ride quality. The use of modules also reduces the required body strength and weight since the rigidity is not such an important issue and the suspension force is distributed more evenly along the length of the vehicle. This means a 14.1m long chassis with a fundamental bending frequency of 15Hz maybe used.

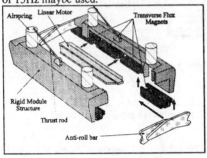

Figure 4 HSST 100S Module   **Figure 5** HSST 100S Bogie steering structure

The lateral suspension structure is shown in Figure 5 and the use of the sliding tables mounted on linear bearings and the intermodule hydraulic suspension cylinder allows all the modules to form a curving line while still adequately supporting the vehicle body by having the magnets directly below the track. This arrangement allows the HSST 100S vehicle to follow a minimum radius curves of the order of 25mR [9]. Intermodule hydraulic cylinders are avoided in the 100L and the sliding tables connected to the body via springs. The use of offset magnets to improve the lateral damping has been discontinued due to the success of the secondary suspension steering mechanism. Consequently the magnets are arranged aligned rather than staggered and this reduces the flux leakage and improves magnet efficiency. The HSST 100S vehicle has undergone 4 years of extensive testing at a purpose built 1.5km long test track in Nagoya and has been certified as fit for passenger transport in Japan by the Ministry of Transport. During the period May 94 to March 95 a total of 20000km was travelled and it was found that the lateral suspension and hydraulic system required minor adjustment about twice a month. During the above period the HSST 100S proved able to operate with 140mm of snow on the track and under different conditions travel at 70km/h against 100km/h winds [10].

4.2 Suspension Controller

The HSST Suspension controller is principally a position control loop providing a very stiff suspension with as small an airgap variation as possible, typically 1mm r.m.s in response to the track roughness. Originally the HSST control strategy was developed for higher speed

(>200km/h) vehicles and the target speed has been reduced over the years. The HSST company is based in the aerospace industry where a greater degree of experimental tuning is prevalent. The HSST linearized magnet model is shown in Figure 6 and is very similar to the Birmingham linear magnet model. Flux density is eliminated from the model but the effect of changes in the airgap on the supply voltage are included as required by Faraday`s law.

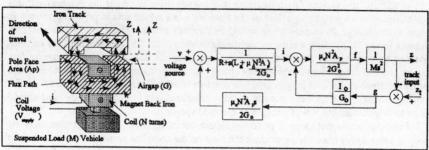

Figure 6 HSST Transverse Flux Magnet and linearized magnet model

The controller design procedure is less transparent than that for the Birmingham controller and has been tuned experimentally, relying on the experience of the HSST engineers.

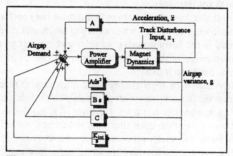

Figure 7 HSST Controller Structure

In the controller A, Ad, B and C are the filter gains and allow a blending of the measurements with a pseudo airgap velocity and airgap acceleration both derived from the airgap measurement. An additional airgap integral feedback loop encloses the whole controller to improve the response to load changes which for a 20% load change is almost zero displacement. The controller has a bandwidth of about 10 to 12Hz providing the necessary guidance. The 4 magnet coils in a module are divided into two magnet systems which are independently controlled by two sets of sensors. However the measurements of airgap and acceleration are generalised measurements for each module and not for each magnet. The positioning of the sensors was decided experimentally and optimising their positions may provide a certain degree of decoupling between, for instance, the pitching and bounce of the module structure.

5.   COMPARISON

From the outset the Birmingham Maglev was designed specifically for the low speed range whereas the HSST Maglev has developed from a design for higher speeds. In both designs the non-linearities are considered to be soft and a linear model and a linear control technique are used to stabilise the system. Both models use a voltage drive to power the magnets and this enables the inherent stabilising effect of the Faraday`s law which acts to prevent a change in airgap flux density. A current drive would nullify this effect. The use of transverse flux magnets in the HSST design reduces the magnetic drag caused by circulating eddy currents produced by the magnetic flux of the poles moving along the rail. However a transverse flux magnet generally require an increased slot size which entails more steel in magnets and track. The Birmingham design uses longitudinal flux magnets which although using less steel do suffer from higher magnetic drag, but since the speed is comparatively low (54km/h) the drag should not be too significant but it will increase at elevated speeds.

The two distinct control loops of the Birmingham controller make for easier commissioning a loop at a time rather than having to close the loop on the complete controller as in the case of the HSST controller design. The flux loop enclosing the non-linearities does make the magnet much easier to control and more robust to magnet parameter changes. After the flux loop has been successfully closed the control designer has a large margin for error when developing a stable suspension and can select the controller main loop bandwidth as necessary. Flux density control is not used on the HSST controller so it is more sensitive to magnet parameter variation such as changes in the leakage reactance.

The HSST magnet controller has the advantage that it must only perform two tasks namely load bearing and track following which are not conflicting and a stiff suspension will perform both tasks. A conflict exists in the Birmingham controller and requires an understanding of the trade off between the requirements of the suspension but in the case of the Birmingham system a satisfactory compromise has been found.

The HSST Controller requires a high quality track alignment to maintain ride quality of about $4.0ms^{-2}$ r.m.s whereas the Birmingham controller can provide ride quality of $4.5ms^{-2}$ r.m.s on a secondary quality track which brings with it savings in alignment and maintenance costs (see Table 1). In any Maglev system the track alignment should be a one off cost as there should be very little wear since there is virtually no contact between track and vehicle.

The use of the airspring and steering mechanisms in the secondary suspension means that the HSST vehicles are able to follow tighter corners, not have a speed limited by canting and ride quality requirements and enables a longer less rigid and cheaper chassis. The rigid and light weight chassis on the Birmingham design is expensive and too small in terms of passenger capacity to compete with more conventional vehicles such as monorails or rubber tyred systems which are generally twice its length. However the secondary suspension mechanism of the HSST is complex and will almost certainly require a greater amount of maintenance throughout its life compared with the Birmingham Maglev.

Analysing the controllers adjusted for the HSST type magnet and track it can be shown although the Birmingham design is extremely adaptable it is difficult to match the performance of the HSST controller simultaneously in categories of airgap variation, ride quality and response to track deflection. Typically the HSST controller gave an airgap variation of 1mm r.m.s with a magnet acceleration of 35%g whereas the Birmingham controller typically gave a higher airgap variation of about 1.5mm r.m.s but a lower acceleration of 23%g when using a 3rd order Butterworth filter at 5.5Hz in the ride filters. Comparison studies are ongoing.

236

6.   CONCLUSIONS

HSST power consumption appears to have been an important consideration exemplified by the suspension airgap and the linear induction motor airgap being half that of the Birmingham Maglev`s. Hence the Birmingham Maglev is less efficient but avoids mechanical complexity required to operate with such small airgaps. For the Birmingham design power consumption was not such a high priority and this approach is validated by the low yearly power costs for the year 1988/89 which suggest power consumption is not significant to low speed Maglev operation. The HSST design will be able to operate at a theoretically higher speed range with greater comfort, tighter curving capabilities and better power consumption but the module system is rather complex and the magnet controller more difficult to adapt to incorporate suspension requirements other than the tracking. The Birmingham Maglev design has proven low maintenance because of its mechanical simplicity but at present the vehicle length is limited by the need to use a rigid and consequently expensive chassis. Birmingham Maglev type control using the flux loop is more robust to alteration and as the advantage that it is adaptable to varying requirements.

ACKNOWLEDGEMENTS

The author would like to acknowledge the European Union for their support of this work under the European Science and Technology Fellowship Scheme.

REFERENCES

1.   Goodall, R.M., The Theory of Electromagnetic Levitation, Physics in Technology, 1985, Vol. 16, No. 5, Pp. 207-213.

2.   Heidelberg, G., Niemitz, K., Weinberger, H., The M-bahn System, Proc IMechE Conf. 'Maglev Transport - Now and for the Future' Paper C413/84, Pp 159-168, 1984.

3.   Tomohiro, S., The Development of the HSST 100L, Maglev'95 14th International Conference on Magnetically Levitated Systems Nov. 26-29, 1995, Pp.51-55.

4.   Taylor, D.R.D, Goodall, R.M., Oates, C.D.M., Theoretical and practical considerations in the design of the suspension system for Birmingham Maglev, Proc IMechE Conf. 'Maglev Transport - Now and for the Future' Paper C394/84, Pp 185-192, 1984.

5.   Walker, J.N., Birmingham's Experience with the Maglev System. Airport Forum, 1989, Vol.5, Pp.15-18.

6.   Goodall, R.M., Dynamic characteristics in the design of Maglev suspensions, Proc. Instn. Mech. Engrs. Part F., 1994, Vol 208.Pp.33-41.

7.   Nenadovic, V., Riches, E.E., Maglev at Birmingham Airport: from system concept to successful operation, GEC Review, 1985, Vol.1, No.1.

8.   Goodall, R.M., Suspension and Guidance for a DC Attraction Maglev Vehicle, IEE Conf. Publication 142, 1976.

9.   Masada, E., Kitamoto, M., Kato, J., Kawashima, M., Present status of maglev development in Japan and HSST-03 project, Proc IMechE Conf. 'Maglev Transport - Now and for the Future', 1984, Paper C394/84, Pp 9-22.

10.   Fujino, M., Mizuma, T., Total test operation of HSST 100 and planning project in Nagoya, Maglev'95 14th International Conference on Magnetically Levitated Systems Nov. 26-29, 1995, Pp.129-133.

Table 1 Birmingham and HSST Maglev Comparison

	Birmingham Maglev Vehicle	HSST Maglev Vehicles	
		100S	100L
Roughness Factor [m]	1×10^{-6}	1×10^{-7}	
Track Quality	secondary quality	primary quality	
Alignment	10mm over10m	5mm variation over 10m	
Horizontal Shape, Min. Curvature [mR]	40	25	50
Gradient standard / maximum	1.5% / 5%	6% / 7%	
Verified cant angle	unknown	8°	
Guideway Natural Frequency [Hz]	7	6 / 7	
Chassis length/width/height [m]	6 / 2.25 / 3.5	8.5 / 2.6 / 3.3	14.1 / 2.6 / 3.3
Chassis Fundamental Frequency (Hz)	40	unknown	9
Total Vehicle Weight Tare/Loaded	5 / 8	10 / 15	15 / 25
Passengers (per car) Standing / Seated	34 / 6	42 / 28	84 / 37
Magnet/Module Pitch	2.8	2.5	
Operating speed kmh^{-1} (ms^{-1})	54 (15)	100 (27.7)	
Secondary Suspension (deflection [m])	None	Airspring(0.35)	Airspring (0.45)
Secondary Suspension Natural Frequency [Hz]	Not applicable	1.8	1.0
Airgap actual [mm]	15	8 (6 mm mechanical clearance)	
Airgap at operating speed [mm] r.m.s	±5 (7.5)	±1 (4)	±1 (3)
Max airgap change due to gradient trans. [mm]	±7.5	±1	
Vertical Acceleration on magnet %g r.m.s	4.5	30	14
Vertical Acceleration vehicle %g r.m.s (empty)	4.5	4.0	5.3 (4.5)
Lateral Ride on magnet %g r.m.s	3.3	35	9
Lateral Ride on vehicle %g r.m.s (empty)	3.3	3.0	1.6 (0.8)
Power con. standstill/op. speed [kW/tonne]	2.5 / unknown	0.6 / 0.9	
Lift/Weight(ratio at Nominal Airgap)	11.7	8.3	8.9
Gap Sensor	Eddy Current	Eddy Current Sensor	
Accelerometer	Piezoelectric	Servo/Piezoelectric/Capacitive Type	
Flux Density Sensor	search coil	none	

C514/069/96

Experimental study on digital brake control for MAGLEV trains

H OSHIMA MJSME
Central Japan Railway Company, Tokyo, Japan
M KISHI
Sumitomo Metal Industries Limited, Amagasaki, Japan
H SAKAMOTO MJSME, MASME
University of Washington, Seattle, USA
B SALAMAT
Crane Company, California, USA

SYNOPSIS

As a back-up brake system for Japanese MAG-LEV trains in case of the normal electric brake failure, carbon-carbon composite brake units are going to be used. In case of train brake, the equalization of absorbed energy into each brake unit needs to be considered from the viewpoints of unit capacity and disc wear life.

This report explains the torque limit control to get the energy equalization among numerous brake units in the train. As the first step, the dynamometer test using two brake units has been done to understand the torque behavior when having multiple brake units. As the next step, simulation study using MAT-LAB has been done to incorporate the experimentally obtained torque behavior appear in a computer calculation. The results show that the torque of two brake units tends to diverge due to the tire radius difference and the simulation can predict this behavior.

INTRODUCTION

Carbon-carbon brake units are going to be used as a back-up brake system for Japanese MAG-LEV trains, when the normal electric brake fails[1][2]. The number of brake units will be numerous in the case of trains. The car load on each wheel is considered to be variable due to car weight distribution, weight shift by pitching motion during braking, and tire radius difference. Such a load variation may cause scatter in absorbed energy in each brake unit. Therefore, consideration of energy equalization is highly important, specially in trains.

The previous study on the digital brake control for MAG-LEV trains presented at '93 S'TECH consists of a simulation study of torque balancing for a 3-car model and a dynamometer study using one brake unit[2]. However, it was not enough to simulate the energy equalization for a whole train, since the torque balancing means to be the same torque value in each brake unit. The torque limit or torque range control is thought to be needed for multiple brake control units in a train. Because each torque value by a brake control unit for a truck should be different due to the need for primary deceleration control demand according to different weight trucks.

The existing dynamometers that belong to brake unit or tire makers normally utilize the testing of only one brake unit. Therefore, dynamometer test of multiple brake units has not been previously performed. Sumitomo Osaka has a wheel testing machine, which has been used for a wheelset and a single wheel or a single brake unit. After modification of the brake testing stand, the testing for two brake units became possible to be performed. The test using two independent brake units is expected to give the information of brake performance for multiple brake units.

In this report, the tests using two brake units that will be the first case in the world are introduced. They are expected to be helpful for future study on the brake performance of multiple brake units in a train. The simulation analysis that is expected to explain the experimental results is also described based on the brake characteristics from the experiments.

NOMENCLATURES

I_d: Moment of drum inertia

P_b: Brake pressure

R_d: Drum radius

S: Operator in Laplace transformation

V: Vehicle Speed

V_s: Slip ratio

μ: Frictional coefficient

ω_d: Drum rotational speed

ω_w: Tire rotational speed

I_w: Moment of tire inertia

P_v: Valve pulse module width

R_w: Tire radius

T_b: Brake torque

V_c: Command voltage

W: Tire load

θ: Rotational angle

ω_n: Natural frequency

ξ: Damping factor

SIMULATION STUDY ON SINGLE UNIT BRAKE TESTING

The brake unit studied is the one for Japanese MAG-LEV trains, which was reported at the last conference[1]. As the brake control system, the control functions are required to be autobrake, anti-skid, and torque limit or torque range control[2]. The idea of torque limit or torque range control was raised to provide the energy equalization for absorbed energy into each brake unit. Figure 1 shows the brake control loop and brake unit.

The disc brake system is going to be used for emergency braking when the normal electric brake fails. Therefore, the initial speed for disc brake application can be 500km/h, and the brake tests from 500km/h were done. The simulation study has been done on the experimental results, and is described for clarity in the following before the dynamometer study with two brake units is reported. For previous experimental results refer to report [2].

Simulation Model

In order to analytically understand the brake performance obtained by the dynamometer tests[2], the simulation study has been done by using the following model.

Valve pulse module width, P_v, is a function of command voltage, V_c, as

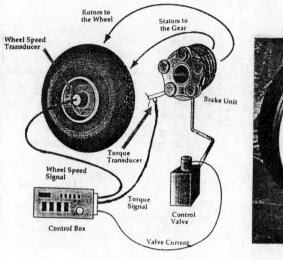

(a) Brake Control Loop   (b) Wheel and Brake Unit

Figure 1 Brake Control Loop and Brake Unit

$$P_V = f(V_C) \tag{1}$$

The characteristic of $f(V_C)$ shown in Figure 2-a was obtained by the experiment of single valve motion. The brake pressure, P_b, is derived from the valve pulse module width by

$$P_b = \frac{\omega_n^2}{S^2 + 2\omega_n \xi S + \omega_n} P_V \tag{2}$$

The values of ω_n and ξ were obtained by response function test for the oil system used.

As for the characteristics of carbon brake disc, the relation of T_b and P_b is

$$T_b = f(\omega_n) \cdot P_b \tag{3}$$

The function of $f(\omega_n)$ was obtained as the average value for the experimental results.

The rotational motions for tire and drum can be described as

$$I_W \cdot \omega_n = -T_b + \mu \cdot W \cdot R_W \tag{4}$$

$$I_d \cdot \omega_d = -\mu \cdot W \cdot R_d \tag{5}$$

The frictional coefficient, μ, is a function of slip ratio, V_S, and is described as

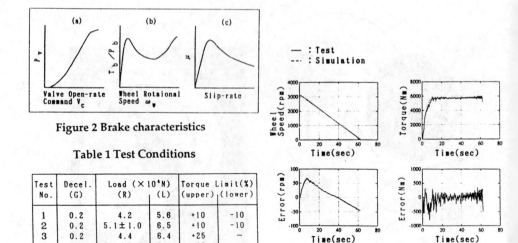

Figure 2 Brake characteristics

— : Test
--- : Simulation

Figure 3 Simulation for Test Result

Table 1 Test Conditions

Test No.	Decel. (G)	Load (×10⁴N) (R)	(L)	Torque Limit(%) (upper)	(lower)
1	0.2	4.2	5.6	+10	-10
2	0.2	5.1±1.0	6.5	+10	-10
3	0.2	4.4	6.4	+25	—
4	0.2	5.1±0.9	6.4	+25	—

$$\mu = f(V_s) \tag{6}$$

$$V_s = \frac{R_d \cdot \omega_d - R_w \cdot \omega_w}{R_d \cdot \omega_d} \tag{7}$$

Simulation Results

The simulation work was done for the brake test results[2] obtained by the dynamometer with a single brake unit. Figure 3 shows the example of simulation in case of 0.25 G deceleration and 5.39×10⁴ N tire load test condition.

Although some differences between simulation and experiment exist, the simulation can be expected to predict the brake performance.

EXPERIMENTAL STUDY USING TWO BRAKE UNITS
Test Method and Condition

The brake testing using the dynamometer with two brake units was carried out. Although there is only one drum, the two brake units are supported by individual dummy landing gear, to which different tire load can be applied. The capacity of the dynamometer is the maximum speed of 300 km/h and the maximum energy of 40MJ, which are not enough for testing under the condition of MAG-LEV trains[1],[2].

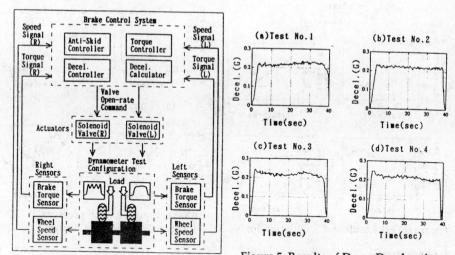

Figure 4 Block Diagram of Test and Control

Figure 5 Results of Drum Deceleration

However, the conventional dynamometers only perform the testing with a single unit. Therefore, the testing with two brake units is expected to give the information of the brake performance for multiple brake units even under such insufficient conditions. Simulation work should follow this experimental work with two brake units to predict the brake performance in service.

Table 1 shows the test conditions. The load conditions shown in Table 1 were applied to right, (R), and left, (L) landing gears. For cases of No.2 and No. 4, the sinusoidal load variation of 0.5 Hz frequency was applied. For each brake testing, the torque limit control with the value shown in Table 1 was conducted. Figure 4 shows the diagram of the brake testing and control.

Test Results and Simulation

Under the brake conditions shown in Table 1 the tests were done, and the results of drum deceleration for each case are shown in Figure 5. The set deceleration 0.2 G was almost achieved for each case, even under the fluctuating load condition.

Figure 6 is the result for test No.3, in which the torque divergence occurred. The torque of the right brake unit goes up and the one of the left unit goes down with time. However, the variation of torque was within the set limit value for each case in this experimental study.

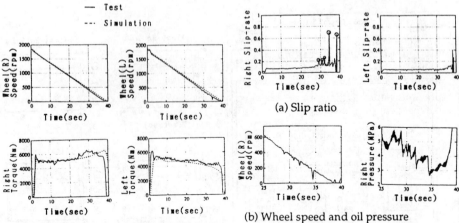

(a) Slip ratio

(b) Wheel speed and oil pressure

Figure 6 Test Result and Simulation for No.3   **Figure 7 Anti-skid Control**

DISCUSSION

Anti-skid Function

In test No. 3, skidding occurred in the right tire. Whenever skid happened during the braking of No.3, the brake controller controlled the anti-skid function. As a result, the tire rotation recovered. The set load of the right tire is lower, and the torque required to get the deceleration was higher than the one of normal frictional coefficient. Figure 7 shows how the skidding happened in the figures of slip ratio, wheel speed, and oil pressure.

In the former tests reported at the last conference, skidding did not occur. The slip ratio calculated for the tests is lower than 0.1. Therefore, skidding may happen if the slip ratio overcomes the value of about 0.15.

Dynamic Tire Radius

Because of the difficulty with measurement of dynamic tire radius, the radius was calculated by tire rotational velocity difference of two tires at the beginning of braking. The time calculated is between touch-down and start of braking. The tire speed ratios of left tire to right tire for cases of No.2 and No. 3 are 1.0058±0.0015 and 1.0089. The average static radius is around 408 mm, and from these values it is estimated that the right tire has a radius of 2 mm larger than the left tire. It is also calculated that the enlargement of dynamic radius is about 0.4 mm per 10 kN.

Torque Unbalance

Through the test of No. 3, there is a tendency that the brake torque of right tire increases and the brake torque of left tire decreases, as is shown in Figure 6. The reason of this tendency was thought to be due to the variation of frictional coefficient for two brake units. The variation in frictional coefficient of the right tire is larger than that of the left tire. However, this is not the reason for the torque unbalance behavior of right and left brake units. This was found by simulation, taking account the frictional characteristics. Instead, 2 mm difference of tire radius was put into the simulation, and the result obtained is the one shown in Figure 6. The experimental torque unbalance was reproduced in the simulation by using the tire radius difference.

As a conclusion of the torque unbalance, it is found that the unbalance is caused by the tire speed difference due to the tire radius variations, as was discussed in the previous paper[2]. This is because the brake controllers for both brake units are independent.

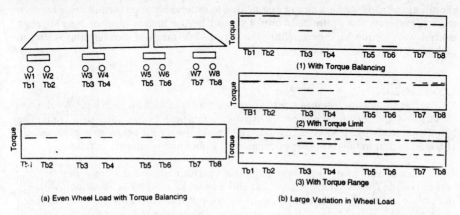

(a) Even Wheel Load with Torque Balancing

(b) Large Variation in Wheel Load

Figure 8 Torque Behavior for Different Cases of Tire Load and Control Method

Energy Equalization

The experiments reported herein did not show whether the torque limit function worked or not. The torque values measured for all cases of No.1-4 were within the set torque limit. The large difference of torque value for two brake units is not admitted, because the energy into two brake units comes from the same drum inertia. In a real train, the inertia might be different in each car or truck, and the difference can be large if the weight of car varies due to passenger occupancy.

In Yamanashi Experimental Line, the first proto-type MAG-LEV train is going to have a running test in Spring, 1997. The train consists of three cars and four bogie trucks. If the train has some variation in car weight, the torque required to obtain the set deceleration of the train will be different for each bogie truck. Figure 8 shows the torque for the case of even car weight load and the case of different car weight load. For the torque limit control or torque range control, the figures are just presumptions. If the torque limit or torque range control makes the variation in torque values smaller, the deceleration for each truck may not be the same as the set value. This causes the zigzag movement in back and forward. Equations of (8) and (9) express the relationship between torque and energy and between torque and deceleration. If the deceleration is the same, the torque depends on the weight and tire radius. If the torque is different, the energy is not the same.

$$\text{Absorbed energy} = \frac{1}{2} \cdot \frac{W}{g} V^2 = \int T_b d\theta = T_b \cdot \theta_{total} (T_b \text{ is constant}) \tag{8}$$

$$\text{Deceleration Force} = \frac{W}{g} \cdot (\text{deceleration rate}) = T_b / R_w \tag{9}$$

As a conclusion for this discussion of energy equalization, the optimized control including torque balance and torque limit or torque range is needed under the condition without having a zigzag motion due to different limited torque value for each truck. Therefore, the optimized control should be the torque limit or torque range control with torque balancing within the permissible range of deceleration scatter in each truck from a riding quality.

CONCLUSION

The dynamometer test using two wheels with a brake unit for each has been done to understand the torque behavior when having multiple brake units in a train. The simulation study using MAT-LAB followed to make the torque behavior experimentally obtained appear in the computer calculation. The obtained results are as follows.

(1) The brake tests using two brake units were performed, and the brake performance was obtained. The functions of anti-skid and autobrake worked as expected. Although the torque limit and torque range control were put into the controller, the functions were not explicitly shown in the experiments. The torque values measured were all within the set torque limit or range values.

(2) In test No.3, the torque divergence due to tire radius difference appeared more clearly than in the other tests. The torque behavior was also demonstrated by the computer simulation. The simulation is expected to predict energy equalization needed in a train.

(3) Further studies will be needed to get the optimized control taking into consideration of torque balance, torque limit or torque range, and constant deceleration without uncomfortable movement due to different limited torque values for each truck.

REFERENCE

1) Igarashi, M., Oshima, H., Sakamoto, H., and Takakuwa, T.; Carbon-carbon Composite Brake System for MAG-LEV Trains, '93 S'TECH, Yokohama, Nov. 1993, 430-434

2) Igarashi, M., Oshima, H., Sakamoto, H., Qi, K., and Salamat, B.; Digital Brake Control System for MAG-LEV Trains, '93 S'TECH, Yokohama, Nov. 1993, 223-227

Aerodynamic and Environmental Issues

C514/038/96

The reduction of excessive aerodynamic lifting force and the performance on current collection at 300km/h with low noise current collectors

Y FUJITA, K IWAMOTO, S MASHIMO, H NORINAO, and M HAMABE
West Japan Railway Company, Japan

SYNOPSIS

West Japan Railway Company, JR-West, has been developing a low noise current collector for operating trains at 300km/h. The most challenging problem in this development was to reduce and stabilise aerodynamic lifting force without damaging its characteristic of low noise. In this paper, major experimental results of reducing aerodynamic lifting force and the performance on current collection at 300km/h are described.

1. INTRODUCTION

For commercial service at 300km/h, it is necessary to satisfy the anti-noise criteria required by the Environment agency. Current collectors on the train roof are the most significant source of noise. To meet this requirement, West Japan Railway Company, JR-West, has been developing a low noise current collector.

To obtain lower aerodynamic noise level, the total number of the parts was reduced, and the appearance became very simple. Also the collector head was converted from conventional two rectangular pipes to one elliptical head.

However the elliptical shape produced considerable amount of aerodynamic lifting force compared with the conventional pantograph. For the excessive lifting force causes damages or wear in contact wire, it should be reduced to a suitable level.

Through wind tunnel tests, it was found that the lifting force was influenced by the presence of support pillar of the collector head, which produced turbulent aerial flow. By modifying the cross-sectional shape, the lifting force could be reduced to a suitable level and in the actual train operation test, the effect was confirmed. To reduce noise level more, two types of horn were provided. The horn also made an influence on the lifting force.

Since then the experimental train, WIN 350, has been running at

300km/h using two low noise current collectors in the long-run tests. Fig.1 shows the appearance of the low noise current collector. The collector head and upper cover moves straight up and down by piston-cylinder mechanism in the cover. The constant pressurised air yields static uplifting force. To reduce the aerodynamic drag and noise, the piston and cylinder was covered with cylindrical case at first, and later, changed to elliptical one.

In this paper, the major experimental results and the performance of current collection at 300km/h are described.

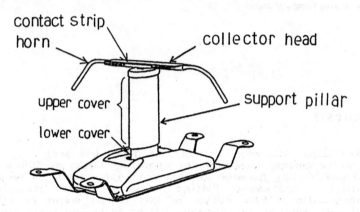

Fig.1   The appearance of the low noise current collector

2. INFLUENCE OF SUPPORT PILLAR

In order to confirm the effect of the aerodynamic lifting force due to support pillar, wind tunnel test was conducted using a real size model. In the test, as shown in Fig 2, both ends of the collector head were connected with load cells which measured the lifting force of the head, F. Both of the horns were not attached to the collector head because the wind was blown only at the central part of the collector head. The support pillar of the current collector was substituted to a cylinder in this wind test. Both the side bars and the cylinder were fixed to the base, but the height of the top of the cylinder was changeable. In Fig.1, ϕ is the diameter of the cylinder and δ is the distance between the top of the cylinder and the bottom of the collector head. The wind speed was 27m/s and not changed. Aerodynamic lifting force, F, was measured by changing ϕ and δ. Table 1 shows the results.

Table 1   Aerodynamic lifting force changing with ϕ and δ

F [N]	-1.47	1.47	2.45	2.84	4.70
ϕ [mm]	—	40	40	145	145
δ [mm]	∞	55	13	60	15

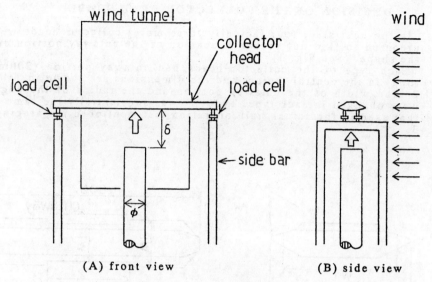

Fig.2 Conditions of the wind tunnel test

When $\delta = \infty$, which means there was no cylinder, aerodynamic lifting force of the collector head itself showed minus. But when there was a cylinder, aerodynamic lifting force increased by the turbulence of aerial flow. From the results of this wind tunnel test, following features were obtained.

(1) aerodynamic lifting force increases as diameter ϕ increases.

(2) aerodynamic lifting force increases as distance δ decreases.

Therefore, it is necessary to avoid the influence of cylinder or support pillar causing the increase of aerodynamic lifting force.

3. EFFECT OF CUTAWAY PORTION

When there is a centre pillar, upward air flow may occur at the connection with the collector head and cause lifting force. To avoid the influence of this effect, the collector head in the central region was cut away. See Fig 3, the cross section of the central part became rectangular and the rest of it remained smooth. The effect of this cutaway was admitted in the actual train operation test to measure aerodynamic lifting force under the condition that the contact strip was not contacted with the contact wire.

However, in the current collection test, the uplift of the contact wire was measured about 80mm at 260km/h in a tunnel. This value would be beyond the criterion of 100mm at 300km/h. By using more steep angle of contact strip showed less uplift of the contact wire, but it would come also near the criterion. Therefor, reconsideration was necessary. The target for aerodynamic lifting force at 320km/h was set below 98N.

4. DECISION OF THE COLLECTOR HEAD SHAPE

In the next step, four (actually three more) collector heads were prepared to find out proper dimension of the cutaway portion and the shape. (See Fig. 3)

Each type of the collector heads had cutaway portion (300mm long) in the central region, and the dimensions are shown in Table 2. The width of the cutaway portion and the radius of the edges were changed in each type. Type "i" was substantially the same as that used in the actual train operation test mentioned in paragraph 3.

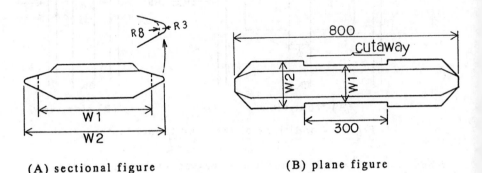

(A) sectional figure   (B) plane figure

Fig. 3   Tested collector head

Table 2   Aerodynamic lifting force with the changes of the cutaway width and the radius of the edges

| type | dimension | | | lifting force | | ratio |
	W1 [mm]	W2 [mm]	R [mm]	measured at 275km/h [N]	calculated at 320km/h [N]	
i	135	165	3	65.2	90.5	1
				79.5	114.0	1
m	120	165	3	44.9	60.2	0.69
				58.4	76.9	0.73
n	135	150	8	54.1	72.9	0.83
				84.5	114.3	1.06
o	120	150	8	15.0	20.0	0.23
				31.4	41.1	0.39

1. upper column : values in open-air
   lower column : values in tunnel
2. the bases of the ratio are the values of type "i"

Aerodynamic lifting force was measured in the same manner as that for actual train operation test in paragraph 3. Table 2 shows the test results at 275km/h and also the calculated aerodynamic lifting force at 320km/h. Type "i" was used as the basic shape and the ratios were calculated.

It was possible to reduce the aerodynamic lifting force below 98N in open-air section at 320km/h. However, excessive force in tunnel had to be considered. When the cutaway portion was enlarged as type "m", it was possible to reduce the force by approximately 30%. By enlarging the curvature radius of each edge, type "n" enabled to reduce the aerodynamic lifting force in the open-air section.

In the case of type "o", which was changed in both the cutaway width and the radius, a remarkable reduction in aerodynamic lifting force was achieved.

Thus, these two types, type "m" and "o", satisfied the criteria that aerodynamic lifting force at 320km/h had to be below 98N.

5. HORN SHAPE AND AERODYNAMIC LIFTING FORCE

Through the tests, it became clear that the horn also made an influence on the aerodynamic lifting force. Fig. 4 shows the shapes of horns. Continual holes were cut out to reduce the noise level from the horn itself. This effect of holes had been confirmed in other low noise wind tunnel tests before.[1]

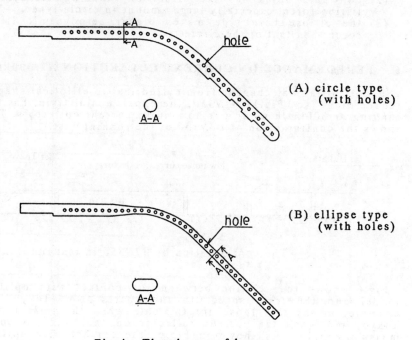

(A) circle type
(with holes)

(B) ellipse type
(with holes)

Fig.4   The shapes of horns

The measurement conditions of the aerodynamic lifting force are as follows:
(1) horns were attached to the common collector head of a single-arm type current collector (see Fig.8)
(2) in wind tunnel test, wind speed was 50m/s
(3) in actual train operation test, the values were estimated at 300km/h from those measured at 275km/h

Table 3   The effect of the cross-sectional shape and the holes of the horn

	uplifting force [N]			
	circle type		ellipse type	
	no holes	with holes	no holes	with holes
wind tunnel test	13.9	9.2	18.6	18.0
train operation test	49.0	41.2	not measured	55.9

Table 3 shows the results of the tests, and these conclusions are obtained:
(1) the cross-sectional shape of the horn influences aerodynamic lifting force
(2) continual holes has the effect of reducing aerodynamic lifting force, especially large amount in circle type
(3) the ellipse horn type makes bigger aerodynamic lifting force than that of the circle type

6.   PERFORMANCE ON CURRENT COLLECTION AT 300km/h

Support pillar was changed from cylindrical to elliptical shape to reduce drag. (see Fig.1) *WIN350*, the experimental train, has been running at 300km/h using two low noise current collectors. Fig.5 shows the configuration of *WIN350* at the running test.[2][3]

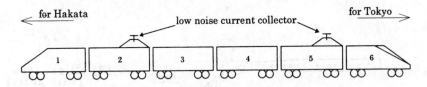

Fig. 5   The configuration of *WIN350* in test run

Fig.6 shows the relation between the contact wire uplift in tunnel and the train speed. In the chart, "No.2-Tokyo", for example, means that those are the data when the train ran for Tokyo, and when the current collector on car No.2 passed the measuring point. As shown in Fig.6, the contact wire uplift at

300km/h was about 50 mm, which is clearly below the criterion of
100mm. Fig.7 shows the relation between the interruption rate of
contact and the train speed. The interruption rate of contact at
300km/h was also bellow the criterion of 30%.

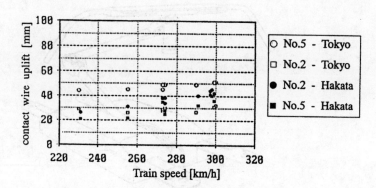

Fig.6   Contact wire uplift

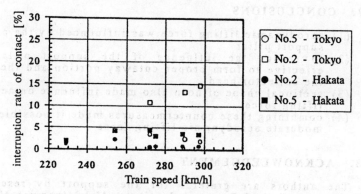

Fig.7   Interruption rate of contact

Another type of low noise current collector has been on the test.
Fig.8 shows the appearance. This single-arm type current collector
shows a distinctive rod shape to simplify the outline. The test
data will be present in near future.

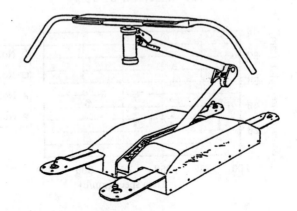

Fig.8   Single-arm type current collector

7.   CONCLUSIONS

(1) aerodynamic lifting force was influenced by the presence of support pillar
(2) to reduce the influence of the support pillar, it was effective to form proper cutaway portion and the radius of the collector head
(3) sectional shape of horn also made influence on aerodynamic lifting force
(4) combining these countermeasures made it possible to obtain moderate aerodynamic lifting force

8.   ACKNOWLEDGEMENT

The authors are grateful for the support by researchers at Railway Technical Research Institute, Mr. Miyamura M., and Mr. Yajima S., engineers at Toyo Electric Mfg. Co., Showa Corporation, and many others who co-operated in this development.

REFERENCES

[1] Aso T., et. al., *"Reduction of Current Collecting Noise by Wing-shaped Current Collector"* (in Japanese), The 4th Transportation and Logistics Conference, Japan, 1995, pp.342-344.

[2] Higashi A., et. al., *"Aerodynamic Noise from Car Bodies and Pantographs of WIN350"*, S-TECH '93, 1993, vol.2, pp.59-64

[3] Okamura, Y., et. al., *"Solving Aerodynamic Environmental Problems Arising from Train Speedup"*, S-TECH '96, 1996, Submitted Papers

C514/047/96

The aerodynamic sizing of tunnel cross-sections for train operation

R G GAWTHORPE BSc, MSc, CEng, MRAeS, FIMechE, T JOHNSON BSc, MSc, MAFIMA, and G I FIGURA-HARDY BSc
B R Research, Derby, UK

Synopsis

New routes require the straightest alignments to take best advantage of high-speed services. Except in the flattest terrain, this will necessitate the costly construction of numbers of tunnels. Since tunnel construction costs are mainly dependent on the amount of earth and rock to be removed, there is a need to minimize the cross-sectional area of the tunnel. For speeds above about 200km/h, it is the aerodynamic pressure disturbances from the train which fix the cross-section. The Paper describes the basis of a procedure developed by BR Research for assessing the appropriate aerodynamic area depending on other features of the tunnel, and the geometry and operating features of the trains. Though emphasis is given to the design of plain tunnels used by modern unsealed trains, other alternatives are also examined.

1. INTRODUCTION

When a train passes through a tunnel, it generates a succession of compression (positive) and expansion (negative) waves in the tunnel (ref 1) which propagate along the tunnel at approximately the speed of sound. For train occupants, these rapid pressure changes unless limited may be felt as an excessive and unacceptable sensation on the eardrum. Although the limiting pressure change that people will tolerate varies considerably from person to person (ref 2, 3) the general limit of acceptable values is known (ref 4). Modern day operations produce pressure changes that are up to the limiting values, and could exceed them unless appropriate action is taken.

The amplitude and time dependent nature of the pressure transients are the important parameters in this context, and they depend on a number of geometric features of the tunnel and train, and on the operating conditions. In particular, they are dependent on train and tunnel cross section (together defining the blockage ratio), tunnel length, train length, and the geometric shaping and configuration of the train. In addition, the pressures are strongly dependent on train speed (approximately to the square of speed).

For double-track tunnels, the possibility exists of two trains passing each other within the tunnel. As well as each train producing its own family of sonic pressure waves, it will also produce an additional pressure pulse on the side of the other train as it passes it. This pressure is felt in addition to the sonic wave pressures and therefore adds to the total sensation of pressure discomfort.

Thus, a number of different situations, or operating scenarios, can occur, each giving rise to a particular level, and time history, of pressure in the tunnel. These pressures will be felt by passengers within the trains.

When a new railway route is being planned which has tunnels along its length, it is clearly important from a construction cost point of view and also from a passenger comfort point of view that its cross-section area is optimized. Put into simple terms, it is important that the tunnel is large enough to avoid excessive discomfort from the pressure fluctuation yet not so large as to incur unnecessary construction cost.

Thus, a procedure is required so that the known details of the tunnel configuration, the trains to be used and train operation can be input in order to assess the cross-sectional size of the tunnel which produce pressures which fall within the maximum allowable level. So as to provide a practical method which can be widely adopted for general use, the procedure should be relatively simple and straight forward.

2.   PROCEDURE

The two essential elements needed to set up the procedure are, firstly, a computer-based method which will predict accurate pressure histories of the conditions existing on the train for a generalized set of circumstances and, secondly, a chosen criterion defining the limiting pressures that are acceptable. It is then a matter of identifying the size of tunnel which will just meet the pressure criteria over the required full range of conditions envisaged for future operations in that tunnel.

Thus, an iterative procedure is undertaken which consists of the following steps:

a)   for the tunnel under study, list down the required geometric and configurational data that are known or can be assumed. This also includes essential empirical aerodynamic input data for the tunnel roughness and portal shape coefficients.

b)   list down the geometric and operating data (such as maximum speed) for the service trains envisaged for use in the tunnel. This also includes empirical aerodynamic data for the surface roughness and train shape coefficients.

c)   run the pressure history prediction program for the worst case (e.g highest speeds, 2 trains passing case) envisaged for future operations in the tunnel, assuming a realistic tunnel area (as a starting value).

d)   choose a limiting pressure change criterion for the service (see section 2.2).

e)   compare the most severe pressure change predicted by the calculation programme to occur on the train with the pressure comfort criterion.

f)   re-run the prediction program with a new value of tunnel area until a value is found for which the pressures just satisfy the criterion. This iterative process can be quite brief as a specially devised pressure scaling program is used to adjust the tunnel area rapidly to the final value. Typically, two to four runs of the computer program are sufficient, depending on the successful choice of the initial starting value. For each iteration, new train and tunnel aerodynamic coefficients are calculated for the input data (since these are themselves dependent on tunnel cross-sectional area).

2.1   Pressure Prediction Program

The use of a practical reliable computer-based calculation method is clearly essential. It must be based on a representation of unsteady compressible flow. A number of such methods (refs.5,6,7,8,) now exist each having its own slightly different features and refinements.

The main elements of the prediction method used by BR Research were developed by Vardy (ref 6) and use a one-dimensional flow model in which the underlying equations are solved by the method of characteristics. Three dimensional viscous flow effects around the train and tunnel ends are modelled using pressure loss coefficients and friction factors as is common practice in the analysis of steady pipe flow problems.

The program is very comprehensive and is capable of modelling the airflows in multiple tunnel complexes with cross-passages and airshafts. Predictions can be furnished for trains running singly, in flights or passing in opposite directions.

The program has been used extensively by BR Research for calculating the flows in mainline railway tunnels and has been validated against experimental data. Fig 1 shows a comparison of calculations undertaken using the AEROTUN program (which is an earlier but essentially the same aerodynamic calculation as used in the THERMOTUN program), and experiment for the transit of two trains through a tunnel over 1km long. Fig 1a shows a comparison between measured and predicted pressures at fixed points in the tunnel (ref 9). Train-borne pressures are compared in Fig 1b for the two trains involved in the same test. There is a slight over estimate in the magnitude of the later reflected waves and they occur slightly in advance of experiment. However the general standard of agreement between theory and experiment, for both pressures on the moving trains and at stationary points in the tunnel, is good.

The train borne pressures predicted by the THERMOTUN program are those which would be experienced on the exterior of the coach. The internal pressure created inside a train as a result of the generated external pressures depends on the characteristics of the external pressure transient and the degree of pressure sealing of the train structure. For a completely unsealed train the internal pressures felt by the passengers are of similar magnitude to the external pressures. As a degree of sealing is introduced, a pressure difference across the structure of the vehicle is created, the external pressures being attenuated during their transmission internally. In the case of well sealed trains, the rapid external pressures are well attenuated through the coach structure, and therefore very little pressure change is propagated into the coach interior.

In the example of the UIC leaflet (ref 11) discussed in Section 3 of the Paper, the procedure describes the tunnel design for unsealed rolling stock operations appropriate to trains such as the British HST, IC225, the French TGV-PSE, TGV-A, the Anglo French Eurostar, the Italian ETR 460 and the Swedish X2.

2.2   Pressure Comfort Criteria

The choice of pressure limit is not straightforward because of the variation in perception of discomfort felt between one person and another and also between one journey and another depending on the journey characteristics. Though the variations between people can be dealt with statistically, i.e. a trade-off between pressure limit and a given percentile of passengers finding it unpleasant, the choice of a limiting pressure criterion appropriate to a particular route depends on a number of complex factors. Amongst them, the type of tunnels, e.g. single or double track tunnel (that is, the likelihood of being passed by another train), and frequency of tunnels along a particular route have a large effect on the choice of pressure criteria adopted. It also depends on the standard of comfort that the operator wishes to provide. For new prestigious routes, where the public expectation of comfort is high, stricter comfort limits are usually adopted. However, the budgetary constraints on such a choice are clear.

Designers of tunnels on new high speed railways are responsible for the expenditure of vast sums of money, and they need to have precise design parameters. They need to know what pressure histories are likely to be experienced by passengers and what is the limiting level of discomfort, caused by these pressure changes, which people will accept before they turn to alternative modes of transport. Otherwise, tunnels may be built with unnecessarily large cross-sectional areas or with unacceptably small areas giving rise to excessive discomfort. The former error implies uneconomic construction costs whilst the latter implies either lost revenue or subsequent tunnel modification costs to remedy the situation.

Fig 2 shows the approximate variation in pressure change magnitude with tunnel-to-train length ratio for given values of B (blockage ratio) and hence, for a given train speed and train area, it indicates the effect that a change of limiting pressure criterion will have on the tunnel cross-sectional area. Clearly, the choice of appropriate pressure comfort criterion is paramount.

It is considered that the form of the criterion defining a limiting change of pressure occurring within a few seconds correlates best with people's perceptions of discomfort. This has been demonstrated in ref 10 and is also supported by fundamental research in the Aerodynamics Pressure Chamber Facility at BR Research such as that reported in refs 3 and 4.

Very few operators have published criteria from their operations and it is believed that there is still a genuine uncertainty about what is the correct choice for their own operating circumstances and what are the cost implications of that decision. The most recent and widely known criteria are probably those used by Great Britain and Germany.

For existing InterCity routes, BR have used a criterion defined as:
Maximum change of pressure of 4kPa within any time period of 4 seconds.

This is for a worst case situation involving the most adverse passing of two trains in a double-track tunnel. These routes are used by unsealed trains.

For the planned high-speed rail link to the Channel Tunnel from London, where about 20% of the route is in the tunnel, criteria have been recommended of :
    3.5kPa change within any period of 4 seconds for double-track tunnels,
    2kPa change within 4 seconds for single-track tunnels.

Again, these are chosen for unsealed train operations and the most adverse train passing case is assumed for double-track tunnel case.

German Railways (DB AG) have been assessing criteria defined as :
    Maximum pressure change of   0.5kPa within 1 second

0.8kPa within 3 seconds
1.0kPa within 10 seconds

However, these maximum pressure changes refer to single train operation only, even in double-track tunnels. These criteria are presently applied on DB's Neubaustrecke routes, where about 30% of the routes are in tunnel and the majority of trains used have sealed rolling stock.

Due to the lack of international consensus, it is presently necessary in a tunnel sizing method such as that described in the next section to incorporate the comfort criterion as a variable, where it is defined as a maximum pressure change occurring in either 1 second, 4 seconds or 10 seconds.

3. EXAMPLE RESULTS

An illustration of this procedure and its results is given in a recent UIC leaflet 779-11 (ref 11) which adopts the same procedure as that in the Paper and indeed has used the same material. It caters only for unsealed trains and for plain tunnels.

Because tunnel pressures, and hence tunnel sizing, are dependent on a large number of factors, the compilation of a manageable document covering the ranges of all these parameters has, out of necessity, to contain some rationalization of the results. Although project decisions having major cost implications (such as tunnel size) should be made on the basis of specially-conducted calculations specifically for the tunnel operation concerned (as described in Section 2), it is feasible to provide a set of generalized graphical results which gives an invaluable indication for initial design studies of the tunnel size required.

Fig 2 illustrates the way in which the results can be displayed. It is an example case for a double-track tunnel designed on the basis of the worst case pressures generated by two trains (designated as " streamlined high-speed trains "), both travelling at 250km/h, and passing in the tunnel. It shows the variation of the maximum pressure change (defined as a change occurring within any 4 sec period) with the length ratio of tunnel-to-train for a number of given values of B (the "blockage" ratio of train-to-tunnel cross-sectional area). Thus, the required tunnel size may be evaluated by drawing a horizontal line, equivalent to the comfort criterion limiting pressure, to meet a vertical line for the tunnel-to-train length ratio. The point of intersection will be between two of the curved lines of B, and an interpolation between these will give the value of B for the intersection. Knowing the train cross-section area, then the tunnel area is equivalent to train area divided by the value of B, the blockage ratio.

An alternative form of the results is illustrated in Fig 3, where the variation of maximum pressure change with blockage area ratio is shown for the case of two " standard modern trains" passing in a double-track tunnel having a train-to-tunnel length ratio of 0.2 (i.e tunnel-to-train length ratio of 5) for 3 train speeds. S1 is equivalent to both trains running at 180km/h, S2 is equivalent to 200km/h, and S3 to 220km/h.

The category of "standard modern train" as used in ref 11 is a locomotive-hauled train of conventional coaches and is in fact representative of the generation of train design prior to current streamlined trains.

Thus, a series of graphs can be produced for a range of values of train speed, tunnel-to-train length ratio, train type, single or double-track tunnels, etc., allowing an estimate of tunnel area to be determined.

4. OTHER CONSIDERATIONS

The whole intention of this design procedure is to establish the best choice of the cross-sectional size for a tunnel satisfying the comfort requirements of the passengers at the lowest cost to the builder and operators. The procedure has specifically been applied to the general case of a smooth plain tunnel i.e one without airshafts, tunnel junctions or other changes to the tunnel area. It also assumes a conventional modern unsealed train. However, a number of other interesting options exist for the optimization of tunnel operations when other tunnel configurations and train design features are examined.

4.1 Airshafts

These are defined as plain open shafts which rise up from the tunnel to ground surface level. They do not normally contain ventilation plant. Previous work, for example ref 12, has shown the considerable advantages to be gained from the use of airshafts in reducing the amplitude of the strongest waves. Consequently, smaller diameter tunnels can be used for the same limiting pressures. Whilst the greatest advantage is gained from using a number of shafts (for example, shafts spaced at intervals equivalent to the length of the trains involved), Table 1 taken from ref 12 shows that even the use of two shafts offers considerable benefit.

4.2 Modified Tunnel Portal Shapes

The addition of flared and perforated extensions to the tunnel section (see ref 13,14) can also provide substantial advantage in reducing pressure or, alternatively, allowing smaller tunnel cross-sections for a given pressure limit. However, studies have shown that the length of portal necessary is considerable and, generally speaking, needs to be about one-third of tunnel length. Whilst they are particularly advantageous for single-track tunnels, similar benefits apply to double track tunnels although of course portal extensions then need to be constructed at both ends of the tunnel. In general, it can be said that the cost of providing a portal extension solution to a tunnel is rather greater than for the airshaft solution.

4.3 Sealed Trains

The "alternative" solution to the tunnel pressure problem is to transfer attention to train design, the most effective modification being to seal the rolling stock sufficiently to reduce the internal pressure transients to an acceptable level (see ref 15). However, this requires considerable structural and component re-design, and therefore a substantial cost penalty, not only for initial cost but also for maintenance. Clearly, on routes where there are large numbers of tunnels, the sealed train solution has attractions.

Once a decision has been made by the train manufacturer at the project stage to seal the train, then it becomes sensible and indeed quite practical to attain a high degree of sealing, though at a high cost. Put another way, even a relatively modest degree of sealing requires considerable re-design and attention to detail, but then relatively little extra attention is needed to achieve a well-sealed design. Having produced a well-sealed coach means that comparatively high pressures can be sustained external to the train. This suggests that relatively small tunnels become practical. However, the disadvantage that now arises is that a more highly stressed coach structure is necessary, implying heavier and more costly construction. A further problem that occurs is that, in the event of a structural sealing failure e.g window breakage or equipment malfunction, the passengers can then be subjected to the much higher pressures that can exist outside the train, and therefore much more severe pressures than would occur in unsealed train operations. Of course, the risks of this occurring are slight and the train would as soon as possible be subjected to a speed restriction in order to limit the pressure changes.

To offset some of the considerable pressure difference across the coach structure, developments have been proposed, notably by the Japanese RTRI (ref 16), to control the pressure internal to a coach by an actively-controlled fan or other airflow supply or extract device, so as to provide a more acceptable structural pressure loading whilst still maintaining a reasonable comfort environment. Such an optimization procedure theoretically allows the smallest tunnel cross-sections to be adopted for a given operating speed especially when used in combination with airshaft and special portal shapes. However, considerable development is still required in this area.

5. CONCLUSION

BR Research has developed a simple methodology for the cross-sectional sizing of railway tunnels to take account of the aerodynamic pressures generated by train operation. It relies heavily on the use of one of the modern computer programs which have been specially developed for unsteady compressible flow prediction. A UIC leaflet based on the methodology has been produced and enables preliminary estimates of tunnel size to be evaluated for project purposes. BR Research undertake specialist calculations leading to tunnel size recommendations for particular circumstances where required.

The leaflet presently deals with unsealed trains operating through plain tunnels. The Paper identifies other modifications which can be made to the tunnel configuration and also to rolling stock design to help minimise the cross-sectional area of the tunnel. Such effects can however be included within the specialist analysis undertaken by BR Research.

Active or Passive sealing I

ACKNOWLEDGEMENTS

The author wishes to thank the Managing Director, BR Research for permission to publish this Paper. He also wishes to acknowledge his colleagues in the Aerodynamics Team for their major contributions.

References

1. *Gawthorpe R G & Pope C W*, "The Measurement and Interpretation of Transient Pressures Generated by Trains in Tunnels." Paper C3, Proc. Second Int. Symp. on the Aerodynamics and Ventilation of Vehicle Tunnels. (Cambridge, UK, 23-25 March 1976) BHRA Cranfield 1976.

2. *Gawthorpe R G*, "Human Tolerance to Rail Pressure Transients - a laboratory assessment." Paper C4. Procs. 5th Int. Symp of the Aerodynamics & Ventilation of Vehicle Tunnels (Lille, France, 20-22 May 1985) BHRA Fluid Engineering. Cranfield UK 1985.

3. *Gawthorpe R G, Figura G I & Roberston N*, "Pressure Chamber Tests of Passenger Comfort in Tunnels." Proceedings : 8th International Symp on Aerodynamics & Ventilation of Vehicle Tunnels. 6-8 July 1994, Liverpool, UK. pp 227-243. Organised by the BHR Group, Cranfield UK. Mechanical Engineering Publications Ltd, London, UK.

4. *Gawthorpe R G*, "Pressure Comfort Criteria for Rail Tunnel Operations." Procs - 7th Int Symp of the Aerodynamics and Ventilation of Vehicle Tunnels, pp 173-188. (Brighton UK 27-29 Nov 1991). BHR Group, Cranfield, UK 1991.

5. *Fox J A & Henson D A*, "The Prediction of the Magnitude of Pressure Transients Generated by a Train Entering a Single Tunnel." In : Proc. Inst. Civ. Engrs., 49, May 1971, pp 53-69.

6. *Vardy A E*, "On the use of the Method of Characteristics for the Solution of Unsteady Flows in Networks." In : Proc. 2nd Int. Conf. on pressure surges (London, UK, Sept 22-24 1976) Cranfield, UK, BHRA Fluid Engineering, 1976, Paper H2, pp 15-30.

7. *Woods W A & Pope C W*, "On the Range of Validity of Simplified One Dimensional Theories for Calculating Unsteady Flows in Railway Tunnels." In : Proc. 3rd Int. Symp on the Aerodynamics and Ventilation of Vehicle Tunnels (Sheffield, UK : Mar 19-21 1979) Cranfield, UK, BHRA Fluid Engineering, 1979, Paper D2, pp 115-150.

8. *Harwarth F & Sockel H*, "Unsteady Flow due to Trains Passing a Tunnel." In : Proc. 3rd Int Symp on the Aerodynamics & Ventilation of Vehicle Tunnels (Sheffield, UK : Mar 19-21 1979) Cranfield, UK, BHRA Fluid Engineering, 1979, Paper D3, pp 151-160.

9. *Glockle H & Pfretzschner P*, "High Speed Tests with ICE/V Passing Through Tunnels, and the Effects of Sealed Coaches on Passenger Comfort", Proc 6th Int Symp on the Aerodynamics and Ventilation of Vehicle Tunnels, BHRA Fluid Engrg, Durham UK, 23-44, 1988.

10. *Glockle H*, "Comfort Investigations for Tunnel Runs on the New Line Wuerzburg-Fulda" In : Proc 7th Int Symp on the Aerodynamics and Ventilation of Vehicle Tunnel (Brighton, England, 27-29 Nov 1991). Organised by BHR Group, Cranfield, UK. Published by Elsevier Science Publishers.

11. _____ "Determination of Railway Tunnel Cross-sectional Areas on the Basis of Aerodynamic Considerations." UIC leaflet 779-11. To be published by International Union of Railways, Paris, France 1996.

12. *Gawthorpe R G & Pope C W*, "Aerodynamic aspects of train design for operation through the Channel Tunnel." Conference C451/003. Train Technology for the Tunnel, Inst. of Mech. Engrs. 4-5 November 1992, Le Touquet, France.

13. *Fox J A & Vardy A E.* "The Generation and Alleviation of Air Pressure Transients Caused by the High-speed Passage of Vehicles Through Tunnels." In : Procs. 1st Int. Symp on The Aerodynamics & Ventilation of Vehicle Tunnels (Canterbury, England, 10-12 Apr 1973) Organised by BHRA Fluid Engg, Cranfield, UK, 1973. Paper G3, 16pp.

14. *Vardy A E*, "Ventilated Approach Regions for Railway Tunnels." Transportation Engineering Journal, ASCE, Proc ASCE Vol 101, No TE4, Nov 1975, pp 609-619.

15. *Diepen P*, "Deutsche Bundesbahn's Pressure-Sealed Passenger Coaches." ZEV Glasers Annalen 117 (1993) Nr 6. June 1993, pp 180-193.

16. *Kobayashi M, Suzuki Y & Akutsu K*, "Alleviating Aural Comfort of Passengers under Shinkansen Speed-up by Controlling Flow Rate in Ventilating System." Paper B4-2-(3). In : Procs. of The International Conference on Speed-up Technology for Railway & Maglev Vehicles (Yokohama, Japan, 22-26 Nov 1993). Organised by Japan Society of Mechanical Engineers.

Table 1 Estimated areas for single-track tunnels
for a train speed of 225km/h and 400m length
(Reproduced from ref.12)

1.5km Tunnel		3.0 km Tunnel	
No. of Shafts	Area, m^2	No. of shafts	Area, m^2
0	58.5	0	49.0
2	29.5	4	26.8
4	23.9	9	23.8
9	22.9	19	22.8
29	17.4	59	21.21
Criterion 2.5 kPa in 4s			

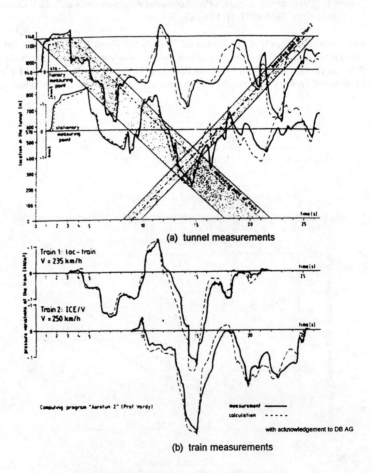

(a) tunnel measurements

(b) train measurements

Fig 1 Comparison of AEROTUN prediction with experiment
(Reproduced from ref.9)

© IMechE 1996 C514/04

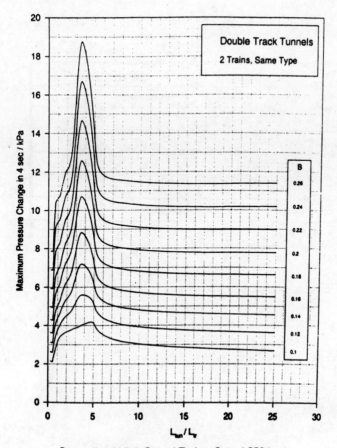

Streamlined High Speed Trains, Speed 250 km/h

Fig 2 Variation of pressure change with length ratio for different blockage ratios

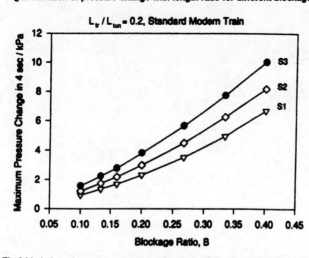

Fig 3 Variation of pressure change with blockage ratio for single trains at 3 different speeds

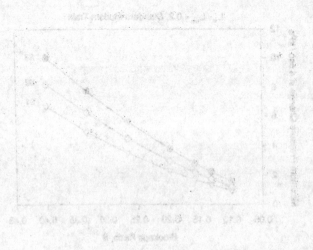

C514/015/96

Optimum nose shape for reducing tunnel sonic boom

M IIDA MJSME, **T MATSUMURA, K NAKATANI, T FUKUDA,** and **T MAEDA**
Railway Technical Research Institute, Tokyo, Japan

SYNOPSIS

This paper describes a train nose shape which is effective for reducing an impulsive pressure wave ('tunnel sonic boom' or 'micro-pressure wave') radiating out of the exit of a railway tunnel. First, the effects of fundamental nose shapes are made clear by numerical simulation of tunnel entry problem using axisymmetric models. Next, an optimum nose shape for reducing the impulsive pressure wave is obtained by nonlinear optimization method coupled with the above axisymmetric simulation. The effect of the optimum shape is confirmed by model experiments. Finally, the effects of three-dimensional models which have the same cross-sectional area distribution as the axisymmetric optimum shape are investigated by three-dimensional simulation.

1. INTRODUCTION

When a train nose enters a tunnel, it generates a compression wave in the tunnel. The compression wave propagates through the tunnel and arrives at the exit. At this moment an impulsive pressure wave radiates out of the tunnel exit toward the surrounding area[1]. The situation is shown in Fig 1. The impulsive pressure wave causes an explosive sound and/or an abrupt rattling of window frames or shutters of houses near the tunnel exit if its magnitude is larger than a certain level (about 20Pa of the peak value at a house). This phenomenon is called a *tunnel sonic boom* or a *micro-pressure wave*. It is necessary to solve this environmental problem for speed-up of the service of high-speed trains in mountainous countries like Japan.

An acoustic analysis using far-field and low-frequency approximations leads us to the conclusion that the maximum value of the impulsive pressure wave is proportional to the maximum pressure gradient of the compression wave arriving at the tunnel exit[2]. Consequently, it is effective for reducing the impulsive pressure wave to decrease the pressure gradient of the compression wave at the tunnel exit. Based on this principle, several measures against the impulsive pressure wave have been proposed[1, (3)].

A typical measure against the impulsive pressure wave is installation of a hood with openings at the tunnel entrance. This measure aims to fully decrease the pressure gradient of the compression wave at the tunnel entrance in the stage of its generation during train nose entry. The decrease in the pressure gradient of the compression wave at the entrance results in its decrease at the exit, for steepening of the propagating compression wave due to nonlinear effect is smaller if its initial pressure gradient is smaller. Thus, we can reduce the impulsive pressure wave by means of the tunnel entrance hood. For instance, a 49m hood which is installed at an entrance of a 6.7km slab track tunnel reduces the peak pressure of the impulsive wave from 300Pa to 20Pa at a measured point[3]. As for the distortion of the compression wave during propagation in the Shinkansen tunnels, see the reference[4].

We can also apply measures to the train in the following manner. If we make the maximum cross-sectional area of the train smaller, the pressure rise of the compression wave becomes also smaller and consequently its pressure gradient decreases. If we make the train nose slenderer, the rise time of the compression wave becomes longer and hence its pressure gradient decreases. However, there are restrictions on those measures from the commercial point of view. Accordingly, it becomes important to find out an optimum nose shape under the conditions that the maximum cross-sectional area and the length of the train nose are given.

There are several works that investigated the effect of the train nose shape for reducing the impulsive pressure wave by using model experiments or numerical simulations[5],[7]. In the paper[5] by Maeda et al. a qualitative discussion is given regarding a nose shape effective for reducing the impulsive pressure wave. Based on this result, we intend to seek an optimum nose shape using numerical simulation in this paper.

2. NUMERICAL SIMULATION OF TUNNEL ENTRY PROBLEM USING AXISYMMETRIC MODELS

To estimate the effect of the train nose shape for reducing the impulsive pressure wave, numerical simulation of the tunnel entry problem is conducted[7]. As mentioned above, the smaller the pressure gradient of the compression wave at the tunnel entrance, the smaller the impulsive pressure wave. Therefore, we use the maximum pressure gradient of the compression wave at the tunnel entrance as an indicator of the effect of the train nose shape for reducing the impulsive pressure wave.

We assume a compressible inviscid flow in the numerical simulation. This is because we pay attention to generation of the compression wave by the train nose and the influence of viscosity is considered to be neglected. In addition, we use axisymmetric models in Sections 2 and 3. The effects of three-dimensional models will be taken up in Section 4.

2.1 Computational methods
To obtain the form of the compression wave generated during train nose entry using numerical simulation, we set a problem in which a semi-infinite train enters a semi-infinite tunnel. A computational model of this problem is shown in Fig 2, where d: the diameter of the tunnel, a: the length of the train nose, b: the radius of the maximum part of the train, V: the running velocity of the train.

The governing equations in the numerical simulation are the unsteady axisymmetric compressible Euler equations. Space discretization is done by the cell-centered finite volume method(FVM). Numerical fluxes at the FVM cells are calculated by the total variation diminishing (TVD) scheme[8] of MUSCL type. Time integration is performed by the two-step explicit Runge-Kutta scheme.

The boundary conditions are treated as follows. On the solid wall, the slip condition is imposed and the value of the pressure on the wall is extrapolated from the values at the centers of the neighboring cells. On the center line of the axisymmetric cells, the numerical fluxes are set to be zero. Non-reflecting conditions are imposed on the far-field boundary and the artificial boundary in the tunnel located at the distance of $8d$ from the entrance.

Relative motion of the train and the tunnel is treated by using multiple grids. A moving grid is assigned for the train, while a stationary grid is assigned for the tunnel. The two grids relatively move each other, sliding on their interface boundary. Information is transferred across the interface boundary by the interface scheme similar to the one by Rai[9].

A close-up of the grid system is shown in Fig 3. The tunnel grid is 178x26 H-grid and the train grid is a combination of 416x23 H-grid and 230x11 O-grid. These grids are generated by Transfinite method.

Computation begins with an impulsive start of the train. The train is initially positioned at a distance of $6d$ from the tunnel entrance and starts suddenly. It runs at a constant velocity and then enters the tunnel. The initial position of the train is so determined that the pressure wave generated by the impulsive start is sufficiently attenuated by the time the train enters the tunnel.

2.2 Computational results

Fig 4 is a typical example of computational results. It shows time evolution of pressure contours in the computational domain and illustrates the generation process of the compression wave by the train nose entering the tunnel. The pressure inside the tunnel begins to rise just before the front end of the nose part enters the tunnel and continues to rise until a short while after the rear end of the nose part enters it. The pressure rise propagates through the tunnel as a compression wave and passes the artificial boundary in the tunnel without reflection.

Computational results of the compression wave form are presented in Fig 5. The results are obtained under the conditions: $R=0.116$, $M=0.29$, and $a/b=5$, where R: the blockage ratio, M: the running Mach number, a/b: the slenderness ratio. Nose shapes of the train for this computation are three fundamental ones: an ellipsoid of revolution, a paraboloid of revolution, and a circular cone. Fig 5 shows the pressure change and its time derivative on the tunnel wall at a distance of $6.8d$ from the tunnel entrance, which expresses the form of the compression wave and its pressure gradient. Here the time is set zero when the front end of the nose reaches the tunnel entrance and is nondimensionalized with respect to reference time defined as d/V. The pressure is expressed as a coefficient defined as $(p-p_0)/\frac{1}{2}\rho_0 V^2$, where p: the pressure, p_0: the ambient pressure, ρ_0: the ambient density. In the figures that follow in this paper, the pressure and the time are expressed in the same way.

From Fig 5, we can see that the maximum pressure gradient of the compression wave in the case of the paraboloid of revolution is the smallest of the three nose shapes. Fig 6 shows the relation between the nose length and the maximum pressure gradient of the compression wave for the three fundamental shapes, which was obtained by computation for different values of a/b. In Fig 6, the maximum pressure gradient is normalized by that in the case of the ellipsoid of revolution whose slenderness ratio a/b is three. Simultaneously Fig 6 shows the results of model experiments[5] under the same conditions as the present computation except the running Mach number. The figure indicates the following: (1) the longer the train nose, the smaller the maximum pressure gradient of the compression wave; (2) the maximum pressure gradient of the compression wave in the case of the paraboloid of revolution is the smallest of the three nose shapes if the nose is not very short, i.e., $a/b>2$; (3) the agreement of computation and model experiments is quite good if $a/b>2$. From the last result, we can conclude that the numerical simulation can be applied for estimation of the effect of the train nose shape for reducing the impulsive pressure wave.

3. OPTIMIZATION OF NOSE SHAPE USING AXISYMMETRIC MODELS

In the preceding section, the paraboloid of revolution was presented as a shape effective for reducing the impulsive pressure wave. In this section, we seek a more effective shape by modifying the paraboloid of revolution. We use a numerical optimization method to this end.

As mentioned earlier, the smaller the blockage ratio and the larger the slenderness ratio of the nose, the smaller the maximum pressure gradient of the compression wave. Consequently, our optimization problem to be solved here is as follows: Using an axisymmetric model, determine the nose shape that minimizes the maximum pressure gradient of the compression wave generated during its entry into the tunnel on the condition that the blockage ratio and the slenderness ratio are given.

3.1 Optimization methods
The numerical simulation described in the preceding section is used as method for estimating the effect of the nose shape. Coupling the simulation with the nonlinear optimization method enables us to optimize the nose shape automatically using computers. As nonlinear optimization approach we use the method of rotating directions proposed by Rosenbrock[10]. Since the method is one of the direct search methods and does not require the derivative of the objective function, we can apply it easily to a problem where the objective function is given as an output of simulation programs.

For the numerical optimization, we must express the nose shape mathematically using design variables. In this paper, we express the cross-sectional area distribution of the train nose $A(x)$ as

$$\frac{A(x)}{\pi b^2} = \left(1-\alpha_2\right)\left[\left(1-\alpha_1\right)\frac{x}{a}+\alpha_1\sqrt{\frac{x}{a}}\right]+\alpha_2\left(\frac{x}{a}\right)^2 \dots\dots\dots\dots\dots(1)$$

where x: the distance from the front end, a: the length of the train nose, b: the radius of the maximum part of nose. α_1 and α_2 are the design variables to control the shape of the nose. If $\alpha_1=0$ and $\alpha_2=0$, the cross-sectional area distribution is linear and the shape of the nose is the paraboloid of revolution. Increasing the number of the design variables makes a finer tuning possible. However, this causes an increase in iteration numbers of optimization process. We can obtain practically effective results by Eq (1) under the conditions treated in this paper.

In the present optimization problem, the design variables are α_1 and α_2 and the objective function to be minimized is the maximum pressure gradient of the compression wave generated by the axisymmetric nose shape whose cross-sectional area distribution is expressed as Eq (1). Determined the design variables, the value of the objective function is obtained by the following procedures: (1) definition of the nose shape, (2) generation of the computational grid, and (3) numerical simulation of the tunnel entry problem. The algorithm of the optimization is the one by the above Rosenbrock's method. The initial nose shape is set to be the paraboloid of revolution, i.e., $\alpha_1=0$ and $\alpha_2=0$.

3.2 Axisymmetric nose shape obtained by numerical optimization
Fig 7 shows the result of the optimization under the condition that $R=0.116$ and $a/b=5$. Fig 7(a) shows the cross-sectional area distribution and the radius distribution of axisymmetric models. Fig 7(b) shows the form of the compression wave and its pressure gradient. In the figure, the results in the cases of the other fundamental shapes are also shown for comparison. From Fig 7(b), we can see that the form of the pressure gradient is flattop shape and the maximum value is smaller in the case of the optimum nose shape than in the case of the other shapes. Consequently, we can conclude that the optimization of the nose shape has been successful in this case.

The characteristics of the optimum shape shown in Fig 7(a) are as follows: (1) the shape of the front end is blunt; (2) the cross-sectional area continues to increase up to the rear end of the nose part. In the case of the nose shape whose front end is blunt, the pressure begins to rise early when the nose approaches the tunnel entrance. In the case of the nose shape whose cross-sectional area continues to increase up to the rear end, the pressure continues to rise longer after the rear end of the nose enters the tunnel entrance. Both of them lengthen the rise time of the compression wave and contribute to decreasing the maximum pressure gradient of the compression wave.

The results of the optimum nose shapes in the cases of different slenderness ratios and different blockage ratios are not shown here on account of limited space. However, those results indicate that the optimum shapes under the condition that R ranges from 0.116 to 0.3 and a/b ranges from 3 to 12 are similar to the shape shown in Fig 7(a) in normalized form, which means the relation between $A(x)/\pi b^2$ vs. x/a. Usually the slenderness ratio of high-speed trains is considered above three, and consequently we can say that the characteristics of the train nose shape effective for reducing the impulsive pressure wave is represented in Fig 7(a). Moreover, the shape in Fig 7(a) agrees quite well with the shape explained qualitatively based on the results of model experiments[5].

Some operators may face a problem of operating in tunnels where blockage ratios are higher than 0.3. The optimum nose shapes in these conditions are also considered effective.

3.3 Model experiments on axisymmetric optimum nose shape

We have done a model experiment to confirm that the optimum nose shape obtained by the numerical optimization is actually effective to decrease the maximum pressure gradient of the compression wave. The results are shown in Fig 8. Fig 8(a) gives the shapes used in the experiment, one being the shape obtained by the numerical optimization and the other being the paraboloid of revolution for reference. In Fig 8(a), we notice that there is a contracting part from $x/b=5.5$ to $x/b=8$. This part is formed for a reason unrelated to the present study.

Fig 8(b) and (c) show the results of the numerical simulation and the model experiment using the nose shapes shown in Fig 8(a). We observe that there is a negative pressure gradient region caused by the contracting part but that its influence is not large. Comparing the result of the simulation with that of the experiment, we see that in both results the maximum pressure gradient in the case of the optimum shape is smaller than in the case of the paraboloid of revolution. Hence we have confirmed that the shape obtained by the numerical optimization is actually effective to decrease the maximum pressure gradient of the compression wave.

4. NUMERICAL SIMULATION OF TUNNEL ENTRY PROBLEM USING THREE-DIMENSIONAL MODELS

In this section using the result on the axisymmetric optimum nose shape, we attempt to construct three-dimensional shapes which can be applied to the nose shapes of real trains.

The three-dimensional shapes are so determined that their cross-sectional area distribution coincides with that of the axisymmetric optimum shape. Although there are innumerable three-dimensional shapes that have the same cross-sectional area distribution, we will consider two typical examples. One is made by cutting an axisymmetric shape in half, which is called a *half-axisymmetric* optimum shape. The other is a *two-dimensional* optimum shape whose longitudinal cross-section does not change in the lateral direction. Fig 9 shows the shapes that have the same cross-sectional area distribution as that of the axisymmetric shape in Fig 7.

As before, the effect of the three-dimensional shapes for reducing the tunnel sonic boom is investigated using the numerical simulation of the tunnel entry problem. The flow model and the computational methods are almost the same as in the axisymmetric case. The computational conditions are the same as in the computation of the optimization of Fig 7. Fig 10 shows a close-up of computational grids on the surface of the train and tunnel. The train runs over the center line of the floor of the tunnel. There is a gap between the bottom of the train and the floor of the tunnel. The computational grids are rather coarse in the present three-dimensional computation. Hence, we will discuss qualitatively about the results in this section.

Fig 11 shows the compression wave and its pressure gradient generated by the three-dimensional nose shapes in Fig 9. For comparison, Fig 11 also shows the result in the case of an ellipsoid of revolution cut in half. The maximum pressure gradient in the case of the two shapes that have the optimum cross-sectional area distribution is smaller than in the case of the ellipsoid of revolution cut in half. This confirms the effect of the optimization.

Next, comparing the cases of the two optimum shapes, we see that the maximum pressure gradient in the case of the two-dimensional model is larger. From the result of the compression wave form, we also see that the final pressure rise of the compression wave in the case of the two-dimensional model is also larger. The reason is considered that in the case of the two-dimensional shape shown in Fig 9, the flow separates to some extent when turning around the corners of the edge lines enclosed by the upper surface and the side surface and the net cross-sectional area of the train increases. Consequently, we can expect that it is possible to further decrease the maximum pressure gradient in the case of the two-dimensional shape by rounding the corners of the edge lines and preventing the separation.

5. CONCLUDING REMARKS

The optimum nose shape for reducing the tunnel sonic boom has been presented. The shape has been obtained by nonlinear optimization method coupled with numerical simulation of a tunnel entry problem using axisymmetric models. The effect of the shape for reducing the tunnel sonic boom has been confirmed by model experiments. The effects of the three-dimensional nose shapes that have the same cross-sectional area distribution as the axisymmetric optimum shape are investigated by three-dimensional numerical simulation of the tunnel entry problem.

Except for very short noses, the characteristics of the optimum shape are as follows: (1) the shape of the front end is blunt; (2) the cross-sectional area continues to increase up to the rear end of the nose part. The shape presented in this paper agrees quite well with the shape explained qualitatively based on the results of model experiments[5].

Lastly, some comments are made regarding the relation between the nose shape and the tunnel entrance hood. Improvement of the nose shape enables us to shorten the hood or even negates the use of it. However, since the effect of the hood depends on the nose shape, the necessary length of the hood changes with the nose shape.

The authors wish to thank Professor Ozawa for his comments on reading the manuscript.

REFERENCES

(1) Ozawa,S., Studies of micro-pressure wave radiated from a tunnel exit, Railway Technical Research Report, RTRI JNR, 1979, No.1121 (in Japanese).

(2) Yamamoto,A., Micro-pressure wave radiated from tunnel exit, Preprint of the Spring Meeting of Physical Society of Japan, 1977, No.4, 4p-h-4 (in Japanese).

(3) Ozawa,S., Maeda,T., Matsumura,K., Uchida,K., Kajiyama,H., Tanemoto,K., Counter-measures to reduce micro-pressure waves radiated from exits of Shinkansen tunnels, Proc. 7th Int. Symp. on Aerodynamics and Ventilation of Vehicle Tunnels, Brighton, UK, 1991, pp.253-266, BHR Group Limited.

(4) Ozawa,S., Maeda,T., Matsumura,K., Nakatani,K., Uchida,K., Distortion of compression wave during propagation along Shinkansen tunnel, Proc. 8th Int. Conf. on Aerodynamics and Ventilation of Vehicle Tunnels, Liverpool, UK, 1994, pp.211-226, BHR Group Limited.

(5) Maeda,T., Matsumura,K., Iida,M., Nakatani,K., Uchida,K., Effect of shape of train nose on compression wave generated by train entering tunnel, Proc. Int. Conf. on Speedup Technology for Railway and Maglev Vehicles, Yokohama, Japan, 1993, Vol.2, pp.315-319, The Japan Society of Mechanical Engineers.

(6) Ogawa,T., Fujii,K., Numerical simulation of compressible flows induced by a train moving into a tunnel, Computational Fluid Dynamics J., 1994, Vol.3, No.1, pp.63-82.

(7) Iida,M., Numerical simulation of compression wave generation in tunnels during train nose entry, RTRI Report, 1994, Vol.8, No.6, pp.25-30 (in Japanese).

(8) Chakravarthy,S., The versatility and reliability of Euler solvers based on high-accuracy TVD formulations, AIAA Paper 86-0243, 1986.

(9) Rai,M., Navier-Stokes simulations of rotor/stator interaction using patched and overlaid grids, J. Propulsion, 1987, Vol.3, No.5, pp.387-396.

(10) Konno,H., Yamashita,H., Nonlinear Programming, 1978, Nikkagiren Shuppan-sha, (in Japanese).

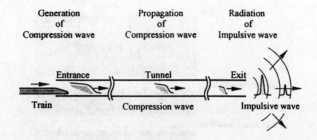

Fig 1 Generation of impulsive pressure wave from tunnel exit

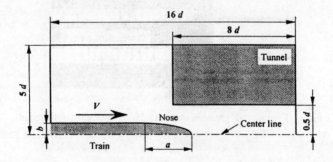

Fig 2 Computational model

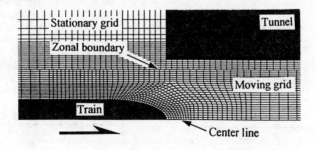

Fig 3 Computational grid system for axisymmetric model

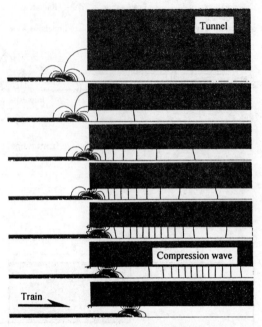

Fig 4   Evolution of pressure contours during nose entry into tunnel

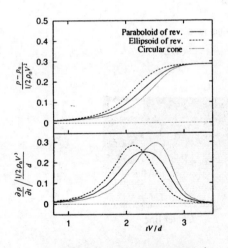

Fig 5   Compression wave generated
during nose entry into tunnel

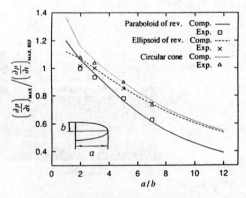

Fig 6   Relation between nose length and
maximum pressure gradient

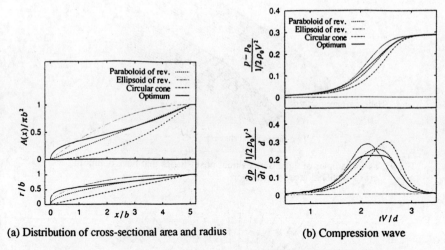

(a) Distribution of cross-sectional area and radius

(b) Compression wave

Fig 7   Axisymmetric optimum nose shape and generated compression wave

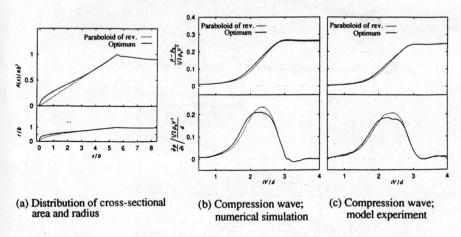

(a) Distribution of cross-sectional
area and radius

(b) Compression wave;
numerical simulation

(c) Compression wave;
model experiment

Fig 8   Numerical simulation and model experiment on axisymmetric optimum nose shape

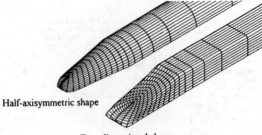

Half-axisymmetric shape

Two-dimensional shape

Fig 9 Two examples of three-dimensional nose shapes that have optimum cross-sectional area distribution

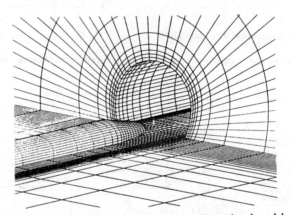

Fig 10 Computational grid system for three-dimensional model

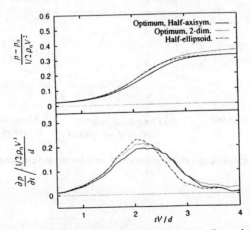

Fig 11 Compression wave generated by three-dimensional nose shapes shown in Fig 9.

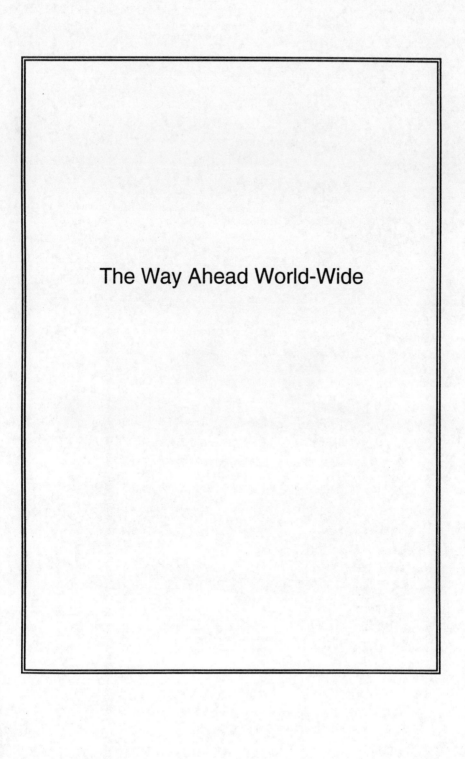

The Way Ahead World-Wide

C514/070/96

Role of super high-speed technologies in modern urban transportation

C D'SOUZA and **A VON GABLER**
ABB Daimler-Benz Transportation (Deutschland) GmbH, Germany

Summary

Since 1969 Germany has been involved in innovative developments of super high speed transportation systems. The new TRANSRAPID-technology is based on on contactless electromagnetic levitation and is designed for operating speeds upto 500 km/h. At the test facility (TRANSRAPID Versuchsanlage TVE) in Emsland analyses under operational conditions have been conducted since 1985. A photograph of TRANSRAPID is shown in figure1. In 1991 the German railways and 7 Universities confirmed the system to be technically mature for operation. Elaborate developments and tests of sophisticated subsystems to determine their reliability as well as their economy are currently being conducted to streamline practical applications.

Figure 1: TRANSRAPID Test Line

This paper describes the TRANSRAPID project to be realized between Berlin and Hamburg. The integration in available transportation infrastructures as well as suitable connections at junctions facilitates short travel times between city centers. This, together with adequate pricing and passenger comfort makes the system attractive for both business and private travellers.

Introduction

The initial ideas, concepts and patents for electromagnetic transportation systems date back to Hermann Kemper in 1922. In fact, the first scientific report of a railway type system was published in 1953. Later on as the need for fast and economic interconnections was realized, one returned to Kempers idea in 1965 and the German Ministry for Research and Development initiated a study for „High Efficiency, High Speed Railways". This study resulted in the funding of Magnetic Levitation Technology.

Private companies have ever since been developing this magnetic concept that caters to travelling speeds of 500 km/h. Table 1 depicts major mile stones over the past 25 years.

1969	Initial Study of Super Speed Railway Travel
1970 - 1978	Alternative Studies for Concepts of Levitation, Guidance and Driving of the Companies AEG, BBC, Dyckerhoff & Widmann, Krauss Maffei, MBB, Siemens and Thyssen Henschel
1977	Government's Decision to Develop Electromagnetic Levitation System based on Longstator Linear Propulsion (EMS)
1979	First Demonstration of Magnetic Levitation Vehicle at International Transportation Fair, Hamburg (IVA)
1979 - 1987	Construction of Test Facility at Lathen (TVE)
1989 - 1991	Certification of Technical Maturity by Central Office of Germany Railways
1992	Official Decision to Connect Berlin-Hamburg (Gov. Plans) by Adtranz
1994	Legislation for Magnetic Leviation Passed by German Court Constitution of 'Magnetschnellbahn Planungsgesellschaft'

Table 1: TRANSRAPID Chronology

In the electromagnetic system motion is caused by magnets attracting the objects i.e. the train, from below to the stator imbedded in the track, while lateral direction is maintained by guide rails on either side. An electronic control system ensures that the train is magnetically cushioned over a gap of 10 mm. The first object based on this principle was exhibited at the International Traffic Fair at Hamburg in 1979.

For the first time, more that 50 000 passengers had the opportunity of „floating" over a magnetic cushion over a distance of 900 m.

The next step was to implement an improved system over a length of 31,5 km at the test facility in Emsland [1]. For this purpose a consortium consisting of 6 members from the private sector (electromechanical and civil engineering firms) was founded. From 1981 this unique facility, operated by MVP (Versuchs- und Planungsgesellschaft für Magnetbahnsysteme) has proven the development results with respect to maturity and reliability. Already we have travelled over 500 000 km and reached operating speeds of 450 km/h.

Institutions such as german railways and Lufthansa have indicated interests to operate TRANSRAPID complimentary and not competitively to existing transportation modes. Among various other reasons, but also due to the environmentally friendly aspects, passengers from road and air will hopefully be diverted to this mode of transportation on this selected stretch.

After intensive tests, the german railways in 1991 certified the system as technically mature. This paved the way for commencing planning procedures for a concrete implementation. In 1992 the consortium members established a financial concept for building the route Berlin-Hamburg with TRANSRAPID.

EMS Concept and Test Facility

Figure 2 depicts the cross section of a conventional train and a TRANSRAPID vehicle. The floor of both bodies are at the same level. By virtue of the larger width of 3900 mm as compared with 2950 mm of conventional trains, the TRANSRAPID can accomodate 5 instead of 4 seats in a row. Alternatively it is possible to design other seating variations. The cell of the magnetic train embraces the „track" from below thus making it secure against derailing. All electronic equipment and on board power supply is accomodated in the space between floor and lower frame work. They are arranged so as to enable easy access from the top or from the sides. The propulsion system itself is embedded in the track and therefore distributed over the length of the train, that does not require extra motors or drives. This intrinsic feature enables installing extra energy at the place required, i.e. at slopes or where particularly high acceleration is called for.

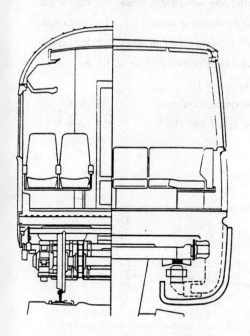

Figure 2: Comparison of Cross Section

The facility for high speed testing consists of a straight stretch with loops at either end (figure 3). The track is basically composed of 25-m long beams that are supported on stilts at approx. 4,5 m above ground level. The switches at the entrances to the loops are steel beam girders approx. 150 m long.

At the southern loop as well as at the test center the track shows a gradient of upto 35 ‰. The northern loop provides conditions for travel upto 300 km/h in curves and also for developing the necessary acceleration for speeds upto 450 km/h in the straight stretch. Before entering the southern loop the train is decelerated to 200 km/h. Passengers are offered a test ride over 41 km (due to routing of the testrun) at 420 km/h.

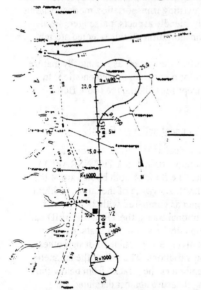

Elevated or At-Ground Level Guideway with:

Beams of Prestressend Concrete or Steel Girders

Length of Beams 12 m, 25 m (Standard Span)

31 m and 37 m

2 High-Speed-Switches, 1 Low-Speed-Switch

- South Loop with 1000 m radius   9,4 km Length
- Straight Guideway about   9,6 km Length
- North Loop with 1690 m radius 12,5 km Length
- Total Length of the Test Line   31,5 km
- Max. Lateral Tilt   12°
- max. Gradients of up to   35 ‰
- max. Speed with TR07   450 km/h

Figure 3: Test Track

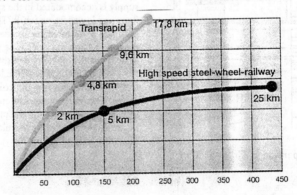

Figure 4: Acceleration Characteristics

The acceleration capacity of the magnetic train is shown in figure 4. After 160 sec TRANSRAPID traverses 9,6 km to reach a velocity of 400 km/h. In comparison, the ICE train requires 440 sec and 25 km to attain a velocity of 280 km/h. The same is valid during deceleration since the linear motor reacts with same forces at a 180° phase lag used for braking purposes.

Environment and Technology.

Due to the contactless mode of motion the magnetic train has good acoustic characteritics. The induction of forces via the wide track area minimize any jerks and contribute to passenger comfort felt in smooth riding. At all places and at all speeds the measured magnetic and electromagnetic values are far below the permitted minimum levels[3]. Figure 5 shows a comparison of noise emission levels measured for TRANSRAPID, ICE, TGV and conventional trains. The graph confirms that at a measuring distance of 25 m and at a speed of 400 km/h, the noise emission of TRANSRAPID at 93db(A) is equivalent to the TGV travelling at 250 km/h. At a speed of 300 km/h it is as loud as a conventional train travelling at 100 km/h.

Distance from measuring point 25m

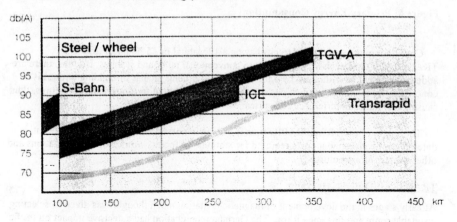

Figure 5: Comparison of Noise Emission

Generally, railway systems consume less primary energy in comparison with automobiles or airplanes. Assuming a 66 % pay load on all systems, energy consumption comparison is shown in figure 6. For an objective judgement, a short distance flight at 400 km was selected (in general the consumption at ground level is even higher).

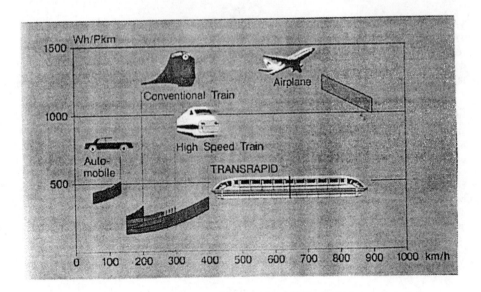

Figure 6: Primary Energy Consumption

Considering its capability on tracks having a 12° lateral or 35 ‰ linear gradient, TRANSRAPID has decisive advantages to conventional railbound systems. Besides, there are added investment advantages where the necessity of drilling tunnels or building bridges is concerned. By building the track on 'stilts', it is easier to adapt to topographical features, thus causing less radical interference to landscapes.

The length of the TRANSRAPID train is also an important economical factor, since it determines the investment costs required for stations, haltpoints, workshops, shuntstations and other related infrastructures.

TRANSRAPID Berlin-Hamburg

At early stages in the development of magnetic levitation much thought was given to selecting a suitable route and financing it too. The German reunification had a decisive impact on traffic flow and showed a growth potential in the East-West axis. It therefore seemed appropriate to connect two dynamic developing cities Berlin and Hamburg with each other. A fast and comfortable connection would certainly attract car and air passengers between the two cities. For intermediate connections or travel the local networks would be better frequented and the existing railway network could be optimised for freight transport.

Thanks to the interest and support of the legislative and executive governmental instances, a company MPG (Magnetschnellbahn Planungsgesellschaft) was founded in 1994. The MPG was alloted the task of providing all planning prerequisites based on legislative law to realize the connection Berlin-Hamburg.

The planning concept is based on the following facts:

- Connecting Berlin-Hamburg with TRANSRAPID.

- Halts at the endpoints as well as peripheral points in Berlin and Hamburg and one intermediate station in the area of Schwerin.

- Travel time less than 60 minutes.

- Travel frequency of 10-15 minutes.

- Peak operating velocities of over 400 km/h and 200 km/h in stop-free and city entry areas respectively.

- An estimated passenger ride capacity of approx. 14 mio by the year 2010.

- Investment totalling to 8.9 milliard DEM according to the financing concept of 1993.

Table 2: Planning cross data

Based on these premises the public hearing and planning procedures for alternative routes were initiated. In June 1996 a preferred route was established which is currently basis for further planning.

Distance	292	km
Speed	450	km/h
Travel Time	<60	min.
Frequency	64	Trains per day and direction
Fleet Size	14+2 Trains, 6 Sections each	
Capacity	500 Seats per Train	
Passenger	approx. 14 Mio./year	
Traffic capacity	approx. 4,5 Milliard Pass. km/year	
Average Load	63 %	

Table 3: Key Data of Berlin-Hamburg Route

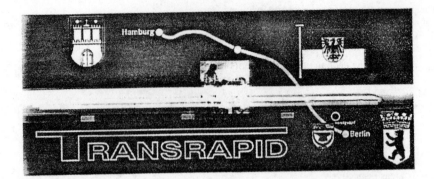

Figure 7: Planned Transrapid Route between Berlin and Hamburg

The preferred route begins at the Main Station in Hamburg, passes the first peripheral halt at Moorfleet leads on to Holthusen before stopping at Spandau-Peripheral of Berlin and then terminating at Lehrter Station. A central control station as well as maintenance and repair workshops will be erected at Perleberg.

The major mile stones of the intended implementation are shown in <u>figure 8.</u> This will be realised in joint efforts by the Private Sector, German Railways, Lufthansa and Public Sector.

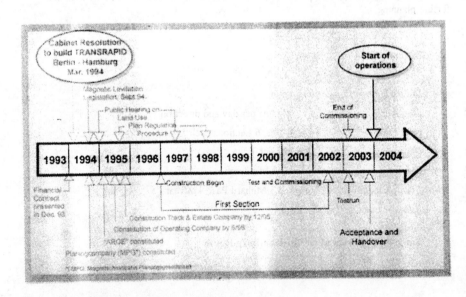

Figure 8: Intended Project Mile Stones

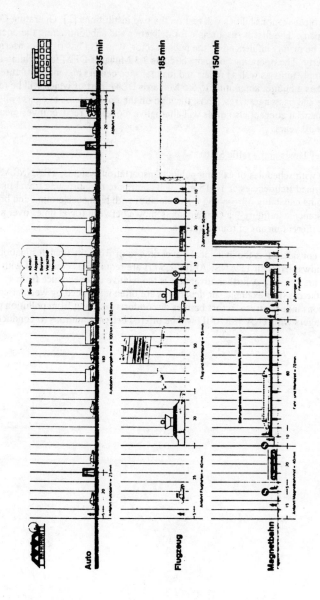

Fig. 9. Comparison of Travel Time over 300 km

In course of building and operating TRANSRAPID between Berlin and Hamburg new perspectives pertaining to economy, traffic flow and employment will be opened.

The major project-responsibilities will rest on the two institutions [2]: Operating Company and Track Company. The first is responsible for deliveries and subcontracting the operation. The second will be mainly influenced by the public sector, who also has rights concerning estate and property. The operating company budgets 3.3 Milliards DEM for building the trains, tracks, powersupply units as well as repair and maintenance centers and infrastructure. The track company has a budget amounting to 5.6 Milliards DEM, part of which will be recovered by operating fee and passenger fares. Thus, the load on the public sector budgets will be reduced. The financial concept also deals will alternative models to recover investment costs over a period of 30 years.

Important Travel Times and Traffic Shift

In coordination with schedules of existing public transportation systems, TRANSRAPID operating at planned frequencies of 10-15 min. will contribute to shortening travel times. This is mainly due to its capability of reaching high velocities with high acceleration and braking over shorter distances. In figure 9 a comparison is shown between travel times over a 300 km distance using different means of transportation.

What is not considered, is the individual travel time to or from the final destinations. The investigation shows that with TRANSRAPID 2,5 hrs are required as compared with 3 hrs 5 min by air or 4 hrs by road. A further advantage is the relative independance from environmental factors. For example, air departures are often influenced by weather conditions, traffic congestions induce stress factors besides discomforts resulting from pollution problems. Business passengers with TRANSRAPID will be offered better conditions to conduct work during travel.

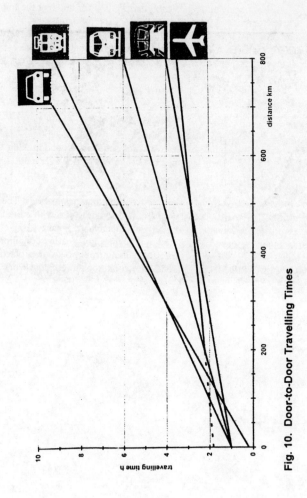

Fig. 10. Door-to-Door Travelling Times

The graphics in figure 10 attempt to objectively characterize transport means with respect to distances of travel. Below 100 km the automobile has advantages based on flexibility and ease of travel preparation. In ranges upto approx. 500 km the TRANSRAPID has plus points while the air plane begins to gain attraction at longer distances.

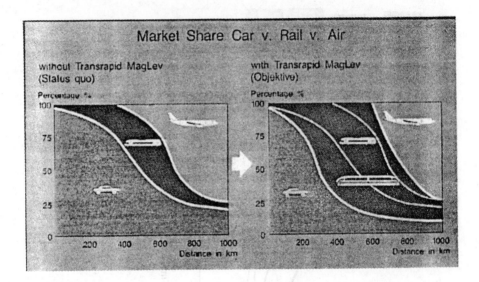

Figure 11: Passenger Shift

A further investigation, - results shown in figure 11: - predicts expected traffic shifts from road and air to TRANSRAPID. For distances below 600 km air traffic will yield to TRANSRAPID while for distances above 1000 km they will retain their customers. This analysis presumes an infrastructure with a well integrated magnetic levitation transportation system.

Conclusion

The TRANSRAPID connection between Berlin and Hamburg signals the introduction of a new means of transportation that will certainly influence future traffic patterns. The acceptance of this system and the readiness to use it instead of using cars or planes will largely depend on the competitive advantages to be established. Some of these are, without order of priority, summarized below:

- shorter travel times and good connections to Park & Ride, tram and other public transportation systems

- adequate comfort with respect to seating, noise emission, passenger information systems, air circulation and design

- fair fare pricing and flexible booking systems compatible with other transportation means.

- well adapted integrated travel schedules

- high safety reliability and low environmental pollution

- modern station design and baggage handling

- on board communication via telephones and facsimiles

- well equipped interior for office work

These features and others that have been or are being designed will provide prerequisites for comfortable travel in shorter times thus making it attractive for both private and business customers.

Literature:
[1]   TRANSRAPID MagLev System
      Hestra-Verlag Darmstadt 1989

[2]   Finanzierungskonzept
      Magnetschnellbahn Berlin-Hamburg GmbH
      München 1993

[3]   The New Dimension in
      Transportation Technology
      TRANSRAPID International
      München 1995

C514/058/96

The Italian tilting train ETR 460

C CASINI, G PIRO, and G MANCINI
Ferrovie Dello Stato SPA, Firenze, Italy

1 - PENDOLINO FEATURES

An increase in the speed of railway system can be reached either by increasing the line speed or by using vehicles capable of negotiating the curves of the actual lines at increased speed but with no extra forces on the rails and without reducing the ride quality: the second solution is represented by the tilting technology.

The tilting trains, nicknamed "PENDOLINO", designed and manufactured by FIAT FERROVIARIA, have been used from the Italian State Railways for about 25 years. They belong to the series of ETR 401, 450, 460, 470, 480 and others running or that will be running on many European networks, like German State Railways, Swiss Federal Ry., Finland State Ry., Czeck Sate Ry., Spanish State Ry., Portuguese State Ry., and others in the future.

The tilting system makes it possible to increase speed in curves without modifying the existing track and without affecting passengers with increased accelerations, so that curves can be negotiated at a speed increased by 20-30 % compared to traditional trains.

The first operational series of this kind of trains is given by the 15 trains ETR 450, in service from 1988 on the main Italian lines; each train is formed by 9 vehicles, including 8 power cars and one trailer and can run at the maximum speed of 250 Km/h; the total running time from Milan to Rome (with no intermediate stops), is reduced to about 4 hours at a commercial speed of more than 160 Km/h.

The operating and commercial results of the first series are satisfying, with about 25 millions Km covered in 8 years with a daily average of the fleet of almost 12,000Km.

The main technical aspects which allow all the series of the "Pendolino" trains to have very high performances can be synthesised as follows:
- a very low axle load;
- traction motors hung to the car-body;
- lateral active suspension which prevents car body from bumping into the bogie frame in curve negotiated at very high cant deficiency ;
- the active tilting system.

Nowadays following the good operation results of these trains, the FS has chosen to go ahead with this philosophy and new electric and diesel Pendolino trains are foreseen in next years, either for domestic lines, or for European connections across the difficult alpine lines.

The second generation of the Pendolino trains, named ETR 460, are provided with many important improvements compared to the first series. In the ETR 460 the tilting system makes it possible to tilt the body by 8°, corresponding to a compensation of the centrifugal acceleration of 1.3 m/s^2 (200 mm of cant deficiency); this allows the train to run with a non-compensated acceleration equal to 2 m/s^2 (300 mm of cant deficiency), while the centrifugal acceleration affecting passengers will be equal to 0.7 m/s^2 (110 mm of cant deficiency). The body control and actuation system is made of an electronic control unit which receives information from the various transducers installed on the first vehicle front bogie and calculates the car body angle inclination and the tilting speed. The calculated angle signal is used to drive the hydraulic cylinders making the body tilt; the measured angle signal is used to control the tilt. The tilting system is located entirely under the body, so that more space is available for the passenger compartment area. To improve the bogie stability, the bogie wheelbase is increased to 2.7 m (against 2.45 m of ETR 450). The main section is widened by 8 cm, so that the adoption of four seats per row is possible in the 2nd class. The passenger comfort is improved with a new highly performing air-conditioning equipment, better insulation, and most of all, total air tightness of the coaches. Also the electrical traction equipment is improved, with GTO converters with asynchronous three-phases traction motors.

The main characteristics of this series, (the trains are formed by 9 vehicles, 3 traction units of 2 vehicles each and 3 trailers) are as follows :

Supply voltage	3 kV d.c.
Gauge	1435 mm
Bogie wheelbase	2700 mm
Max. axle load	13 t
Top speed	250 km/h
Trainset composition	9 car (6 motor coaches)
Total length	237 m
Total weight	435 t
Power at wheel-rim	6000 KW
Available seats on 1st and 2nd Class	460
Motors	Asynchronous three-phases
Motor output (each)	500 kW

Following this series, another one with multisystem traction equipment has been developed, for the connections with the European Countries, through the Alpine links: 9 trains with 3 and 15 KV supply will be delivered this year for a Italian-Swiss Company, that operates the services between Milan and the most important Swiss Cities.

Some other trains are entering in service from Italy and France through the difficult alpine lines, on which a reduced time-running is very important. Others series will follow in order to connect the other European Countries to Italy; all these trains will be equipped by electrical traction with the different tension supplies used in Europe; as a matter of fact in Europe the power supply is different for the different Countries as follow:

Italy, Belgium, Spain (normal lines), Poland, Czeckia	3 KV d.c.
France (normal south-lines), Nederland	1,5 KV c.a.
South-East England	750 V d.c.
Germany, Switzerland, Austria, Scandinavia	15 KV c.a.
France and Spain (new lines), Portugal, Finland, England	25 KV c.a.

The fact that practically all these trains have a top speed of 250 Km/h (with the exception of the trains running on the Alpine routes, where gradients until 3 per cent are present and for that reason a 200 Km/h top speed is sufficient), has sometimes caused a little misunderstanding, because tilting trains are always thought as High Speed Trains and not "all-purpose trains", as they are in reality, with the possibility to adopt them also for regional services; as a matter of fact the reduction of running time may be more significant at lower than at higher speed.

For this reason the FS have planned the construction of some diesel tilting trains for the operation on non-electrified lines, beginning from the main Sardinia line to other important lines like the Bari-Reggio Calabria on southern Italy. In this case the train may be formed by two or three elements, with a top speed of 160 Km/h, rated output of about 1000 KW, air conditioned and with the possibility to run on lines with a maximum gradient of 3 per cent.

2 - ETR 460 TESTS

Before starting commercial service in 1994, a 6 month period was dedicated to the acceptance tests of the ETR 460 from the point of view of safety, track fatigue and quality of ride. These tests were carried out according the standards required by UIC (Leaflet 518).

The tests were carried out on the Direttissima line between Florence and Arezzo where speed up to 300 km/h can be reached and on the Old line between Chiusi and Orte where several small radius curves can be negotiated.

Transducers were placed on the train to measure a large number of parameters:

1. wheel to rail forces ;
2. acceleration on car bodies ;
3. acceleration on bogies ;
4. displacement of the primary and secondary suspensions
5. relative displacement between bogie and car body
6. pressure variations inside and outside the train
7. angle and speed of tilt of the car body
8. electrical parameters

The tests conditions include all combinations of speed, cant deficiency and curve radius required by Leaflet UIC 518. Maximum speed of 275 km/h and maximum cant deficiency of 300 mm was reached during the tests.

According Leaflet UIC 518 the assessment of safety and vehicle dynamic behaviour is performed on the basis of criteria relating either to directly measured parameters or to quantities which are calculated from measured parameters. For the acceptance of the train a statistical analysis of the following quantities is required by UIC standards:

1. track shifting forces ;
2. derailment coefficient ;
3. vertical force of the outer wheels in curves ;
4. lateral force of the leading wheel in curves ;
5. lateral acceleration in car body ;
6. vertical acceleration in car body ;

The statistical evaluation requires to split the line into homogeneous (that is straight line, full curve or curve transition) elementary sections of line no longer than 500 m each. The values of the following statistical parameters:

- value at 99.85% ;
- average value ;
- root-mean-square value ;

of each of the previous quantities were calculated on each of these elementary sections according to the methods indicated in the leaflet UIC 518. Finally, these values are compared to the limit values indicated by Leaflet UIC 518. If no values exceeds the limits (covering all the test conditions and requirements) the train can be accepted.

In such analysis the real conditions of test tracks have to be taken into account, since track geometry has a significant impact on the vehicle dynamic behaviour. In the UIC 518 method the geometric quality of the track is described by the following parameters:

1. longitudinal level
2. alignment

The standard deviation and the peak values of such track defects are to be calculated on each of the elementary sections of the test-line in order to define a track geometric quality level of test-line (and to exclude from the analysis the sections where the peak values of the track defects exceed a fixed value). What is important, however, is that such classification of track defects makes it possible to assess the relationships between the values of measured quantities and the value of the track defects on each of the elementary line sections. In fig. 1 is shown the peak values and root-mean-square values of longitudinal level and alignment on each of the elementary sections of the Direttissima line between Arezzo and Florence.

According Leaflet UIC 518, another fundamental parameter to be taken into account is the geometry of the wheel-rail contact. The ETR 460 is provided with wheels with an ORE S 1002 type profile. Before carrying out the tests, the geometry of each wheel has been measured and performed an analysis and calculation of the equivalent conicity. In fig. 2 the characteristics of a contact between wheels with a new ORE S 1002 profile and a theoretical FS track having a gauge of 1435 mm and an inclination of 1/20 is shown.

To have a complete view of the evolution of the dynamic behaviour of the train according to the wear of wheel tread , the test campaign was divided into two phases: in the first phase the tests were carried out with new wheels (with the ORE S 1002 profile freshly turned); in the second phase, the wheels were turned with an ORE S 1002 profile in operation for 400,000 km (copying the profile from the wheels of an ETR 450 with that amount of kilometres covered).

In the next paragraphs some of the main results of the tests are presented.

2.1 - Wheel to rail forces

Though the axle load of ETR 460 is very low (12.5 kN), negotiating curves at very high cant deficiency (until 300 mm) can generate significant track forces. As already noted, the main quantities to be assessed about safety and track loading in the UIC method are the track shifting forces, the derailment coefficient, the vertical force of the outer wheels in curves and the lateral force of the leading wheel in curves.

The values of the statistical parameters of the previous quantities, calculated on each elementary section of the line, were subjected to two types of analysis:

⇒ one-dimensional analysis as a function of the distance covered (performed in real time during the tests);

⇒ two-dimensional analysis as a function of the non compensated acceleration and the characteristic of the test area (curves with 250 m < R < 600 m, curves with R > 600 m, straight sections).

Figures 3 and 4 show the graphs of the values at 99.85% of the track shifting forces and derailment coefficient on the first and second axis of the train in each elementary section of the Arezzo-Florence and Chiusi-Orte lines (one-dimensional analysis). These values are normalised with respect to the limit value according to the Leaflet UIC 518. The first figure refers to a test run in which a speed of 275 km/h (maximum commercial speed of 250 km/h + 10%) was reached; the second refers to a test run in which non compensated acceleration of 2 m/s² was reached.

Figure 5 shows the graphs of the values at 99.85% of:
- track shifting forces of first axle ;
- track shifting forces of second axle ;
- derailment coefficient on the outer wheels in curve of 1st axis
- vertical load on outside wheels in curve of 1st bogie

as a function of the non compensated acceleration in curves with a radius between 250 and 600 m (two-dimensional analysis). The figure also shows the limits according to the Leaflet UIC 518.

As it can be seen in the figures, even in critical conditions, all the values do not exceed the limits.

2.2 - Ride quality

As already noted, the main quantities to be assessed about ride quality according to Leaflet UIC 518 are the lateral acceleration on the vehicle body and the vertical acceleration on the vehicle body.

The method of analysis of the quality of drive is similar to the method used to evaluate wheel to rail force. That is, the statistical parameters of the accelerations are calculated on each elementary section of the line and then subjected to one-dimensional and two-dimensional analysis.

Figure 6 shows the graphs of the values at 99.85% of the vertical and lateral accelerations on the first two car bodies of the train in each elementary section of the Chiusi-Orte line during a test run in which non compensated acceleration of 2 m/s^2 was reached (one-dimensional analysis). Figure 7 shows the graphs of the root-mean-square values of the vertical and lateral accelerations on the first two car bodies of the train in each elementary section of the Chiusi-Orte line (normalised with respect to the UIC-518 limit of 0.5 m/s^2) during a test run in which non compensated acceleration of 2 m/s^2 was reached.

Figure 8 shows the graphs of the values at 50% of:
- lateral and vertical acceleration on the 1st car body
- lateral and vertical acceleration on the 2nd car body

as a function of the non compensated acceleration in curves with a radius between 250 and 600 m. The FS limit shown in the figure is 1.225 m/s^2. In the figure, it will be noticed that the tilting system of the ETR 460 is designed in such a way that the average acceleration in the car body corresponding to the maximum tilt is very low (approx. 0.65 m/s^2); it will also be noted that the maximum tilt is reached for a non compensated acceleration of about 1.67 m/s^2; beyond that value of the lateral acceleration the amount of compensated acceleration does not change and acceleration in the car body increases directly proportional to the non compensated acceleration.

DIRETTISSIMA LINE - AREZZO - FIRENZE SECTIONS

TRACK DEFECTS (JULY 18, 94)

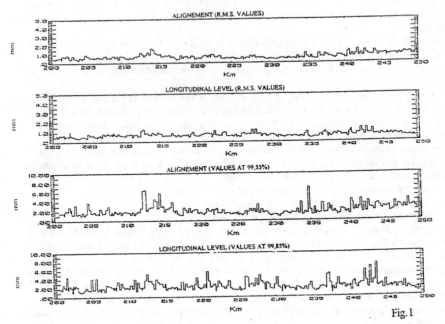

Fig.1

ETR 460 BAC 001 - AXLE 3 (DECEMBER 18, 94)

F.S. TRACK UNI 60 - INCLINATION = 1/20 GAUGE = 1435 mm

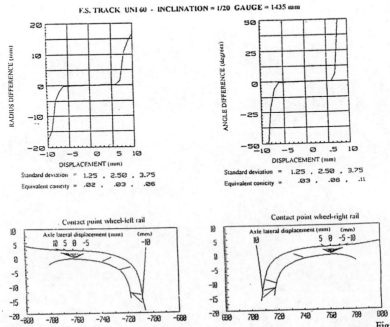

Fig.2

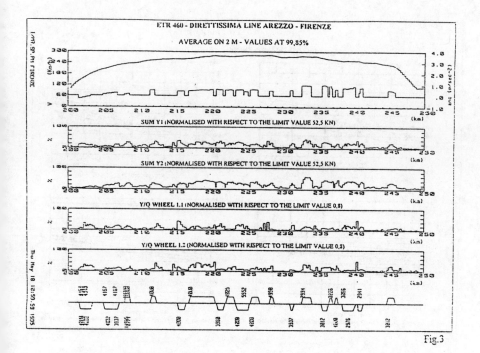

Fig.3

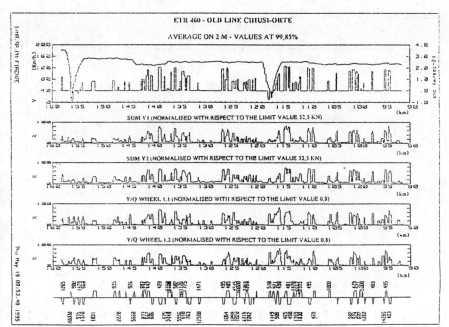

Fig.4

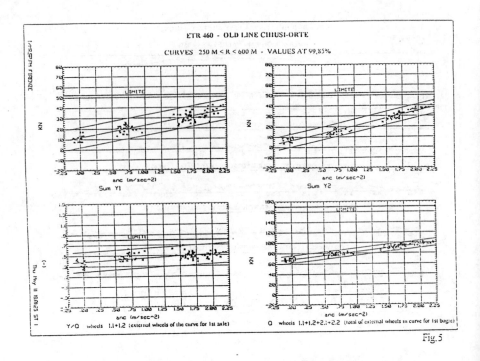

Fig.5

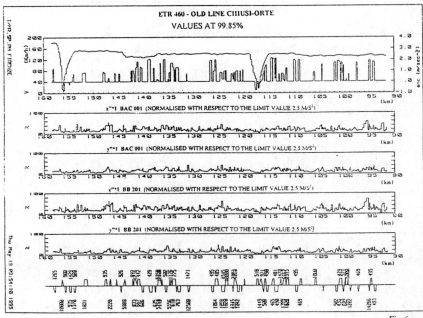

Fig.6

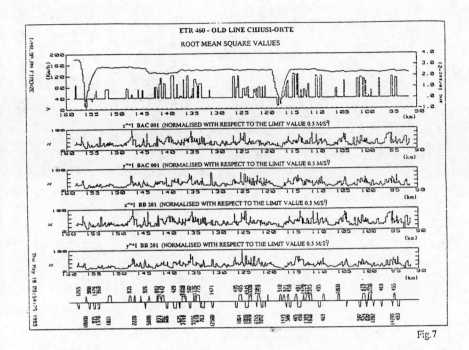

Fig.7

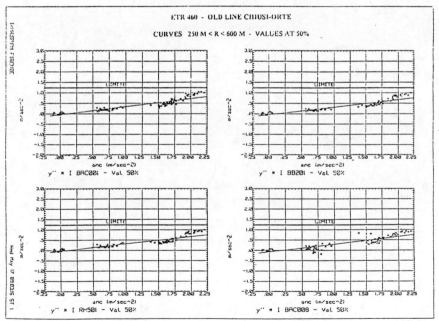

Fig.8

C514/079/96

UK freight development

J WORTH BA, MCIT
English, Welsh, and Scottish Railway, London, UK

Rail freight in the UK is currently going through a period of dramatic change. Ownership and commercial strategy have been transformed by privatisation. Some traditional markets, notably coal, are in decline and open access operation is producing further challenges. At the same time, road congestion and environmental concerns about pollution and land use are leading to greater interest in rail than at any time in the past three decades. New intermodal technology is enabling rail freight to penetrate markets previously exclusively served by road transport.

1996 looks likely to be a momentous year in the development of rail freight in the UK. This paper considers the nature of the changes and where they may lead in the years to come. Firstly, however, it is necessary to review briefly the background to the current situation.

Starting in the early 1980s, British Rail's freight activities were increasingly run as a business, focussed on key markets and industries, working to a clear profit-orientated remit. By the late 1980s this policy had evolved to include responsibility for and control of assets, culminating in 1992 with the freight business assuming full line management responsibility for its operations and personnel. The bulk freight business - Trainload Freight (TLF), consisted of roundly 100 million tonnes and £500 million turnover per annum and was managed through four commodity-based Profit Centres: Coal, Metals, Construction and Petroleum. The remainder of BR's freight activity - containers, non-bulk traffic and European business - was grouped into Railfreight Distribution (RfD), with around £100 million turnover.

The structure which was put in place in 1992 was thus the natural conclusion of the BR policy of "sectorisation", i.e. the change from a geographical/functional structure to a business-led organisation. However, the ink was barely dry on the first week's results for the new organisation when Government announced its proposals for the privatisation of British Rail, to a template which bore no resemblance to that which had just been put in place. Instead, the track and other infrastructure was vested in a new body - Railtrack - and freight train operations and maintenance split into five freight operating companies. TLF's bulk business, together with RfD's domestic UK traffic, was split into three competing geographically-based companies, roughly equal in turnover, but of differing size, complexity and profitability, while RfD's container arm, known as Freightliner, was to be hived off as a separate company. European traffic through the Channel Tunnel was to be the remaining focus for RfD, receiving heavy investment in locomotives, wagons and terminals.

Having managed the business to normal commercial practices, in competition with a highly efficient road haulage industry, for the previous 10 years, the prospect of privatisation held few fears for freight managers, but the Government's chosen structure was contrary to advice given by industry professionals. A further round of reorganisation was most undesirable and the notion of rail-against-rail competition seemed unnecessary given rail's total market share in the UK of just 6% of tonne miles.

Nevertheless, the same professional managers duly complied with Government's wishes and through 1993 planned and prepared the new structure which, in the case of the three TLF units, went live in April 1994. Operating initially as divisions of BR, the three units prepared their own Safety Cases and obtained the necessary operator's licenses to become independent legal entities and subsidiaries of BR in April 1995.

Concurrently, work was underway on the sale of the three TLF companies, now known as Transrail, Loadhaul and Mainline. An Information Memorandum was issued in July 1995 with indicative bids submitted in September. Short listed bidders then undertook due diligence prior to submission of final bids in December. It was during this period it emerged that Government were having second thoughts about the proposed structure and that, after all, were prepared to sell all three companies to one purchaser.

Thus it was that in the first week of 1996 that a consortium headed by Wisconsin Central Transportation Corporation became preferred bidder for all three TLF companies and duly completed the purchase at the end of February. The same consortium had, three months earlier, purchased BR's mail train operation, Rail express systems (Res). The four companies are now being reintegrated into one entity - English Welsh & Scottish Railway Ltd (EWS), retaining an industry/commodity focus within marketing but with common operations, engineering and support functions.

After a protracted three year sale process, the Freightliner container business was finally sold to a management consortium in May 1996. The sale of Railfreight Distribution was announced in June with completion targetted for the end of the year. Thus, 1996 is likely to see the sale of all BR's freight businesses into the privatised sector.

At the same time that major structural change was being implemented within the industry, fundamental changes were occurring in the core market for rail freight - coal for electricity generation. Heavy investment in gas-fired generation led to a substantial reduction in the use of coal in power stations leading to numerous pit closures and serious loss of profitable traffic for TLF. Production and transport of coal will have virtually halved over a five year period.

This was exacerbated by National Power's decision to set up an open-access operation to cater for the heaviest concentration of coal movements - from Selby mine to Drax and Eggbough power stations in Yorkshire. Only one other such operation has appeared - the government-owned British Nuclear Fuels Ltd set up a subsidiary Direct Rail Services (DRS) to move its traffic in north west England. Provision for open access to the Railtrack network was introduced under the 1993 Railways Act.

Nevertheless, on the plus side, road congestion and increasing concern over environmental issues, notably air pollution and road building, have led to a renewed interest in rail for the movement of freight by industrialists and policy makers. It is clear that there is much public and political goodwill for rail freight and industries that have used road transport exclusively for the last 20-30 years are beginning to return to rail. Milk, for example, is likely to be moving by rail later this year for the first time since the early 1970s.

Contrary to the view expressed in some quarters rail freight in the UK is very much alive and providing key links in the supply chain for British industry. At current levels, it equates to 10 million lorry journeys per annum kept off Britain's already congested roads. To take just one example, the M4 in south Wales, there would be an extra lorry every 40 seconds, day and night, 365 days per annum, in the absence of rail freight. The Department of Transport's own projections indicate that nearly 40% of the UK motorway network will be "in distress" by the early decades of the next century as a result of traffic volume, i.e. 40% of motorways will experience the current level of congestion encountered at the M6/M5 junction near Birmingham. In contrast to the congested motorways there is substantial capacity available on Britain's rail system. Other than in the commuter peaks around the major conurbations it is possible to operate additional freight trains on virtually all parts of the network.

Railtrack plays a crucial role in the future of rail freight in the UK. It owns the track, signalling, bridges and most operational land (although the freight companies have the freehold of their maintenance depots). Railtrack timetables and controls the trains and is responsible for the overall safety of the railway. As such, it controls the acceptance regime for new and/or modified equipment that operators may wish to operate on the network. Most important of all, Railtrack levies access charges for use on the network. For freight, these charges are negotiated on a flow by flow basis and, on average, account for around 30% of the cost of rail transport, although this figure can be much higher in certain cases. The level of access charges is obviously a key component of the overall price of using rail.

Mention should also be made of the important role played by other service providers in the rail freight sector, notably the wagon hiring companies and terminal operators with whom we work closely to offer a full distribution service to the customer.

We in EWS believe strongly that by building on existing customer partnerships and establishing partnerships with new customers there is substantial potential for growth of rail freight in the UK, to the benefit of customers, suppliers and the country at large. There are a number of key building blocks to the strategy.

First and foremost, a focus on serving the customer and providing a more flexible, responsive and competitive service. For example, if new or additional wagons or containers are required, we will seek to provide them. If storage and road distribution from railheads are required, we are able to provide them as part of a 'one-stop-shop' package. Furthermore, and of particular significance to many potential customers, we offer a service for lorry size consignments between the main regions of the UK. It is no longer the case that a customer has to provide a full trainload of product. The Enterprise service, developed by Transrail, provides an overnight service for wagonload consignments over long distances (e.g. Kent to Aberdeen) and has attracted over 350 000 tonnes of new business from road in 18 months. The service has already been extended to the far north of Scotland, conveying freezers for Norfrost and further extensions in eastern England are under investigation.

Secondly, to be able to offer competitive prices, EWS has initiated a major cost reduction programme. In part savings will come from cutting overheads and exploiting the synergies generated by combining the four companies and in part by improving operating efficiency. All 7500 employees have been offered a special severance package, to which the initial response has been encouraging. This mirrors the pattern followed elsewhere in the world by Wisconsin Central: initial manpower and cost reductions have enabled new business to be attracted, in turn generating new jobs. In both the USA and New Zealand, the number of employees is now greater than at the outset, in spite of a dip in the interim as efficiency was improved and new business developed. We in the UK now have the freedom to adopt world-wide best practice in our operations and our commercial policy: stark contrast indeed to life in the public sector as part of a passenger-orientated British Rail dominated by Treasury dictat.

Thirdly, investment. British Rail's investment programme was, inevitably, heavily skewed to the passenger businesses and Channel Tunnel related projects. While RfD has seen investment in new equipment for European services, the domestic freight business has seen only 100 Class 60 locomotives in the last 10 years and virtually no new wagons. This is about to change. EWS has placed an order with General Motors for 250 state of the art locomotives for use on both freight and mail trains. These will replace 30 year old machines which are small, inefficient and unreliable. Availability for traffic will be substantially better as a result of much lower maintenance requirements. In addition, work is underway to develop an investment programme for wagons to serve new markets and replace some of the ageing fleet at present in use. In designing the new wagons, we shall be working very closely with our customers to ensure that their needs are incorporated from the outset. Just such a process was followed in conjunction with British Steel some two years ago in designing and fitting cradles to the steel-carrying fleet based in south Wales to meet new handling and transport requirements.

Fourthly, new technology. It has now been conclusively proved that lorry trailers can move by rail in the UK, in spite of low bridges, narrow platforms and other constraints. Since January, the RoadRailer system has been used to carry high quality paper products from Aberdeen. The box or curtain-sided trailer is loaded in the normal way at the mill and driven to the Aberdeen railhead where a rail bogie is placed under the trailer. The trailers then move 450 miles overnight by rail on the Enterprise service to Northampton where they revert to road mode for delivery to paper merchants locally and as far away as Romford. This demonstrates why we cannot afford to be anti-road: we need lorries for local collections and deliveries, leaving rail to concentrate on trunk haulage.

In addition, late summer will see the roll-out of the prototype Piggyback wagons which will carry virtually standard lorry trailers. Initially, low height trailers (3.6 metres) will be used to remain within the existing loading gauge but standard 4 metre high Euro trailers could be running by 1999 if the go-ahead is given for Piggyback gauge clearance on the West Coast Main Line. Piggyback clearance would also permit the movement of 9' 6" containers on standard rail wagons. At a cost of under £100 million a route could be created from the Channel Tunnel to central Scotland and an Irish Sea port, i.e. a 550 mile rail superhighway for about the same price as 15 miles of M25 widening between the M3 and the M40. This would be one of the most strategic transport investments that an incoming Government could make.

The new intermodal systems such as Piggyback and RoadRailer open up substantial potential for rail in markets such as food and drink and consumer goods within the UK. These are demanding markets for transport operators but ones in which we believe we can compete on service as well as price. For example, Transrail's reliability runs consistently at around 99.95% and in the 1995 customer survey, 92% of respondents rated the service in the range acceptable to excellent. Rail already has a foothold in this market with a nightly train from Scotland to south east England conveying Spillers' petfood from their factory to the main distribution centre.

Greater use of rail in the consumer and manufactured goods sector will, in the main, happen by our assembling the necessary logistics packages in association with our partners who operate rail distribution centres. Many of these centres offer a wide range of services including warehousing, stock control, order picking and 'just in time' delivery onto production lines or the High Street. So, whilst you will not see a rail wagon delivering freezers or catfood in the High Street or milk to the doorstep, rail can and will provide the primary long distance trunk haulage to regional distribution centres. In some cases, bi-modal equipment such as RoadRailer will, indeed, drive up to the back door of the supermarket and feed directly onto the shelves.

So, cost reduction, investment, new technology and links with professional distribution centres are providing the means to offer a competitive alternative to road transport in both existing and new markets. We intend to use these advantages to build partnerships with customers across a wide range of commodities and using the full range of equipment now available. The combination of the four companies gives us complete national coverage and a substantial resource base to exploit. Most important of all, though, is a determination from Ed Burkhardt downwards to grow the business and a clear recognition that this will only happen by understanding and delivering what the customer wants - and more

This paper has attempted to demonstrate that, in spite of the policy swings of recent years, rail freight in the UK is poised for a renaissance. With a committed owner, substantial capital investment and growing problems facing road transport, it may indeed be that 1996 proves to be the year that sees the turning point for rail.

C514/031/96

Development of a new Shinkansen vehicle, 'Series 500' for realization of 300km/h commercial operation

K SAKURAI BE, MJSME **M UTSUNOMIYA** BE **E YAGI** MS, and **Y YOSHIMOTO** BE
West Japan Railway Company, Osaka, Japan

West Japan Railway Company (JR-West) is planning to raise the maximum speed of the Sanyo Shinkansen Line to 300 km/h by introducing the Series 500 vehicle. At this speed, aerodynamic phenomena largely influence on both environmental and ride comfort levels. Our researches show that these problems could be solved by newly developed technologies such as a wing-shaped current collector, a super-long front nose and actively controlled suspension systems. This paper describes out efforts for the speedup and the main features of Series 500 adopting these technologies.

1 PROBLEMS AGAINST SPEEDUP

1.1 Sanyo Shinkansen Line and its surroundings

JR-West operates the Sanyo Shinkansen Line (Shin-Osaka - Hakata, 554 km) in western part of Japan. This line is fiercely competing with both airlines and highways (1). Since JR-West earns more than 40 % of traffic receipts in this line, it has been making efforts to strengthen its competitiveness. In 1987, JR-West raised the maximum speed of the line from 220 km/h to 230 km/h by introducing "Grand Hikari" (Series 100N) with four double decked cars. Again in 1993, JR-West raised the speed to 270 km/h by "Nozomi" (Series 300). Its competitors in this area, however, improved their services in the meantime. It has been required to offer better services in speed and ride comfort.

It is difficult to speed up furthermore the Series 300's vehicle. In Japan, the toughest obstacle against speedup is environmental problems along the line. Since the Shinkansen lines run through many residential areas instead of connecting most of major cities efficiently, they cannot avoid these environmental problems. The levels of noise and ground vibration from Shinkansen trains are strictly regulated in residential areas. Actually, Series 300's maximum speed is determined by its environmental levels.

1.2 Environmental problems

The environment along the high-speed railway line is exposed mainly to three phenomena, noise, pressure wave and vibration. The noise from a train consists of four kinds of noises, current collecting noise, aerodynamic noise from car body, rolling noise between wheel and rail and noise from structure vibration. At a high speed like 270 km/h, the current collecting noise, especially aerodynamic noise from current collector's components, occupies in a dominant position, because it increases exponentially in proportion to speed. Other noises can be lowered by a sound-proof barrier along the track, installed in most residential areas. The Series 300's current collector has a large surrounding cover. It lowers the noise level by decreasing the speed of air flow lashing directly the current collector's components. At a higher speed, however, the cover is no longer effective, since it becomes another noise source.

When a train enters into a tunnel, a micro-pressure wave occurs at the other end, and causes an impulsive noise. A long tunnel with a slab track promotes this phenomenon. Since the 50% of Sanyo Shinkansen Line is tunnel section, and its track structure is mostly slab, it is a very serious matter for JR-West to solve this problem. Though some tunnels have special hoods at their entrances to ease the influences, they are too expensive to install for all the tunnels. The fundamental countermeasures against them are needed in the vehicle side. Besides, it turned out there is another phenomenon concerning pressure wave. When a train passes by a house in an open-air section, pressure waves followed about the train rattle the doors and windows.

A train causes ground vibration through rail and track structure. The ground vibration level depends upon train weight, speed and stratum. Car weight reduction is said to be most effective to ease the vibration level. The Series 300 car has only 75% of Series 100 car's weight to run faster by 50 km/h. The maximum axle load is already reduced to 11.4 tons. So further reduction, keeping safety and service qualities, requires hard work. Where the vibration level locally exceeds the standard, elastic sleepers or ballast mat are used. However, they are just a kind of symptomatic therapies, not general ones.

Fig 1 High speed testing train, "WIN350"

1.3 JR West's WIN350 project

In such a situation, JR-West formed a project team to research the problems in the range of 300 km/h, and started test runs using a newly designed experimental train, "WIN350" (Fig 1) in 1992 (2). After it recorded a maximum speed of 350.4 km/h, JR-West concentrated on development to solve environmental problems. In the process of the project, aerodynamic phenomena turned out to largely affect not only environmental levels but also ride quality. In a tunnel section, pressure waves caused by a train sway its car body. In the field of railway researches, however, aerodynamic matters have rarely been discussed. Then JR-West started to tackle with them by getting valuable advice of aerodynamic engineering experts outside the company. Beside the WIN350 test runs, wind tunnel tests and numerical simulation were performed.

Based on the results from this research, JR-West could get a prospect of realizing a high-speed train satisfying environmental and ride comfort requirements, and decided to develop a new Shinkansen vehicle for 300 km/h commercial operation. That is a next generation Shinkansen vehicle, "Series 500."

2 DEVELOPMENT OF SERIES 500 VEHICLE

Fig 2 New Shinkansen vehicle, Series 500

2.1 Concepts of design

The Series 500 vehicle (Fig 2) is carefully designed mainly from four viewpoints, that is, aerodynamic characteristics, weight control, ride comfort and maintenance cost (3).

1) The shapes of car and its parts are drastically changed from conventional ones. It largely contributes to improving environmental levels and ride comfort.
2) The weight of each part is thoroughly reduced and well balanced in a car. Its maximum axle load remains almost the same as Series 300's, though its total power is 1.5 times as large. It minimizes increase of ground vibration.
3) Lateral vibration and inside-car noises are minimized by modification of body and bogie and introduction of new technologies to them. It enables to offer a more comfortable passenger's space.
4) The change of under-floor apparatus layouts and introduction of a monitoring system make easier daily inspection and maintenance.

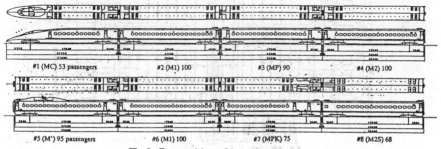

Fig 3 Composition of cars (first half)

2.2 Main feature

2.2.1 Composition and running performance

Car composition of Series 500 is basically the same as Series 300's. It consists of 16 cars, and its passenger capacity is 1324. Car's #8,9 and 10 are first class, and car's #7 and 11 have a snack shop space. Fig.3 shows the first half of composition.

On the other hand, for 300 km/h operation, it has enough power. All cars are motorized, and each motor has power output of 285 kw (constant rated). Total power output per train is 18 240 kw, approximately 1.5 times as large as Series 300's. Fig.4 shows the characteristic curve of traction and speed. Balancing speed is 365 km/h, and starting acceleration is 1.6 km/h/s.

Adoption of the all motorized cars largely contributes not only to power increase but also to weight reduction. It eliminates additional braking systems in trailer cars of Series 300 by fully utilizing regenerative braking systems in all cars.

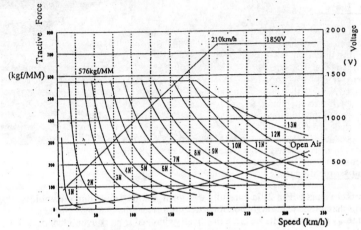

Fig 4 Characteristic curve of traction and speed

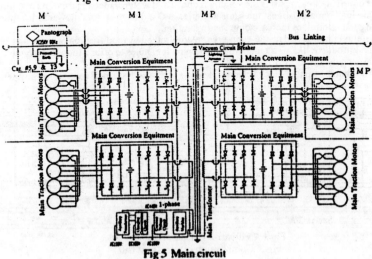

Fig 5 Main circuit

2.2.2 Main circuit system

Main circuit (Fig 5) adopts a VVVF control system with asynchronous main motors. One unit consists of four cars. A main transformer in MP car and two main conversion equipment in M1 and M2 cars drive all the motors in the four cars.

By positively introducing advanced technologies, the parts of the system are modified from Series 300's, which contribute to further weight reduction. As for the main transformer, its capacity is increased to 5400 kVA, and the secondary voltage is raised to 1100 V. Its primary and secondary windings are made of aluminum coils. Main conversion equipment uses large capacity GTO devices, 4500V-4000A for a converter and 4500V-3000A for an inverter. The main motor adopts a frameless structure with aluminum brackets.

To reduce the harmonic influences on catenary voltage, the phases of the systems in a train are controlled to be different. When cutting off a system in a train, the phase control condition is automatically adjusted.

2.2.3 Car body

The shape of car body is drastically changed from aerodynamic viewpoints. The front nose shape is sharpened keeping as constant as possible its cross-section area transformation rate in the progressing direction. As a result, front nose length becomes approximately 2.5 times as long as Series 300's. The cross-section area at cabin is reduced to 10.2 m^2 by rounding its shape, approximately 10 % smaller than Series 300's 11.4 m^2, shown in Fig 6. The underfloor apparatus is covered with a body-mounted structure. These changes in shape are expected to improve the environmental levels caused by pressure waves from a train and the ride comfort in a tunnel, as well as the running resistance.

On the surface of body including the underfloor, protuberances are eliminated or streamlined as much as possible. On the roof, the large between-car insulators of a high voltage bus cable are eliminated by directly stretching the cable. On the sides, it adopts plugged doors and flush surfaced windows with less bumps at the edge of them.

The body is structured using brazed aluminum honeycomb sheets to realize light-weight and high-rigidity at the same time. This structure is also expected to lower the noises in the cabin, particularly those from wheel-rail contacting and from main transformer. Natural frequency of the body structure is set to higher than 120 Hz to prevent from the chattering vibration by the VVVF system.

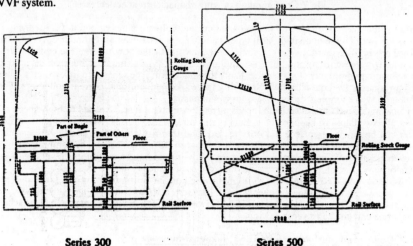

Series 300   Series 500

Fig 6 Comparison of car cross-sections

2.2.4 Wing-shaped current collector and insulator cover

The Series 500 vehicle has two current collectors in car's #5 and 13. This position is determined considering both current collecting ability and noise reduction. At 300 km/h, the distance between two current collectors should not be 75 m or its multiples to avoid the resonance of trolley wire and current collectors. As for the train noise, the front nose and current collector are the two largest sources. To lower the noise level, they should be apart at least the distance corresponding to the time constant of noise meter, 83 m for 300 km/h.

To further reduce the noises from a current collector, JR-West developed a completely new type of current collector, wing-shaped current collector (4). Since this current collector makes much less noise when it is exposed to high-speed air flows, the large cover to surround the conventional current collectors is no longer needed. The wing-shaped current collector has less components than conventional ones, and the shape of each component is aerodynamically designed. Basically, it consists of two parts, a collector head and a mast. While a conventional current collector has two prism-shaped bows on the collector head, the wing-shaped one has only one wide bow with a special wing-shaped cross-section. The wing shape is generally unstable in lift characteristics instead of making less noises. Therefore, the wing shape of bow cross section is chosen after a number of wind tunnel tests and test runs.

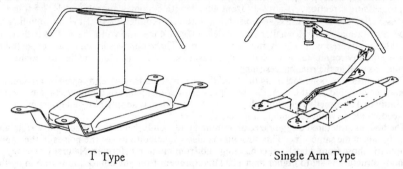

T Type   Single Arm Type

Fig 7 Configuration of wing-shaped current collector

As for the mast, two types of wing-shaped current collectors have been developed. Fig 7 shows the two, T type and single-arm type. The T type's mast is an air cylinder covered with an oval protector. When collecting, it stretches the mast until the bow reaches the contact wire, and otherwise, it shortens and put down the mast. The T-type's simple shape realizes a good aerodynamic performance. On the other hand, the single-arm type's mast is a link which pushes the bow against the contact wire by a spring force like conventional current collectors. The single-arm type's simple structure needs less maintenance.

The wing-shaped current collector needs a small insulator cover. Exposed to high-speed air flows, the insulators to support the collector still make large noises. However, the insulator cover can be small because it has only to surround the insulators, and can be streamlined because laminar flows are the better for the wing-shaped current collector. The combination of a wing-shaped current collector and an insulator cover reduces not only noise levels but also pressure waves by minimizing the maximum cross-section area at the cover.

Besides, a small cover enables to improve ride comfort in tunnels, too. A large current collector cover has to be attached over two cars like Series 300's. The unbalanced aerodynamic force at the cover causes yawing of the body. However, a small insulator cover can be attached to only one car, and causes less yawing.

2.2.5 Bogie and actively controlled suspension system

Out of three types of bolsterless bogies tested in the WIN350, a guide-arm type is chosen as Series 500's bogie from the viewpoint of running stability and maintenance. Based on the results of many test runs and numerical simulations for better ride comfort, the spring-damper system is thoroughly modified, and between-car dampers are attached to connect the adjoining car bodies.

Beside these passive measures for ride comfort, two kinds of actively controlled suspension systems are introduced to lessen lateral vibration, shown in Fig 8. The ride comfort in the end car and the cars with an insulator cover is inferior to that in others, due to their large aerodynamic influences. While car's #1 and 16 have a full-active suspension, car's #5 and 13 have a semi-active one. Car's #8, 9 and 10 also have semi-active one, because they are first class.

The full-active suspension (5) is attached between the car body and bogie. It moves under pneumatic power supplied separately from the main air system. Since the pneumatic actuator is used in parallel with the primary oil dampers, the system is safe from failures. Its control is based on the H-infinity theory, and it calculates the necessary pneumatic power from the observed lateral and vertical accelerations. It targets at the lateral vibrations of 1-3 Hz, and is expected to reduce them approximately to the half of passive condition.

The semi-active suspension (6) uses a variable oil damper instead of actuator. Since it does not need external power supply, it has advantages in space and cost. Its control is based on the sky-hook damper model, and it calculates the necessary damping coefficient from the observed lateral acceleration. Damping power is adjusted by the high-speed electromagnetic valves switching the number of opened orifices. The semi-active suspension is expected to reduce the lateral vibrations approximately to the two thirds.

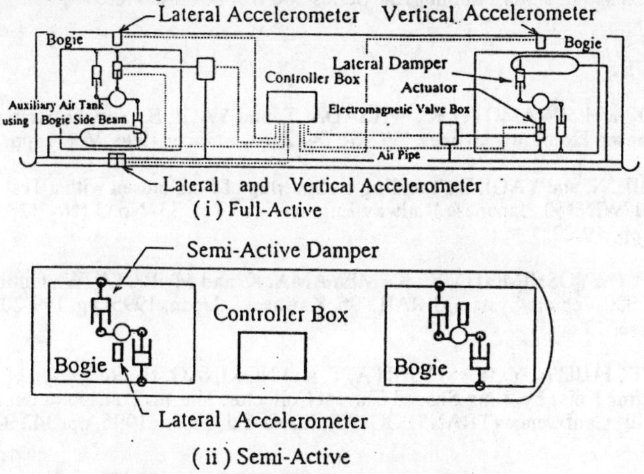

(i) Full-Active

(ii) Semi-Active

Fig 8 Actively controlled suspension systems

2.2.6 Improvement for easy maintenance

The Series 500 vehicle is also designed with a review of the conventional inspection and maintenance processes by introducing advanced technologies. A monitoring system enables an introduction of inside-car inspection. Devices to be inspected are put together as much as possible so that important equipment can be checked quickly in only one side of the body. Bolts on covers of apparatus are changed to detachless ratchets. Instruments easily affected by dust, such as electronic devices, are thoroughly made airtight.

3 TEST RUNS OF SERIES 500

In January, 1996, the first trainset of the Series 500 for commercial operation, W1, designed upon the WIN350 (W0), was built, carried to the rolling stack base in Hakata and started to serve a series of test runs. This scrutiny could have begun earlier by three months unless the Great Hanshine Quake had hit Kobe area

Soon after its delivery, the designed performance of the vehicle was tested in the high-speed test section, between Tokuyama and Shin-Shimonoseki, 106 km, to confirm the newly introduced technologies. With a delightful certification of that, W1 started the whole line test and finished it by the beginning of July, where vehicle's specifications have been fully checked from the aspect of confirmation of the ground facilities. The environmental data has also been collected for evaluation over the whole line. A long run to the amount of several hundred thousand kilometers is being executed to secure its reliability for commercial operation.

4 NEW GENERATION SHINKANSEN

No trainset other than the Series 500 vehicle has ever completed 300 km/h commercial operation in Japan, considering the environmental affairs along the line. Since it can run at the maximum speed of 300 km/h in two thirds of the Sanyo Shinkansen Line, its average speed reaches approximately 240 km/h between Shin-Osaka and Hakata. As well as the vehicle shortens the traveling time, it must be remarkable for the environmental friendliness and ride comfort. The WIN350 has finished

As of July, we have finished most of tests and been satisfied with the acquired data. After the long run test and operator's training, the Series 500 will be on service.

REFERENCES

(1)   HIDAKA, H., IWAMOTO, K., WASADA, T. and YAGI, E., Service Strategy of Shinkansen Excelling Airlines, WCRR '94, Paris, France, 1994, Vol. 1, pp. 103-109.

(2)   YOSHIE, N. and YAGI, E., Tackling on Speedup of Shinkansen with a Test Train Named 'WIN350,' Japanese Railway Engineering, Vol. 33, No. 3 (No. 128, 129), 1994, pp. 19-22.

(3)   YANO, H., YOSHIMOTO, Y., KAMEYAMA, K. and MURATA, W., Outline of Series 500 Vehicle System, J-RAIL '95, Kawasaki, Japan, 1995, pp. 199-202, (in Japanese.)

(4)   ASO, T., FUJITA, Y., YAMASHITA, T. and NORINAO, H., Reduction of Current Collecting Noise by Wing-Shaped Current Collector, The 4th Transportation and Logistics Conference (TRANSLOG '95), Kawasaki, Japan, 1995, pp. 342-344, (in Japanese.)

(5)   TAKENAWA, S., HAMABE, M., SHIMIZU, S. et al., An Active Suspension System for 500 Series Shinkansen, The 4th Transportation and Logistics Conference (TRANSLOG '95), Kawasaki, Japan, 1995, pp. 112-113, (in Japanese.)

(6)   SASAKI, K., SHIMOMURA, T., YAMAGUCHI, H. et al., A Development of Semi-Active Suspension System of Shinkansen Vehicle, The 4th Transportation and Logistics Conference (TRANSLOG '95), Kawasaki, Japan, 1995, pp. 104-107, (in Japanese.)

© IMechE 1996 C514/03

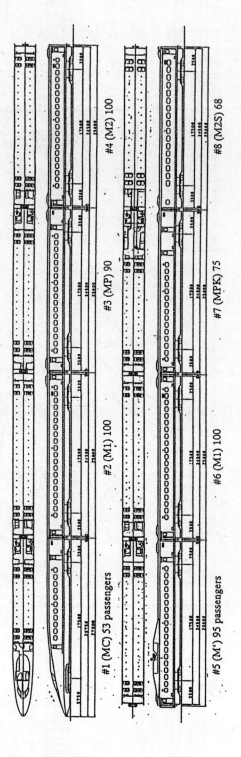

Fig. 3 Composition of Cars

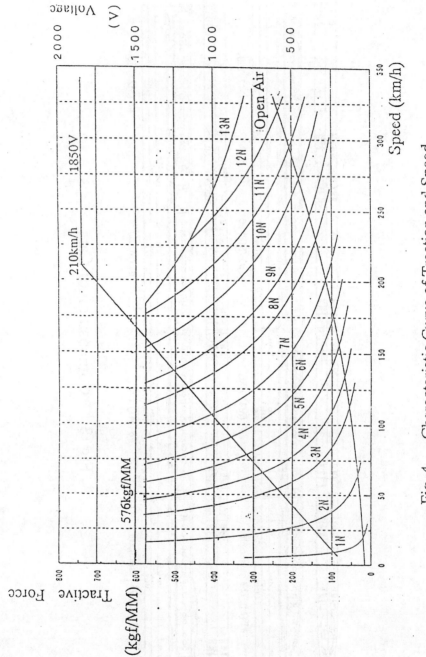

Fig. 4   Characteristic Curve of Traction and Speed

322

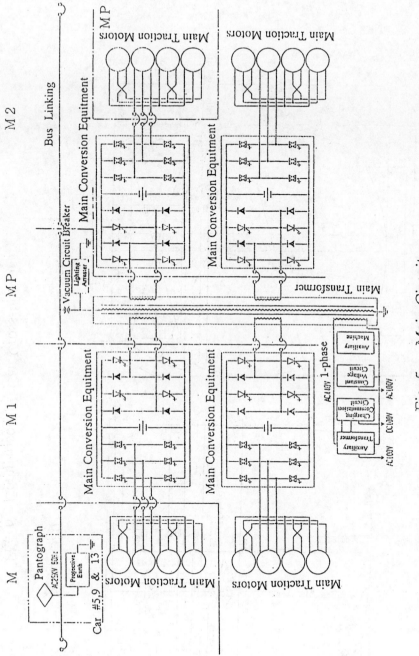

Fig. 5   Main Circuit

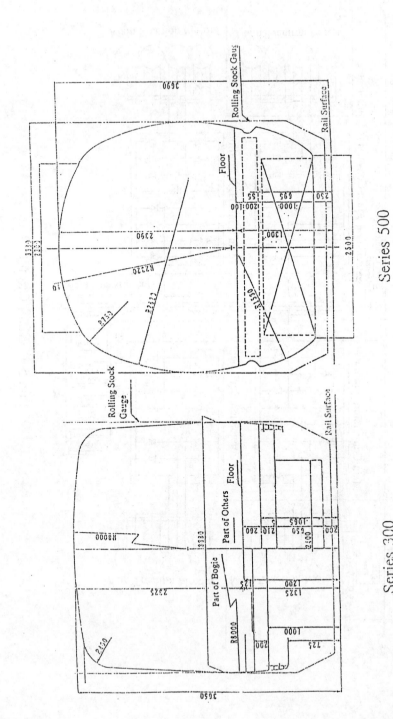

Series 500

Series 300

Fig. 6 Comparison of Car Cross-Sections

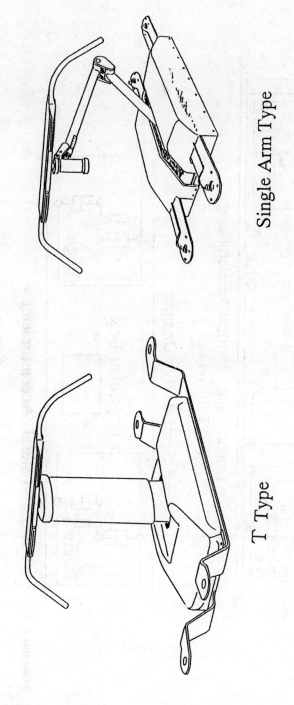

T Type

Single Arm Type

Fig. 7   Configuration of Wing-Shaped Current Corrector

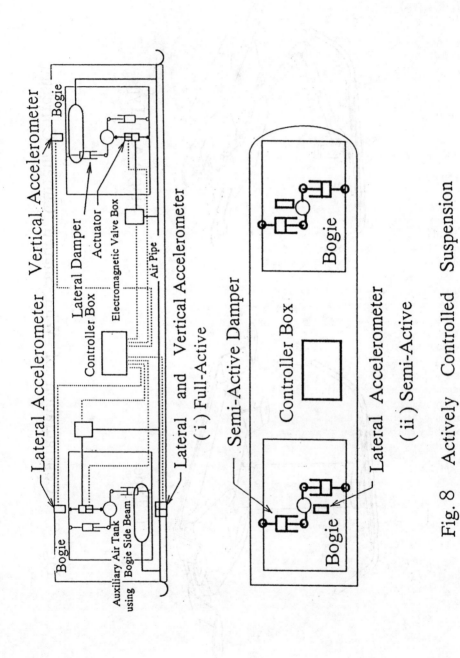

Lateral Accelerometer   Vertical Accelerometer

Bogie

Lateral Damper

Actuator

Controller Box

Electromagnetic Valve Box

Air Pipe

Lateral and Vertical Accelerometer

(i) Full-Active

Bogie

Auxiliary Air Tank
using Bogie Side Beam

Semi-Active Damper

Controller Box

Bogie

Bogie

Lateral Accelerometer

(ii) Semi-Active

Fig. 8   Actively Controlled Suspension

C514/081/96

West Coast Main Line route modernization

N HIGTON BSc, PhD, CEng, MICE
WCML Route Modernisation, Birmingham, UK

SYNOPSIS

Our conference is all about better journey times and better business, and this paper describes how the Modernisation of Railtrack's West Coast Main Line has been approached from a similarly commercial perspective. A potential upgrade within the modernisation of the West Coast Main Line could provide opportunities for High Speed Tilting Trains to significantly reduce journey times between Key centres.

The purpose of my paper is to decribe briefly the totality of the modernisation of the West Coast Main Line and within that context explain the part which the upgrade options including reduced journey times will play.

INTRODUCTION

As I will describe later reduced journey times are potentially significant, as the economic indicators are that they would provide significant scope for attracting more passengers onto the railway, and increase the competitiveness of rail against alternative means of travel, in particular the airlines for business travellers. Modernising the West Coast Main Line is however about very much more than simply decreasing journey times and it would not do justice to the modernisation or to the project to simply confine myself to this issue, important as it is.

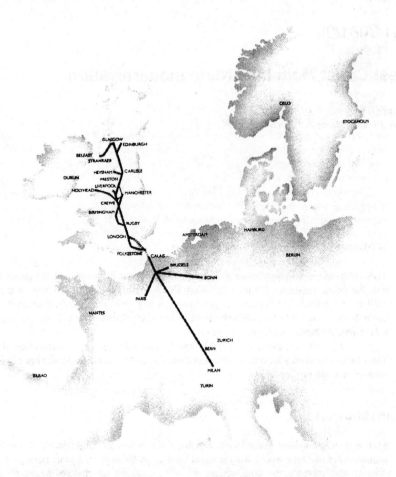

BACKGROUND

The West Coast Main Line is the major arterial rail route linking London with the conurbations of the West Midlands, North West and Merseyside and South and Central Scotland. It is the busiest strategic railway in the country, carrying more than 2,000 trains a day – a mix of passenger (InterCity and commuter) and freight.

It was originally built in the 1830's and 40's and was last modernised and electrified to the south of Manchester in the 1960's and a decade later to the north. The infrastructure is now reaching the end of its life and becoming increasingly unreliable and expensive to maintain.

The signalling system is in a particularly poor state. Deteriorating reliability is compounded by the fact that components are becoming obsolete and such is the condition of the infrastructure that remedial works to solve one problem can cause others.

FEASIBILITY STUDY

There have been a number of proposals to modernise the West Coast Main Line over the past decade or so which have failed to come to fruition. These involved the utilisation of the conventional British railway technology.

When Railtrack became responsible for the infrastructure in 1994 it decided a fresh start was needed including the consideration of innovative technology used in other parts of the world. Therefore, Railtrack in partnership with a private sector consortium (West Coast Main Line Development Company) undertook a feasibility study to look at the best way to modernise to meet the business needs of the route over the next 30 years.

The study recommended a two tier modernisation. This comprised a Core Investment Programme which included the introduction of a new signalling system and delivering improved reliability, lower operating costs and enhanced safety. Funding for the Core Programme comes from the existing Track Access charges levied against train operators.

In addition to this there were a range of upgrade options to improve performance, including shorter journey times through the use of high speed tilting trains, and new freight opportunities. These upgrades require additional investment and are subject to successful negotiations with train operating customers, OPRAF and HM Government.

COMPANY WIDE ISSUES

The modernisation of the WCML cannot be undertaken in isolation and has to be considered within the overall company strategy because innovations developed as part of the modernisation, as well as the implementation techniques, have the opportunity to be utilised on other parts of the network in the future.

Subject to successful development and implementation, the radio-based in-cab signalling system envisaged may well be applied in subsequent modernisations of other high speed routes. Likewise, the proposed centralisation of train control management into a single or small number of centres could have application across the network.

THE EUROPEAN DIMENSION

The WCML has been designated as a priority TENs (Trans European Network) project as part of the European high speed rail network. This is because of its strategic importance linking not only the major conurbations of western Britain via the Channel Tunnel into mainland Europe, but also via the Irish Sea ports offering a link to Ireland.

TENs status brings both opportunities and obligations. There is potential access to significant funding support both through the TENs scheme and other European sources but there are also requirements to comply with various evolving technical standards with regards to Interoperability.

CORE INVESTMENT PROGRAMME (CIP)

This is made up of the following key elements:

- Train Control
- Management Control Centre
- Track and Structures
- Power Supply
- Site Specific Schemes

Train Control

At the heart of the Core Programme is the replacement along much of the route of the existing colour light signals with a radio based in-cab system in which information is transmitted by digital radio directly to the driving cab of the train.

On sections of the route where significant numbers of trains come onto the WCML for relatively short distances (mainly in the conurbations) it is unlikely to be economically viable to fit all the trains with the necessary equipment and in these locations islands of dual signalling (in-cab and conventional) are envisaged.

However, most of the trains using WCML will need cab fitting and this has to be agreed and phased in conjunction with Railtrack's train operating customers and the rolling stock providers.

Whilst in-cab signalling is used in a number of parts of the world there is nothing available "off the shelf" which meets the needs of a busy mixed traffic railway like the WCML. Therefore, significant development work is needed and two parallel development contracts were let in Spring 1996. Subject to successful development, proving and safety validation installation is anticipated to start early in the next century.

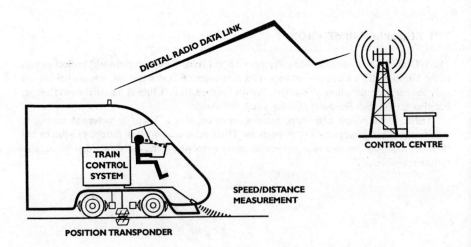

Management Control Centre

The WCML is currently operated from a matrix of 56 signal boxes and four electrical control rooms ranging from some large and relatively modern "power boxes" to single person operated "heritage" lever frame boxes.

It is proposed to manage the whole route from a single or small number of state of the art control centre(s) similar in concept to those used by the power supply industry and Civil Aviation Authority.

Track and Structures

The CIP envisages limited work on track and structures to bring them to a good state of repair and there will also be some realignment and removal of redundant assets.
Works will also include renewals of the numerous minor junctions on the route.

Power Supply

Since the route was electrified there has been an increase in the number of trains, higher speeds and heavier loads. Therefore, the electrical supply system is being severely over stretched, especially in the south. Reinforcement of the power supply system in certain key locations is now needed.

Site Specific Schemes

There are a number of locations along the WCML where consideration needs to be given to redesign and remodelling of the track lay-out. These are at "pinch points" where the existing layout cannot cope with the volume of trains or because of particular operational difficulties.

Sites being looked at include Euston Station and its approaches, Proof House junction on the approach to Birmingham New Street, and Weaver – Wavertree to the south of Liverpool.

Benefits

- Cost Savings
- Enhanced reliability and contractual performance
- Safety

By investing in the modernisation Railtrack will reduce its operating costs on the WCML. This will be achieved through more efficient route control, the replacement of unreliable equipment that is expensive to maintain, and removal of redundant assets.

Modernisation will result in major improvements in reliability. The proposed new signalling system will enable the removal of a lot of lineside infrastructure which is vulnerable to, for example, the effects of weather and vandalism.

The CIP will enhance safety because inherent in the proposed train control system is Automatic Train Protection (ATP) which overrides the train driver should he attempt to exceed his control limits.

By removing much of the track side infrastructure there will be much less need for work on or near the running line.

UPGRADE OPTIONS

Passenger

Incremental to the CIP are a range of upgrades which would improve the performance of the WCML by reducing journey times between key centres. For example, a high speed tilting train running up to 225 kph would take about 30 minutes off the two-and-a-half hour journey time between London and Manchester thus increasing the competitiveness of rail against alternative forms of transport.

The lucrative business market between Manchester and London is of particular significance because of "shuttle" air travel. An upgrade with reduced journey times could make a contribution to "emptying the shuttle."

These higher speeds would involve Railtrack in additional expenditure on the infrastructure than under CIP, in particular with the need to further reinforce the electrical power supply system and additional track works to maintain and improve ride quality and mitigate some of the speed restrictions on the route.

These costs would be recovered through higher track access charges to reflect the incremental investment. To achieve significantly higher speeds new rolling stock would also be needed.

This could include tilting trains which are able to negotiate curves at higher speeds than conventional stock. Originally developed in this country in the 1970's but then abandoned, tilting train technology is now proven and in use in many parts of Europe and the world.

TYPICAL JOURNEY TIME SAVINGS				
ROUTE	STOPS	EXISTING	225 KPH TILTING	TIME SAVING
London - Birmingham	3	1 hr 37 mins	1 hr 22 mins	15 minutes
London - Manchester	4	2 hrs 32 mins	2 hrs 08 mins	24 minutes
London - Liverpool	4	2 hrs 40 mins	2 hrs 11 mins	29 minutes
London - Glasgow	3	4 hrs 59 mins	4 hrs 05 mins	54 minutes

Freight

There are a range of new opportunities available to freight operators in addition to the benefits from the Core Programme. Again these would involve additional investment by Railtrack in particular modifying bridges, tunnels and other structures and reinforcing the electrical supply system.

Structures work to increase the loading gauge would enable 'piggyback' freight movements in which lorry semi-trailers are loaded onto the back of special rail wagons and would enable transportation of the larger European size high cube containers.

By constructing a number of new passing routes at strategic locations the longer European size 750 metre freight trains could operate on the WCML.

With the link to the Channel Tunnel the opportunity to move large volumes of freight between key centres and the heart of Europe is an exciting one. The environmental benefits of moving freight from our congested road network onto rail is widely accepted.

However the difficulties in securing long term contracts with freight customers mean there are significant issues to be resolved with regards to the funding of any freight upgrade.

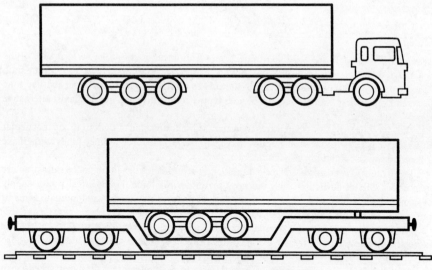

ILLUSTRATION OF 'PIGGYBACK' OPERATION

The Environment

The modernisation offers major environmental benefits through the promotion of rail as an environmentally friendly mode of transport.

Working closely with the new Environmental Agency there will be an assessment of environmental impact with a view to managing the impact of construction works and ensuring compliance with statutory requirements.

SUMMARY

The importance of the WCML to Railtrack's business demands a proactive and planned strategy for the future of the route.

Options ranging from continuing with 'patch and mend' to constructing a new TGV style railway were evaluated and the best commercial modernisation for the company is a Core Programme to restore reliability, reduce costs and enhance safety.

The Core Programme involves the introduction of new technologies for signalling and train control not so far seen in this country.

Additional to this are the upgrade opportunities for both passenger and freight customers. These need to be negotiated with train operators and OPRAF and talks are on-going.

Further benefits to the company include the development of innovative and cost saving technologies with wide scale future applications on other routes.

The successful modernisation of the WCML will enhance the reputation of the company and provide an important demonstration of its commitment to the development of the GB rail network.

C514/078/96

Inter-operability

J A ARMITT MFICE
Union Railways Limited, Croydon, Surrey, UK

INTEROPERABILITY

In recent years, Interoperability has taken up an increasing amount of time of railway engineers and managers throughout the European Community.

So, just what is Interoperability?

The development of railways in Europe has essentially been a national process. Whilst railway engineers have always exchanged views and information, each country has developed its own network and its own rolling stock. At a recent high speed rail conference in Lille, six European countries displayed their high speed trains. The characteristics of each were clearly different and reflected their origins. So the development of national railways has also been different. The result is that trains cannot easily cross borders or, if they do, the drivers instantly find themselves exposed to different rules, signalling systems, electrical power systems, even a different construction gauge.

Interoperability is an attempt to look forward to new high speed railways where infrastructure differences are minimised and trains are economic to purchase.

The initiative has not come from the railway companies. It has come from the European Commission's development of a common railway policy resulting in its report on the European High speed Train Network in 1989.

Studies of future transport needs in Europe resulted in the Trans-Europe Network study projects or TENS as they are commonly called. These projects are not just rail, but also air and road, designed to improve the movement of people and goods throughout Europe. There are many of them with a total forecast cost of ECU100bn.

Cars and planes can cross national frontiers with ease. Trains cannot and for the TENS projects, many of which are new railways between countries, to be successful the Interoperability of trains must be addressed.

The E.C. has developed an Interoperability Directive.

The Directive defines Interoperability as:

- *The ability of the trans-European high speed rail system to allow the safe and uninterrupted movement of high speed trains which accomplish the specified levels of performance.*

The E.C. sees three principal benefits arising from the Directive:

- *The ability of trains to cross borders*

- *Open access for different operators*

- *Increased competition with resultant benefits to travellers and the railway industries in competing world-wide.*

- *Reduced cost of trains and equipment.*

In effect, a single market for high speed rail services.

How is Interoperability to be brought about?

As I said, a Directive has been formulated, debated between the Commission and the various Governments and a final draft approved by the Council. It is now being finalised and could well become Law this summer.

Individual nations will then have to amend their own legislation in order to implement the Directive. Clearly, cohesion is required in this process.

Let us look in more detail at the Directive.

It is essentially a statement of requirements. It identifies the various aspects of rail travel which are generically seen as necessary to be addressed in order to achieve Interoperability.

It breaks the high speed rail system into 6 sub-systems. These are:

- *Infrastructure*
- *Energy*
- *Control, command and signalling*
- *Rolling stock*
- *Operations*
- *Maintenance*

Of these, four are regarded as structural, the other two are operational - maintenance and operations.

Each of these sub-systems will be covered by a TSI or Technical Specification for Interoperability. The key feature of a TSI is that it should only focus on those factors necessary to ensure Interoperability, e.g., the gauge of a tunnel or the height of an overhead electrical power supply, but not whether a train should carry telephones.

In order to establish the TSI's, a joint industry body known as AEIF has been set up by the UIC and UNIFE. The UIC is the representative body of the major national railways and UNIFE the railway industries' manufacturers and suppliers body. The AEIF body has set up separate working groups, each responsible for developing a particular TSI. To ensure compatibility between the work of the groups, there are Economic Evaluation Conformity and Co-ordination groups. The interaction between the various TSI's is complex. Chart A shows some of the main interactions. Wherever there is an interface, there is a choice to be made as to how the requirements of interoperability are to be met. For example, in order to make the rolling stock and infrastructure compatible, we could have:-

- *A single standard platform height and a single vehicle step.*

- *A range of platform heights, and a more complex, adjustable step mechanism on the vehicle.*

In order to ensure that the costs of Interoperability are fully understood, an Economic group will evaluate the additional cost arising from meeting Interoperability requirements and, if necessary, assess the financial and economic benefits. This is an essential but difficult area of assessment. In overall terms, the major investment is that in the infrastructure. There is a tendency, e.g. Talis and Eurostar, to install more complex equipment on trains, leaving national infrastructure standards alone The current Eurostar, with its multiple power and signalling configurations, is a good example, but the result is very expensive and inevitably less reliable, with higher maintenance costs. A challenge will be to use economic evaluations to arrive at a sensible compromise. As we saw on Chart A earlier, the choices will often not be simply bilateral. Signalling rolling stock, infrastructure will all interface, and the overall system balanced with the power supply to minimise electromagnetic interference. An overall economic assessment will not be easy. These six Groups have now been meeting for about a year, and it is likely to be 1997 at least before they have agreed proposals for submission to the Commission. The Commission is to establish what is referred to as the Article 21 Committee, who will receive and, it is expected, approve the TSI's for adoption.

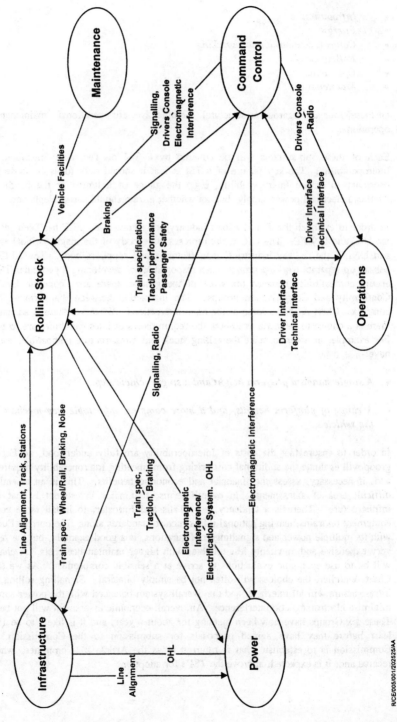

INTERACTION OF TSIs
(Not all minor links shown)

Maintenance

Command Control

Rolling Stock

Operations

Infrastructure

Power

Vehicle Facilities

Signalling,
Drivers Console
Electromagnetic
Interference

Braking

Drivers Console
Radio

Driver Interface
Technical Interface

Train specification
Traction performance
Passenger Safety

Driver Interface
Technical Interface

Line Alignment, Track, Stations

Train spec. Wheel/Rail, Braking, Noise

Signalling, Radio

Train spec,
Traction, Braking

Electromagnetic
Interference/
Electrification System. OHL

Electromagnetic Interference

Line
Alignment

OHL

R/CE/005/001/202525/AA

Now the TSI's are specifications and, clearly, if manufactured items which seek to fulfil the specification are to be recognised across Europe, they will need to be designed and manufactured to agreed standards. So in addition to writing the TSI's the groups will also need to specify the Euro-norms which must be met.

Lastly, in the delivery process there needs to be independent bodies who can confirm that the various elements do actually conform to the standards laid down. Such bodies will be called 'Notified Bodies'. I am not aware that any have yet been established. A key issue will, of course, be that the mutual recognition of such bodies is adopted throughout the community.

What is the timetable for delivery of this whole process?

The Directive is due to become Law during the summer of 1996. It is expected the TSI's will be adopted progressively by 1997/98 and the standards set during 1998/99.

Clearly, this timetable has an effect on the delivery of high speed lines currently being designed and built. It is worth mentioning here that the Directive does not only cover new railways but also upgraded lines. So in the UK both the CTRL and the WCML are candidates for meeting Interoperability criteria. Strictly speaking, the fact that they will be well underway by the time the particular TSI's become effective means they do not have to comply. However, both lines are being designed to meet what are expected to be the key criteria.

As with most European Directives member, states can seek derogation or exemption from the requirements. In the case of the Interoperability Directive the following are the key criteria which would allow derogation:

- *When a project for a new line or upgrading an existing line for high speed is at an advanced stage when the relevant TSI's are published.*

- *For upgraded lines where the loading gauge, track gauge or space between the tracks of the line are different from those on the majority of the European network and where the line does not form a direct connection with the high-speed network of another member state.*

- *In the case of new lines or upgrading existing lines where the rail network is not linked to or is isolated by sea from the high speed rail network of the rest of the Community, e.g., Republic of Ireland.*

- *In the case of a project for a new line or upgrading existing lines, where application of the TSI's compromises the economic viability of the project.*

So far I have talked about the theory or philosophy of Interoperability and the expected process and timetable. However, what are the real issues which are being grappled with on the ground, with which I am sure you will understand is a very complex matter.

Firstly, there are the TSI's themselves. So far most of the groups have had quite a difficult time agreeing the key specification matters **essential** for Interoperability.

As you can imagine, put any group of engineers together to specify the ideal system and they will have great difficulty in restricting themselves to the minimum requirement. Naturally, different countries will wish to ensure that their existing systems are given every chance of complying so there is plenty of room for debate and the Co-ordination Group has a key role in ensuring a consistency and commonalty of approach.

I mentioned just now the need for sensible compromise on economic grounds. The separation of trains and infrastructure into different organisations is a trend not only in the UK which will make such compromises difficult to achieve and we must be careful not to allow economic assessment to stifle rather than help develop and refine the concept of interoperability. Also, UK needs to consider problems of operating mixed traffic (HST/Domestic/Freight) on the HS network.

Another key issue is ensuring that the new systems are not cut off from existing systems. The interfaces will have to be addressed, particularly at stations - for example, platform heights, and over existing routes on the approaches to city centres, which will always be difficult to renew. The interface issues will therefore be important, acting in a way as transitions from old to new, but generating consequences which cannot be ignored in preparing economic evaluations. Part of the Directive identifies the need to ensure that a result of Interoperability is not to create cost-benefit barriers to the preservation of the existing rail network of each State. At the same time, there can be little doubt that there will be extensive costs to be met at these interfaces in modification to existing infrastructure or additional complexity on the rolling stock.

A further important matter is that by setting specifications and standards, we do not hinder the development and improvement of new products and systems.

This is clearly very important in such areas as communication and signalling.

At present, development is taking place by Germany, the UK, France and Italy on a new train control system, ETCS. Care will need to be taken here that the standards for new system do not prejudice future investment and development of improved systems because the process of changing TSI and creating. standards, for example as a result of ETCS, is just too bureaucratic.

There is a parallel issue in how safety should be treated. Should every railway be allowed to set its own specifications with the interopearable standards being the sum of the total or can we bear to move to a position where a lower common standard is adopted by all? The most economic approach is likely to be a risk-based quantitative approach. This is however a relatively new approach to railway safety.

Looking forward, Interoperability presents a combination of opportunities and problems.

The opportunities of open access across the new, high speed network with resultant competition and better services for passengers. The opportunity of creating a larger and stronger rail market in Europe which in itself will be able to compete for international business and a reduction in equipment and train costs.

To achieve these goals will however require more regulation on a European scale, albeit it will hopefully reduce the extent of national regulation rather than duplicate or add another tier. We must be very alive to the potential extra costs of Interoperability. The goal of improving and increasing the demand for rail services will be at risk if extra cost because technically complex TSI's reduce the return on investment and causes investors both Governmental and private to shy away. This process has been driven from the centre, the national railways did not develop the idea. To maintain the momentum it will be important that the Commission maintains its strong interest and lead. It is an expensive process. At present companies are putting valuable engineering and management time into creating the TSI's. The Commission is not fully meeting the cost. If the pressure is relaxed the momentum will undoubtedly slow and possibly be lost altogether. Even more important, is the consequent capital costs in one way or another which the consumer will pay, and he must be aware of sufficient benefits to justify the price.

At another level within the Commission there is already talk of wanting to extend the Interoperability concept to normal lines. This would be a major undertaking and not welcome by most railways. Railways have lost the battle with the motor car during the last 50 years. It will not recover if too much bureaucratic overload and consequent cost in search of the perfect railway is added too quickly. Interoperability , as you can see, is a simple concept whose delivery is complex.

The UK is fully involved in the whole process. It is represented at all levels in the process and has had success, for example, in ensuring that economic evaluation is a key feature of the process. It may be argued that disconnected as we largely are from the European network, do we really need to participate.

The answer is undoubtedly yes -

- *Yes, because the UK is part of the TENS.*

- *Yes, because it will allow European funding to support our new railways.*

- *Yes, because it will improve the services of UK passengers.*

- *Yes, because it will provide opportunity for UK business.*

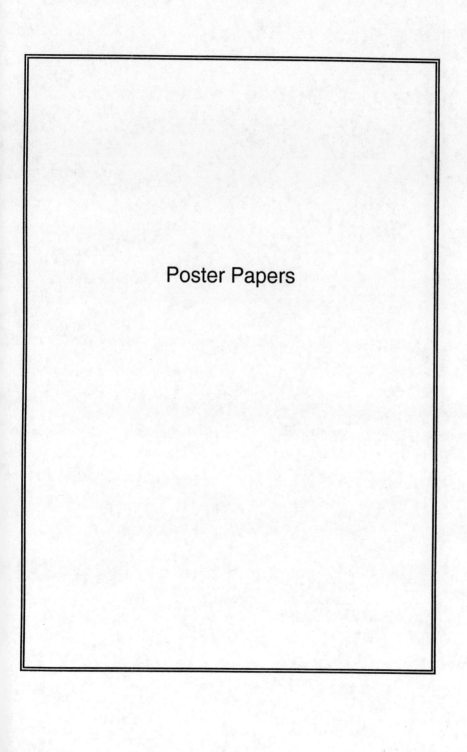

Poster Papers

C514/006/96

Development of observation system for the vibration of a wire-pantograph system with a super-anamorphiclens

A ICHIGE MIEEJ and **K MANABE** MIEEJ
Railway Technical Research Institute, Tokyo, Japan
M AOKI
Central Japan Railway Company, Japan

ABSTRACT

The vibration of overhead wire is one of the most important performances for a wire-pantograph system. It is said that the overhead wire displaces with as long wavelength as several tens of m in the longitudinal direction and as small amplitude as 0.1 m in the normal direction. We successfully developed a super-anamorphiclens with a focal length ratio of 76:1 that permits simultaneous televiewing of its vibration over several spans. [1]

1 Introduction

Until now, we have measured the vibration of overhead wire with a displacement-meter, an accelerometer or a strain-gauge, but in this way we can measure the behavior of the vibration only at several places and cannot make clear the whole behavior of it. It is said that the overhead wire displaces with a long wavelength of several tens of m in the longitudinal direction and a small amplitude of 0.1 m in the normal direction. Because of the difference of the amplitude from the wavelength, we cannot gain a clear image of the wire deformation by an ordinary lens. In order to get a clear image of a vibrating wire, we developed a super-anamorphiclens with a focal length ratio of 76:1. Fig.1 shows a concept of observation system with this lens.

2 Characteristics of a super-anamorphiclens

Table 1 shows the characteristics of the super-anamorphiclens. This observation system uses 1/3 inch CCD camera that has the smallest size of any mass-produced one. If the wire is 50 m distant away from this lens, for example, we can get with the lens an image of the wire covering a length of 52.2 m and a height of 514 mm. The magnification ratio of height to length of the wire appearing on the monitor is about 80:1. Fig.3 illustrates an image of a wire-pantograph system over one span (50 m), which enables the vibrational behavior of the wire to be observed distinctly. Fig.4 illustrates an image of an overlap section, which shows the pantograph motion and the wire waves in the section clearly.

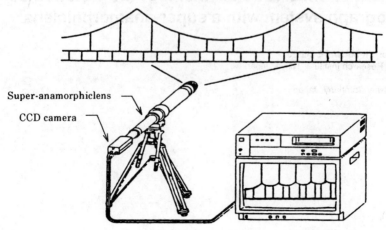

Fig.1 Concept of observation system with super-anamorphiclens

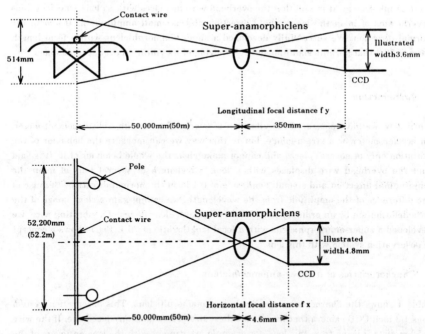

Fig.2 Horizontal and vertical section in the optical system

Table 1 Characteristics of a super-anamorphiclens

Longitudinal focal distance	350mm
Horizontal focal distance	4.6mm
View size in the longitudinal direction(distance 25m)	26m
View size in the normal direction(distance 25m)	240mm
Max aperture ratio	F8.0
Viewing distance range	20m~infinity
CCD resolution in the longitudinal direction	20 lines/mm or more
CCD resolution in the normal direction	20 lines/mm or more
Lens aberration	less than 14%

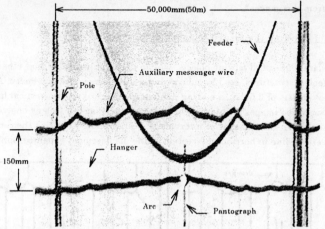

Fig.3   Image of a wire-pantograph system over one span(50 m) observed with super-anamorphiclens

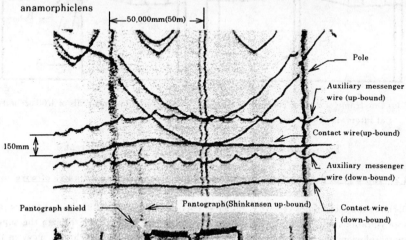

Fig.4   Image of a wire-pantograph system over three spans(150 m) observed with super-anamorphiclens

3 The analysis of the contact wire vibrations under pantograph passage observed by video images through the super-anamorphiclens

Much information can be gained by the analysis of the contact wire vibrations under pantograph passage observed by video images through the super-anamorphiclens. And this method provides a means to get not only the vibration at some discrete points but also its distributions along the wire, which can provide various quantities to determine the propagation of the wave. In particular, the gradient and velocity of the displacement give the wave intensity that would closely influence the contact force. The contact forces are also estimated from the images.

3.1 The way of deciding the displacement of pantographs

As pantographs are high-speed moving objects, we can detect only moving ones other than standing ones (poles, structures, etc.) by subtracting a next image from one image. As one frame of a video consists of 2 fields, a moving pantograph can look as two vertical lines in a frame. If we see video images at intervals of 1/60 seconds, that can look as one vertical line. Then by analyzing video images at intervals of 1/60 seconds, we decide the center of gravity of one vertical line in horizontal direction as the displacement of pantographs.

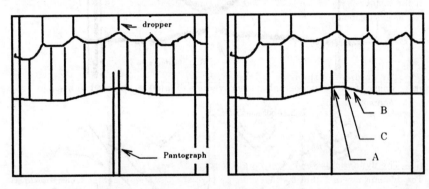

(a) One frame of a video consists of 2 fields at intervals of 1/30 seconds

(b) A field at intervals of 1/60 seconds

Fig.5 The way of deciding the displacement of pantographs

3.2 The way of deciding the displacement of wires on the coordinates of wire

Deciding the coordinates of wire is the most important thing in the analysis of the contact wire vibrations under pantograph passage observed by video images through the super-anamorphiclens. As a contact wire and an auxiliary messenger wire hardly move in the horizontal direction, we decide the center of gravity of wires in vertical direction as the displacement of wires by analyzing video images at intervals of 1/60 seconds.

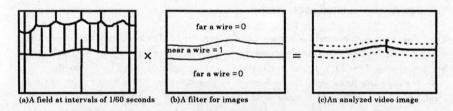

| (a)A field at intervals of 1/60 seconds | (b)A filter for images | (c)An analyzed video image |

far a wire =0
near a wire = 1
far a wire =0

Fig.6 The way of deciding the displacement of wires on the coordinates of wire

3.3 Calculation of various quantities to determine the propagation of the wave

By using the displacement of wires, we can calculate the velocity and the gradient of the displacement, the dynamic power and kinetic energy. As we use a video which can get 30 frames (60 fields) per one second ,we need to analyze 60 images per second. The sampling time($\triangle$t) is 1/60 seconds. The coordinate of image on TV multiplied by scale factors in the vertical and horizontal direction is the actual coordinate.

(1) The velocity of the displacement
If the displacement of wires of i image is named $Y_i(X)$, the velocity of the displacement can be calculated by

$$V_i(X)=(Y_{i+1}(x) - Y_i(x)) \diagup \triangle t .\tag{1}$$

(2) The gradient of the wire θ, the curvature r

$$\theta_i(x)=Y_i(x+\triangle x) - Y_i(x-\triangle x) \diagup 2\triangle x \tag{2}$$

$$r_i(x)=\{Y_i(x+\triangle x) - 2Y_i(x) + Y_i(x-\triangle x)\} \diagup (\triangle x)^2 \tag{3}$$

(3) The dynamic power P, the kinetic energy E
The dynamic power P that is conveyed on a wire and the kinetic energy E per unit length can be calculated by expressions (4) and (5), where T is the tension of a wire and ρ is the density of it.

$$P_i(x)= -T\theta_i V_i \tag{4}$$

$$E_i(x)=0.5\rho V_i^2+0.5T\theta_i^2 \tag{5}$$

4 The experimental result

This chapter describes the analysis of the contact wire vibrations under pantograph of Tokaido Shinkansen 0 type-train with 8 pantographs by changing analogue signal of 768 frames into digital signal.

4.1 The displacement of pantographs

Fig.7 shows the displacement of pantographs in every field. The horizontal axis shows the number of fields. The vertical axis shows the location of pantographs as the number in the horizontal axis(0 ～ 511), 400 on horizontal axis being equivalent to 6.6666 seconds.

The slanting lines show the movement of pantographs. The slanting lines are not straight but a little curved because of the optical strain of super-anamorphiclens. If its strain is corrected, the displacement of pantographs will be measured with good accuracy. The speed of the train is equivalent to the average gradient of the slanting lines. The speed of the train in Fig.7 is found to be 207km/h.

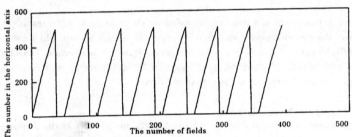

Fig.7   The displacement of pantographs in every field

4.2 The vibration of the contact wire

By analysis of the contact wire vibrations under pantograph passage observed by video images, the displacement, the gradient, the curvature, the velocity of the displacement, the dynamic power and kinetic energy of place A, place B and place C are obtained as shown in Figs 8. From these figures , the following two things are made clear.

(a)   The deformation of overhead wire can be observed with fairly high accuracy.

(b)   The amplitude of the vibration of B point is greater than A point in the area of 10Hz under pantograph passage. That is to say, a span between hangers without dropper vibrates more easily than a span with dropper

5   Conclusions

The images observed through the super-anamorphiclens show the deformation of overhead wire with fairly high accuracy. And we can calculate the gradient, the curvature, the dynamic power and kinetic energy and so on. We will promote the study on wire-pantograph system dynamics by using this system.

References

(1)Manabe, Ichige, Aoki, Development of Observation System for the Vibration of a Wire - Pantograph System with a Super-Anamorphiclens, Quarterly Report of RTRI Vol.37, No.1, MAR. 1996.

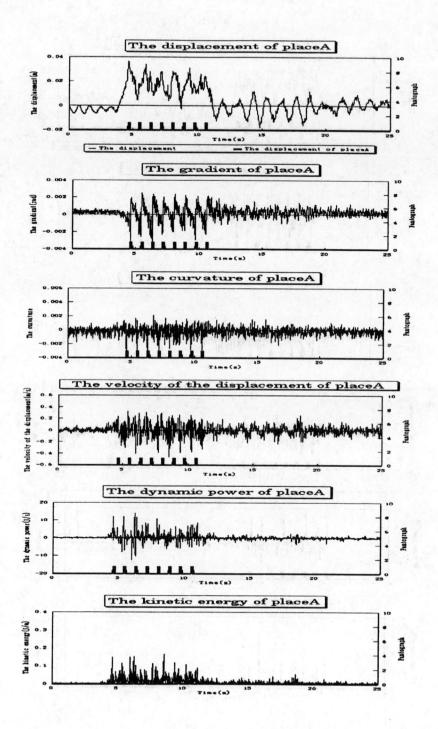

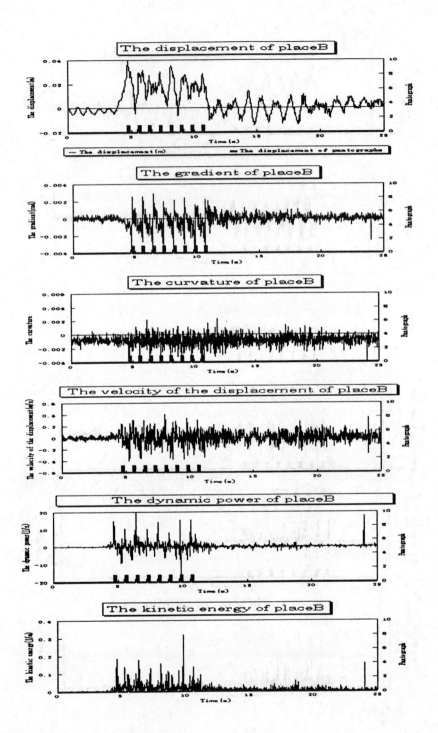

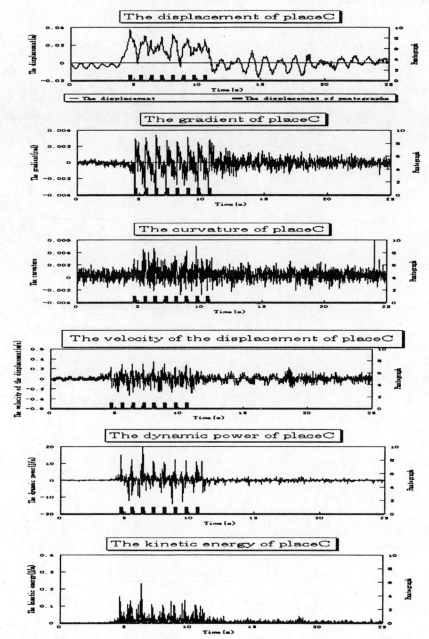

Fig.8 The displacement, the gradient, the curvature, the velocity of the displacement, the dynamic power and the kinetic energy of place A, place B and place C

C514/025/96

Development of the superconducting MAGLEV vehicles on the Yamashi test line

K TAKOA MJSMEand **Y MATSUDAIRA**
Railway Technical Research Institute, Japan
A INOUE
Central Japan Railway Company, Japan

SYNOPSIS

The construction of the superconducting Maglev system on the Yamanashi Test Line started in 1990. The vehicles for the first train set, named "MLX01" type, were completed in the spring of 1995. It consists of three cars, the head car for Kofu (Mc1), the standard middle car (M1) and the head car for Tokyo (Mc2). The train set is an articulated bogie system. The vehicles can run levitated at over 500 km/h (experimental maximum speed 550 km/h) by active track linear synchronous motor. The service brake system is a regenerative brake, and there are two types of emergency brakes: a wheel disk brake and an aerodynamic brake . An important feature of the exterior is the nose shape of the head cars. One is named the "Double Cusp" style, and the other the "Aero-wedge" style, each having superior aerodynamic performance. The test runs will start in the spring of 1997.

1. Introduction

We have been developing the superconducting magnetic levitation (Maglev) systems since 1962. Many experiments on the Miyazaki Test Track with Maglev vehicles of ML500, MLU001 and MLU002 provided much basic and valuable data. And the latest experiments are now continuing with MLU002N.

Meanwhile the construction of the superconducting Maglev system on the Yamanashi Test Line was started in 1990. In the final plan, the length of this test line will be 42.8 km. Now, 18.4 km of this length is being constructed as a priority section. And the vehicles for the first train set of MLX01, which is a 3-car unit, were completed in the spring of 1995 (Figs.1,6). We have two train sets on it, and the second one, which is a 4-car unit, is being designed this year. The Yamanashi Test Line is the final stage for confirming the possibility of commercialization of Maglev system. The beginning of test runs is scheduled for the spring of 1997.

Mc1 ("Double Cusp" style)   Mc2 ("Aero-wedge" style)
Fig.1. Outside view of the "MLX01" Yamanashi Maglev Vehicles

2. Testing Items on the Yamanashi Test Line

The following are main items to be tested on the Yamanashi Test Line;
(1) High-speed stable run at over 500 km/h with safety and comfort.
(2) Reliability and durability of vehicles and ground facility/equipment including SCM (superconducting magnets).
(3) Structural standards specifying a minimum radius of curvature, the steepest gradient, etc.
(4) Distance between track axes taking into account two trains passing each other.
(5) Vehicle performance related to tunnel cross section and the pressure fluctuations in tunnels.
(6) Turnout performance.
(7) Environmental conservation.
(8) Multiple train traffic control system.
(9) Operation and safety system, and maintenance standards.
(10) Control system between substations.
(11) Economy of construction and operation.

3. Design Concept and Details of Development for Maglev Vehicles

3.1 Development of the super lightweight body
(1) Reduction of body weight
We have decided cross-sectional area and body structure for the super lightweight body. A quadratic conical curve, which is nearly a circle and different from the box shape of ordinary trains, is adopted to make the stress flat against the airtight load (Fig.2). The cabin section adopts the semi-monocoque structure to share the vertical load and compression load and mainly to be effective against airtight load (Table 1). It is different from usual trains as a base frame structure.

Table 1. Details of body load

Items	Mc1	M1	Mc2
Vertical Load	205 kN	151 kN	210 kN
Compression Load at Body End	392 kN		
Airtight Load	Maximum: Outer Pressure -20 kPa ~ + 13 kPa		
	Repeat: Outer Pressure -17 kPa ~ +11 kPa		
Correspondence of Bending Force	Over 1.4 GN·m^2		

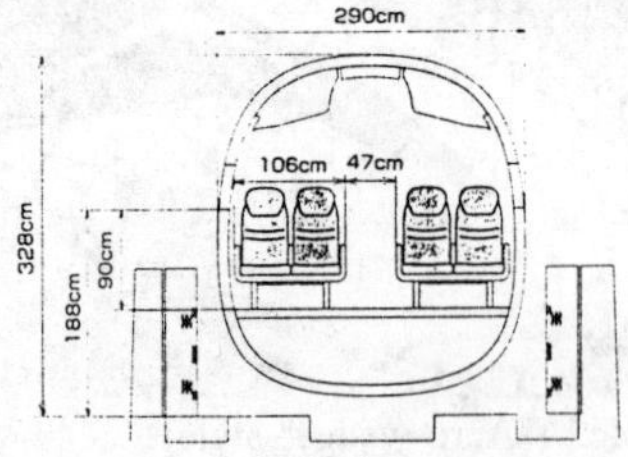

H1-body

H2-body

Fig.2. Cross-sectional area

H3-body

H4-body

Fig.3. Trial bodies

(2) Manufacturing of trial bodies

Various trial lightweight bodies were manufactured to perform various load tests including airtight and fatigue load tests. And then the sufficiency of durability was verified before designing the Yamanashi Maglev vehicles (Fig.3).

3.2 Improvement of aerodynamic characteristics
(1) Development of nose shapes

The nose section length was set up 9.1 m to reduce the cross-sectional area distribution, which reduces aerodynamic drag, aerodynamic noise, and micro-pressure waves in tunnels. Simulating by CFD (Computational Fluid Dynamics) and performing many tests in wind tunnels, a formal shape of the Maglev vehicles has been developed. Finally two nose shapes are selected out of many proposed shapes as the most ideal ones for aerodynamic performance. The air drag coefficients (C_D) of both shapes are under 0.1. Especially, the "Double Cusp" style is superior in controlling the separation of boundary layers when coupled at the extreme tail end of the train, and the "Aero-wedge" style is superior in reducing micro-pressure waves in tunnels (Fig.4).

(2) Fairing of bogies

Reducing the resistance from pressure at the bogies, especially articulated parts, is important in reducing the aerodynamic drag. Therefore, a fairing was installed to make a smooth transition of cross-sectional area from the body to the bogie, and the gap between the body and bogie was filled with a movable fairing (Fig.5).

Air flow around rear car
"Double Cusp" style

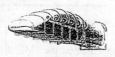
"Aero-wedge" style

Fig.4. Simulating by CFD of nose shapes

Outer hood
fairing

Fig.5. Flatness of body and bogie

3.3 Countermeasure for aerodynamic noise

The causes of noise are the steps, projections, gaps, etc. on the body surface. The flatness of the body surface was examined as follows;

(1) Outer hood: the outer hood which is made of a urethane block is installed between the coupling of bodies to smooth the body surface from front to end.

(2) The flush-surface installation of the window: the window glass is installed on the body with a step difference within 1 mm to the body surface.

(3) The steps of the door: the door is an upward sliding type, with a step difference within 2 mm between door leaf and body surface.

3.4 Magnetic shield

Magnetic shields were installed on the body to shield the passengers from the magnetic fields of the superconducting magnets. The materials and arrangement of the magnetic shields were examined and analyzed to reduce the weight and to make shield performance effective.

Industrial-pure steel and electromagnetic steel were selected for the magnetic shield. These materials are superior to other materials in saturated magnetic flux density, magnetic penetration rate, processing, costs, etc. The most suitable thickness and arrangement of the shields were decided from various analyses and a full-scale mock-up. It was then possible to reduce the strength of the magnetic field to under 2 mT in the cabin, and the weight of the shield to under 1.4 ton in the standard middle car.

3.5 Other components

We have tried to reduce the carbody weight using the following components.

(1) Abolition of cab: a cab which looks like an ordinary train is not necessary for Maglev vehicles, because Maglev systems are active track systems. Then a front glass and a magnetic shield for the cab are abolished to reduce the weight.

(2) Small window: the window glass is made as small as possible without causing passengers a feeling of occlusion, by securing the vision through a window area of 400 mm height x 300 mm width and by adequately setting the seating pitch.

(3) Light weight of equipment: an optical fiber sendings various information, aluminum for the body of devices and the type of air pipe, a complex honeycomb for interior panels, seat, etc.

4. Outline of the First Train Set on the Yamanashi Test Line

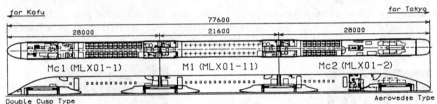

Fig.6. Outline of the First Train Set named the "MLX01" type

The vehicles on the Yamanashi Test Line are named the "MLX01" type (Fig.6). The first train set consists of three cars, the car No.1 is the head car for Kofu (Mc1), the car No.2 the standard middle car (M1), and the car No.3 the head car for Tokyo (Mc2). The train set is an articulated bogie system (Fig.15). The vehicles can run levitated at over 500 km/h (experimental maximum speed 550 km/h) by linear synchronous motor with primary side on ground. The service brake system is a regenerative brake, and there are two types of emergency brakes: a wheel disk brake and an aerodynamic brake.

An important feature of the exterior is the nose shape of the head cars. Mc1 is the "Double Cusp" style, and Mc2 the "Aero-wedge" style, each having superior aerodynamic performance (Fig.1). Both head cars are equipped with seats, baggage racks, etc. for passenger trial riding, and M1 is designed exclusively for use as an experimental car (Figs.8,9,13).

4.1 Features of Vehicles

(1) The Head Car for Kofu (Mc1)

Mc1 is designed as a controlling car and trial riding car. Therefore many controllers and a monitoring system are installed in the head section (mainly in the crew cabin), for example the on-board central controlling system, battery for controllers, antennas and devices for the radio system, converters, inverters for power supply to the bogie system, etc. There are 46 seats, stowage bins resembling those in airplanes, and an air conditioning system in the cabin section for trial riding (Figs.7,8).

(2) The Head Car for Tokyo (Mc2)

Table 2. Technical details of MLX01

Maximum Speed		500 km/h (Maximum Test Speed 550 km/h)		
Model		Mc1	M1	Mc2
Passenger Capacity		46	—	30
Max. Gross Weight		29 ton	20 ton	30 ton
B o d y	Length	28,000 mm	21,600 mm	28,000 mm
	Width	Body Part 2,900 mm, Bogie Part 3,220 mm		
	Height	Levitated Run 3,280mm (Wheel Run 3,320 mm)		
	Cross-Sectional Area	8.9 m ^{2} (Vehicle/Tunnel Ratio 0.12)		
	Construction	Aluminum Alloy Semi-Monocoque Structure		
	Joint Method	Rivet	Spot Weld	Rivet + Weld
	Nose Shape	Double Cusp	—	Aero-wedge
Distance between the Bogie Center		21,600 mm		
Bogie Type		SCM-Rigidly Mounting Type 4-Point Support Bogie		
Superconducting Magnets		4 Coils per Side Same Poles on Both Sides (700 kA)		
Auxiliary Power Unit		DC 600 V		
		—		Ni-Cd Battery Gas Turbine Power Unit
Power Unit of Control		DC 100 V		
		Ni-Cd Battery		
Air Compressor		SF-JF-C1000LA	—	SF-JF-C1000LA
Air Conditioner		Ram Air Ventilation System		Ram Air and Fan V.S.
Brake Systems		Regenerative Brake (Primary) Wheel Disk Brake, Aerodynamic Brake		
Etc	Seat Arrangement	2 x 2, Seat Pitch 880 mm Reclining Seat with Rotation		
	Entrance Door	Upward Sliding Door (Sky Door)		
	Service Equipment	Information Display for Passenger Announcement System		

Fig.7. Crew cabin of Mc1

Fig.8. Passenger room of Mc1

Fig.9. Passenger room of Mc2

Fig.10. Equipment room of Mc2

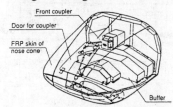

Fig.11. Nose cover of Mc2

Mc2 is designed to serve as an electric power supply car and a trial riding car. Therefore much electric equipment are installed in the head section (mainly in the power source room), for example battery cells, an electric power generator using gas-turbines, a main DC/DC converter, inverters for electric power supply to the bogie system, etc. The power source room has no windows. There are 30 seats, racks, and an air conditioning system in the passenger room for trial riding (Figs.9, 10).

Fig.12. Outside view of M1

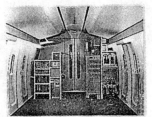

Fig.13. Cabin layout of M1

(3) The Standard Middle Car (M1)

M1 will be used only as an experimental car. Therefore it is not equipped with passenger seats, racks, etc. But an air conditioning system is installed in the cabin section for the measurement operator (Figs.12,13).

4.2 Carbody structure

(1) Mc1

The carbody of Mc1 consists of 3 sections: the head section, the cabin section (passenger room), and the articulated section. The passenger room, gangway at the articulated section, and crew cabin are airtight structures (Fig.14).

The material of the body structure except for the magnetic shield is duralumin, of types 2024 and 7075, which are the same materials as in aircraft bodies. The carbody of

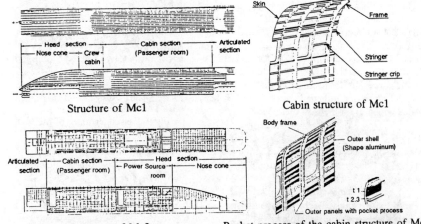

Structure of Mc1

Cabin structure of Mc1

Structure of Mc2

Pocket process of the cabin structure of Mc2

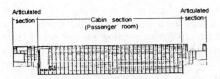

Structure of M1

Articulated part of M1

Fig.14. Carbody structure of Mc1, Mc2 and M1

Mc1 is fabricated using rivets which are the same as in aircraft bodies.

The cabin section structure consists of 3 cylindrical blocks. Each block consists of a top panel, a bottom panel, and 4 side panels. The frame pitch is 440 mm, which is half the seating pitch. The stringer pitch is 170 - 220 mm. The head section consists of the nose cone and the crew cabin. The nose cone is composed of framework to form the "Double Cusp," and of another skin with a thickness of 4.06 mm (0.16 in) to withstand birds and pebbles that may strike the body. The crew cabin is of the same structure as the cabin section. A coupler, a buffer, antennas, marker lights (head and tail lights), and CCD cameras are housed in it. Mc1 has two aerodynamic brake systems, which are installed on top of the crew cabin and articulated section.

(2) Mc2

Mc2 consists of 3 sections: the head section, the cabin section (passenger room), and the articulated section. All sections are built by several methods, and they are connected to each other by riveting or welding. The passenger room and gangway at the articulated section are airtight structures. The power source room is not airtight (Fig.14).

The material of the body structure except for the magnetic shield is an aluminum alloy. The aluminum alloy is of types 5083, 7N01, and 6N01, and the outside shells of the cabin section and power source room are made of 6N01 aluminum extrusion.

The cabin section structure consists of 6 cylindrical blocks: a top panel, a bottom panel, and 4 side panels. The frame pitch is 440 mm and the stringer pitch is 100 mm. The panels of the cabin section are extrusions with stringer. These panels with 2.3 mm in thickness are planed by a skin miller to 1 mm, in order to reduce the weight by more than 15%. This is called the "pocket process." Panels and frames are connected by riveting . The head section structure consists of the nose cone and the power source room. The head section is constituted of frames and stringers, which are joined by arc welding to form the "Aero-wedge." The top of the nose cone is made of FRP for use as a nose cover (Fig.11). The nose cover was confirmed for strength in the event of hitting birds at 500 km/h. A coupler, a buffer, marker lights (head and tail lights), and CCD cameras are housed in it.

The power source room is of the same structure as the cabin section, but the outside shells and frames are joined by spot welding. The articulated part is constituted of frames and plates which are joined by welding.

(3) M1

M1 consists of 3 sections: the cabin section and the articulated sections at both ends of the body (Fig.14). The material of the structure except for the magnetic shield is an aluminum alloy, of types 5083, 7N01, and partly 7075 duralumin.

The cabin section structure consists of 4 longitudinal blocks: a top panel, a bottom panel, and 2 side panels. Each block is constituted of outside skin with frames and stringers which are made of aluminum shapes. The blocks are joined by spot welding or riveting at the outside skins and frames. The M1 body structure features a seamless outside shell, 16 m long. The articulated section structure of M1 is just like Mc2. It is constituted of frames and plates. The aerodynamic brake system is installed on the top of this section on the Tokyo side.

4.3 Bogie

The vehicles are an articulated bogie system (Fig.15). Therefore a 3-car train needs four bogies. The bogie is composed of the SCM, the bogie frame, and several systematically installed on-board devices. The bogie frame mainly consists of side and cross beams of welded or rivet-bonding structure, made of aluminum alloy. The major devices in the bogie

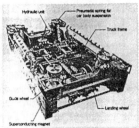

Fig.15. Articulated bogie between Mc2 and M1   Fig.16. Bogie

are the tare weight supporting gears and their actuating power unit, suspension and traction devices, cryogenic equipment, emergency landing apparatus, and on-board mechanical braking system.

4.4 Aerodynamic brake system

An aerodynamic brake system is adopted as one of the emergency brake systems. This system operates when the primary brake system (a regenerative brake) malfunctions during high-speed running. It can act in less than 1.5 sec by the opening command. The power source of this brake is oil pressure which is supplied from the bogies. Therefore it is installed on the top of each car at the bogie positions (Fig.17). There are two types of this brake system as follows:

(1) One-board type: which consists of a main panel with subpanels on both sides of the main panel. This type has two features: a projection area (3.7 m ^{2}) which is a small cut-out area of skin, and a small curvature of the panel which reduces the brake drag difference between forward and backward run.

One-board type   Two-board type

(2) Two-board type: which has two independent panels, each panel projection area being 1.85 m ^{2}. Fig.17. Aerodynamic brake system
The lock mechanism is built into each actuator. This type is very simple.

4.5 Ventilation and air-conditioning system

A ventilation system basically uses a ram air intake at high speed (approximately over 180 km/h) and uses a ventilation fan at low speed. And the passenger ear discomfort from a sudden change in cabin pressure through tunnels is controlled by tuning the intake and exhaust valves while monitoring the cabin pressure against the change of outer pressure.

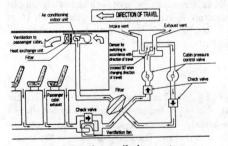

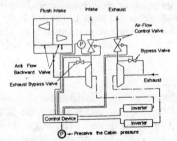

Fig.18. Ram air ventilation system   Fig.19. Ram air & fan ventilation system

There are two types of the ventilation system as follows:

(1) Ram air ventilation system. It is a ventilation of a ram air intake system, which shifts between intake and exhaust depending on the running direction, takes in the running air from a pair of ram air intakes facing each running direction. And this system has a low pressure fan for low speed running (Fig.18).

(2) Ram air & fan ventilation system. It uses both a ventilation fan and a ram air intake at high speeds to introduce fresh air, and uses a ventilation fan when the train is at a low speed and at a stop (Fig.19).

5. Closing Remarks

The carbody weight of the Yamanashi Maglev vehicles is reduced to about 50 % of the Shinkansen Series 300 which is one of the newest trains. But it is necessary to save the weight further to save the running energy and to decrease the burden of the ground structure and ground coils.

It is hoped that the success in the various tests will enable commercialization Maglev operation in the near future.

The development of the superconducting Maglev systems has been subsidized in part by the Japanese Ministry of Transport.

REFERENCES

1. K.Takao, Vehicles for superconducting Maglev system on Yamanashi test line, pp.487 -494, Proceedings of the 4th Int. Conf. COMPRAIL94, Computers in Railways Ⅳ-Vol. 2 Railway Operations, Madrid, Spain, September 1994
2. Kikuo TAKAO, Kiyoshi TAKAHASHI, Vehicles for Superconducting Maglev System on Yamanashi Test Line, pp.150-157, QR of RTRI, vol. 35, No. 3, August 1994
3. K.Takao, M.Yoshimura, N.Tagawa, Y.Matsudaira, K.Nagano, Development of the Super -conducting Maglev Vehicles for Yamanashi Test Line, pp.1210-1215, Proceedings of the Fourth Japan International SAMPE Symposium (JISSE-4) Vol. 2, Tokyo, Japan, September 1995
4. K.Takao, M.Yoshimura, N.Tagawa, Y.Matsudaira, K.Nagano, A.Inoue, Development of the Superconducting Maglev Vehicles on Yamanashi Test Line, pp.233-238, Proceedings of the 14th Int. Conf. on Magnetically Levitated Systems (MAGLEV'95), Bremen, Germany, November 1995
5. Hiroshi SEINO, Ken-ichi KATO, Hiroshi OSHIMA, Masayoshi AZAKAMI, Hiroshi, YOSHIOKA, The Maglev Bogies and its Development, pp.261-266, Proceedings of MAGLEV'95, Bremen, Germany, November 1995
6. Kikuo TAKAO, Yoriharu Matsudaira, Akihiko Inoue, Shirou Hosaka, The Super -conducting Maglev Vehicles on the Yamanashi Test Line, pp.525-531, Preconference Proceedings of the World Conference on Railway Research Conference 1996 (WCRR'96), Colorado Springs, USA, June 1996
7. K.Takao, Y.Matsudaira, A.Inoue, Development of the Superconducting Maglev Vehicles on the Yamanashi Test Line, Proceedings of the 5th Int. Conf. COMPRAIL96, Berlin, Germany, August 1996
8. Kikuo TAKAO, Masafumi YOSHIMURA, Naoto TAGAWA, Yoriharu MATSUDAIRA, Akihiko INOUE, Shiro HOSAKA, Development of the Superconducting Maglev Vehicles (MLX01 type) on the Yamanashi Test Line - Carbody Structure and Equipment of the First Train Set -, QR of RTRI, vol. 37, No. 2, August 1996

C514/007/96

Advanced traction and braking control – an artificial intelligence approach

D A NEWTON MEng, CEng, MIMechE
GEC Alsthom European Gas Turbines Mechanical Engineering Centre, Leicester, UK
H J THOMPSON BSc, AMIEE, MIEEE
GEC Alsthom Traction Limited, Preston, UK

Synopsis

This paper investigates the use of artificial intelligence techniques for maximising the available adhesion for a rail vehicle. A comprehensive non-linear computer simulation of a typical Electrical Multiple Unit is used to demonstrate the techniques and the proposed control scheme. It is shown that there is potential for reducing journey time and energy consumption for variable adhesion conditions when compared with a traditional approach.

1 Introduction

Recent years have seen a resurgence in the use and construction of rail systems. Increasing environmental concerns and road traffic congestion, particularly within city centres and their surroundings has led to a world-wide expansion in the provision of rail based transport. This has presented both an opportunity and challenge to the rail industry with a large potential market, but one that is used to a fast, flexible and reliable service.

Restructuring of the industry has led to a significant cultural change from that of a public service monopoly supported by a state subsidy to a business led organisation existing in a competitive environment. This in turn has changed the emphasis for the suppliers of new equipment where reliability, through life costs and performance are beginning to assume equal importance to the initial capital outlay, and more recent contracts are including severe penalties for failures in these areas.

This paper investigates the use of Artificial Intelligence (AI) techniques to address these issues with the aim of designing a control strategy that will rapidly and safely find the optimum adhesion point for a region of track thus reducing time between stops and vehicle headways. Other benefits could include less vehicle induced track damage, lower energy consumption, smoother ride and better integration of the traction and braking systems.

The new scheme is tested on a comprehensive non-linear model of a typical Electrical Multiple Unit (EMU) and a comparison is made with an existing approach. The issue of safety is also briefly discussed and the potential implications of these methods when presenting a safety case.

2 Electrical Multiple Unit Simulation

To test the proposed algorithm a detailed computer simulation of a typical Electrical Multiple Unit was generated. This is a much more cost effective and rapid means of testing new concepts provided there is some means of validating the model against real data. All major subsystems were modelled in detail and independently verified prior to inclusion in the complete model. The following systems were modelled.

2.1 Motor

The traction industry has undergone something of a revolution in recent years in the change from the well-proven, well-understood d.c. motor to the new and less well-understood a.c. induction motor driven by an inverter. The advantages of the induction motor over the d.c. motor have been documented extensively, namely lower maintenance costs, better performance and smaller size for the equivalent power rating, [1, 2]. The costs of these advantages are more complex control and the necessity of using microprocessors to cope with the calculations, of which the latter has only been possible in recent years. Most contracts currently being placed for traction equipment are for a.c. drives with asynchronous motors and Gate Turn Off (GTO) thyristor or, more recently, Insulated Gate Bipolar Transistor (IGBT) inverters. The model is based on a voltage source inverter with GTO devices driving a standard asynchronous AC motor.

2.2 Brakes

The braking system for a train is usually designed in isolation from the traction motoring system by a separate contractor. There are several different and equally valid design approaches in the industry. For most applications, the motor is used to provide electrical braking to assist the mechanical braking system and the integration of the two systems is known as blending. Usually only a rudimentary interface between the two systems is used, even though both systems use a great deal of common information, partly due to the safety implications of the braking for the train, [3, 4]. A complete pneumatic braking system is modelled based on a typical EMU installation.

2.3 EMU Model

A simulation of a typical a.c. drive Electric Multiple Unit (EMU) has been created on which to study some of these control issues. A schematic of the simulation blocks of the model are shown in Figure 2.4.1.

The mechanical interfaces are modelled using the best available data, taken from measurements on the track. Adhesion was modelled in a macro sense using look-up tabulated data of typical adhesion curves based on vehicle measurements, [5]. The mechanical forces on the bogie, such as that exerted by track curves and the mounting of the motor on the axle, are modelled in addition to equipment weight and passenger loading. The inverter includes a waveform generator for the Pulse Width Modulated (PWM) control of the inverter, rather than pure sinusoidal waveforms. The control algorithm used is as near as possible to the scalar control used in a particular class of EMU.

The wheel slip-slide control scheme incorporated in the simulation is representative of current best practice. When a difference in speed measurement is observed between axles, wheels or the theoretical speed in motoring a 'slip' is detected and torque demand is reduced to zero. After a delay, a percentage of full demand torque is applied and, if no further slip is detected the demand is ramped back up to 100%. If the wheels slip again, torque demand is again reduced to zero and then brought back at a lower percentage than before. This process continues until an operating point is found with no slip. A similar strategy with sliding in braking is adopted. This trial and error process can take time and does not necessarily find the best answer.

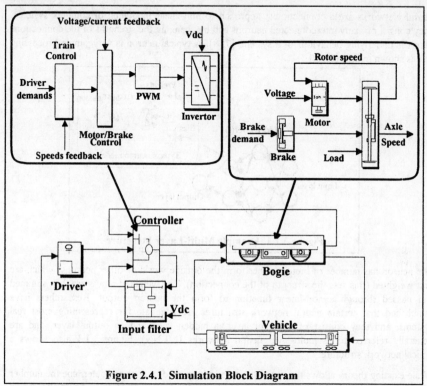

Figure 2.4.1 Simulation Block Diagram

Results from practical tests using the same parameters as those in the model have been compared with simulation results. The data used was taken from normal testing of EMU equipment on the GEC Althsom Combined Test facility, [6]. These tests are conducted on the drive, consisting of control electronics, power converter and motors using real track profile and passenger loading data from the route that the particular EMU would be subjected to in service. When satisfactory agreement between simulated and practical results has been achieved, the model can then be used as a 'benchmark', against which other control system models can be compared with confidence that the results obtained are reasonable predictions.

3 Artificial Intelligence

There are several Artificial Intelligence methods currently used in industry, of which two will be used in this work, Neural networks and Genetic Algorithms. They have the advantage of working with very simple underlying mechanisms that do not require complex mathematics for explanation. However the implication of this subject is one of un-predictability which has significant safety implications. The following sections discuss the two proposed techniques and some issues related to safety and predictability.

3.1 Neural Networks

Neural Networks are a connectionist approach to non-linear modelling of complex systems. They consist of many interconnected neurons and by changing the strengths of the connections learn the underlying behaviour of a system, [7, 8]. A typical neuron is very simple processing unit as shown in figure 3.1.1.

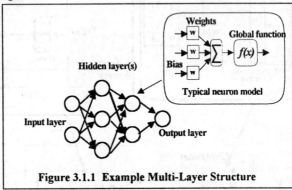

Figure 3.1.1 Example Multi-Layer Structure

The neuron has number of inputs (either form the outside world or other neurons) which are first weighted (this sets the strength of the connection). The weighted inputs are then summed and passed through a non-linear function to form the neuron output. Researchers have established that certain neural network structures are adequate for representing most real systems, and they consist of an input layer, a hidden layer and an output layer, and are generally referred to as multi-layer neural networks (MLNN). Figure 3.1.1 also shows a typical network structure.

While existing theory allows the basic structure to be picked, it does not determine the number of neurons used and at present a trial and error method is used. This is part of the design process and would not be carried out as part of the normal running procedure for the algorithm. The number of inputs and outputs would normally be dictated by the actual system being modelled. One feature often neglected is the need for pre-processing of input data, particularly normalisation. This often greatly improves the convergence properties of the network by enhancing the conditioning of the problem.

The only other variables to be tuned are the connection weights, and a learning algorithm is employed. A trial network is initialised with random weights and some typical inputs are applied to it and the system. The system output is measured and compared with the network output and the error calculated. This error is then "back-propagated" through the layers of the network and a portion of it assigned to each weight, depending mainly on how much the weight was perceived as contributing to the error. A simple update law is used to make a small adjustment to each weight and another test run is carried out. This procedure is continued many time until the NN has accurately modelled the system. The NN is then tested against other system input/output to ensure that it has learnt a general system model (and not simply matched point for point the training data !).

3.2 Genetic Algorithms

Genetic Algorithms (GAs') provide a framework for a robust search mechanism based on biological principles of evolution and "survival of the fittest", [9]. The technique can be used on many design problems to size components, design circuits, select parameters etc. In the context of this work the main interest is gain selection either for a controller, or a neural network.

A GA works by first randomly selecting a population of parameters. The parameters are then encoded into a single number string, the most common approach being to convert each parameter into its binary equivalent and concatenating the binary strings to form what are termed "chromosomes". This is repeated many times to produce the initial population. Each set of parameters is tried in turn on the problem and given a score depending on how successful they were. A selection process picks the better strings from the initial population for modification.

The modification process is called crossover, figure 3.2.1, and consists of randomly selecting two bit strings where the better (or fitter) strings are given a degree of preference. The selected pair swap part of their strings thus creating a two new strings of hopefully better performance. An additional minor bit operator, "mutation", is also very occasionally used to invert randomly chosen bits (generally set to around 1 in a thousand). This continues until a new population is formed and the entire process of experiment and evaluation repeated.

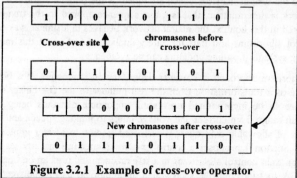

Figure 3.2.1 Example of cross-over operator

The user has several selectable properties to tune, the population size, number of generations, and probabilities of mutation and crossover (typically 0.05 and 0.8 respectively). It is important to note is that each generation is much better than the previous by virtue of the selection and crossover mechanisms. Thus after a few generation the initial very random estimates will have converged to a solution for the problem.

This method has several advantages over more traditional techniques. It requires no gradient data (unlike the majority of classical methods), the fitness evaluation can incorporate rules and other heuristics, empirical measures and is not restricted to a quadratic cost function, and it does not require a model of the system, but uses only input/output data. It is a powerful technique well suited to multi-dimensional search problems, particularly where direct measurement is difficult, for example relative slip on a wheelset.

3.3 AI and Safety

Safety is an important area for new rail vehicle developments with the need to satisfy all aspects of the safety case, particularly for methods potentially using on-line adaption. A MLNN is represented by the following mathematical formula.

$$y_l(k) = \sum_{i=1}^{h} f_i(u(k), w_i, \delta) B_{i,l} \tag{3.3.1}$$

$$u(k) = \begin{bmatrix} r(k) & r(k-1) & \dots & r(k-p) & y(k-1) & y(k-2) & \dots & y(k-z) \end{bmatrix} \tag{3.3.2}$$

$$w_i = \begin{bmatrix} w_{i,1} & w_{i,2} & w_{i,3} & \dots & w_{i,e} & w_{i,e+1} \end{bmatrix}^T \tag{3.3.3}$$

where h is the number of hidden neurons, l is the number of network outputs and e is the number of inputs to the network. The non-linear activation function is defined by:

$$v_n(k) = \Phi_n \left[\left(\sum_{i=1}^{e} u(k) \bullet w_{n,i} \right) + \delta \bullet w_{n,e+1} \right] \quad n = 1, 2, 3, \dots, h \tag{3.3.4}$$

typically

$$\Phi = \frac{1}{1 - \exp(-x)} \text{ or } \frac{1 - \exp(-2x)}{1 + \exp(-2x)} = \tanh(x), \text{ and } 0 < \delta \le 1$$

The variables are the weight matrices which are tuned by the training algorithm. Once trained the neural network is deterministic, that is a test sequence applied to the trained network is repeatable. Viewed in this context the neural network behaves in a similar way to other more traditional control algorithms and thus the safety implications of using this method may be similar to existing systems if on-line updating can be avoided.

The Genetic Algorithm approach is however dependant on randomness for both operators. However the use of a transformation from the real number to the GA binary representation allows restrictions to be incorporated preventing impossible demands being made on the system. These can range from current and voltage limits to limiting speeds and accelerations. Thus the number of allowable controllers is restricted leading to a safer implementation and faster execution. Section 4 proposes a control system architecture that utilises both of these methods to design train control algorithms in a safe manner, and uses an on-board rail vehicle model for the GA to trial controllers providing an additional barrier between the learning process and the actual rail vehicle.

4 Proposed Control System

Various methods for controlled design exist, but the majority require accurate dynamic process models to manipulate and thus design control algorithms using linear algebra. In the absence of such a model, and anticipating highly non-linear variable behaviour, it is proposed that an AI approach may offer a solution by exploring in a restricted manner possible control schemes.

The proposed scheme is shown in figure 4.1. The control system uses two neural networks to model parts of the rail vehicle, a genetic algorithm to search for the best available adhesion control law, and a train controller to determine the demand torque for both the traction and braking equipment. In this example a non-linear control law is used based on a third neural

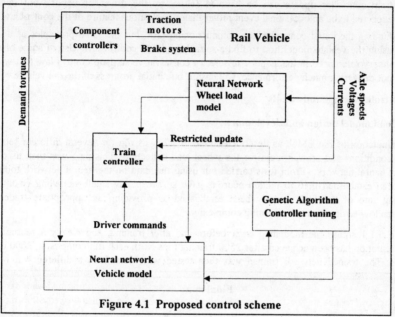

Figure 4.1 Proposed control scheme

network. This approach is proposed because direct measurement of relative slip is difficult to carry out reliably, repeatably and accurately, but by using some form of measured or estimated signal an improvement of existing techniques may be possible.

The NN Wheel-load model uses inputs of estimated applied torque (motor or brake) and rotor speeds to calculate the actual wheel load. Initial training is carried out on a simulation of a typical rail vehicle and, with careful design, a generalised model can be produced. This provides the train controller with an estimate of load which is directly related to the adhesion. The control law is selected to maximise this feedback signal, and to recognise when too great a torque is demanded and the vehicle enters the unstable slip/slide condition.

Selection of the control law using these techniques is potentially hazardous because a poorly set-up GA will probably try a few completely unsuitable controllers. Therefore a second neural model is trained to represent the entire rail vehicle, and the GA designs a controller to best control this model. Using a vehicle model ensures that only high performance, safe controllers are used on the actual vehicle via the "Restricted update". In practice it is envisaged that the control system would identify when a new controller was needed and request driver permission to update the train controller, or alternatively store possible new controllers for checking in more detail at the depot prior to implementation.

The main points to note are

- The train controller uses a non-linear algorithm and thus should be capable of dealing with a wide range of different adhesion conditions <u>without the need for any changes</u>
- Independent checking identifies any unexpected occurrences and tries to find a better controller using an on-vehicle model.
- If a better controller is found that will deal in addition with this new occurrence then the train controller update can be handled (either by the driver or depot), but this is expected to be an occasional event rather than a continuous feature of the control law.
- Finding the initial controller is a commissioning task, but the same principles apply of using the vehicle model first to find a suitable and safe control law. Use of some of the theory referenced in this paper can select a conservative start-up control law that while not delivering much performance, will ensure safe initial progress of the rail vehicle.

5 Controller design and results

5.1 Load model design and development

The simulation of the EMU, as described above in 3.2, was run for several different sets of track conditions and relevant data for the wheel load collected. This data was then used to train a neural network. From tests carried out using this data on the neural network tools it had been established that 10 hidden neurons gave an acceptably small modelling error. A learning rate of 0.1 had similarly been established as providing an appropriate trade-off between low error rate and computing complexity.

Figure 5.1.1 shows the modelling error before and after training, and clearly a reasonable representation has been achieved after 2500 training runs, with each run comprising 2000 data points. The trained network trained was then tested with data from a different run with

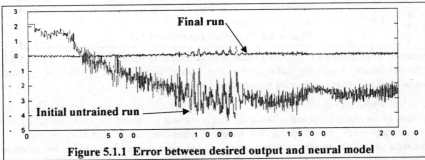

Figure 5.1.1 Error between desired output and neural model

changed gradient and track curvature parameters. Figure 5.2.2 shows this with the same axes scaling as Figure 5.1.1. Clearly despite the change in environment the neural model has accurately mapped the underlying non-linear process, which includes a adhesion model generic to the particular rail vehicle and expected range of adhesion conditions.

5.2 Rail vehicle neural model

The rail vehicle model was developed in a similar way to the load model in Section 5.1, although the application is rather simpler. Clearly a neural model of a simulation is only intended to demonstrate the ability of the network to model this complex system. However the end result is a relatively small computationally efficient network capable of running as part of a

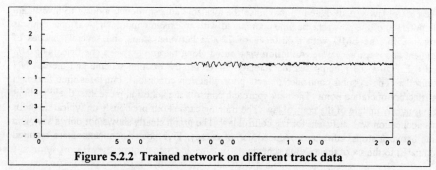

Figure 5.2.2 Trained network on different track data

control algorithm. If the existing simulation were used then to run in real time it would require a very significant amount of processing power, and probably would be simplified by linearisation, losing accuracy in the process.

5.3 Controller design

It has been shown that neural models can be obtained for the wheelset loading regime and overall train. A GA is used to select a control law to maximise the available adhesion. This requires some repeated runs of the two neural models to find the maximum of the load model, the point of maximum adhesion.

The GA population selected was a binary encoding of the weights of a third neural network operating as the train controller. Each population comprised 80 members, with random selection using the stochastic remainder method with no replacement (an enhanced version of the more traditional `roulette wheel' method). The probability of crossover was set at 0.75 and mutation was set at 0.01. The "chromosomes" were over 60 bits long so multi-point crossover was used where each chromosome is split into sub-chromasomes and each sub-portion crosses over with its equivalent. This significantly speeds up the search process.

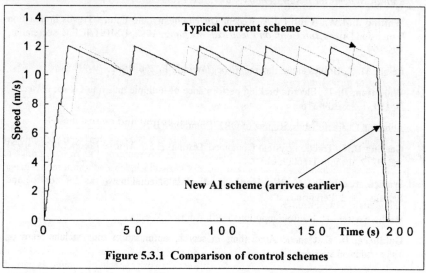

Figure 5.3.1 Comparison of control schemes

The control law was designed to maximise the available adhesion continuously to investigate the potential of the new scheme when compared with an existing approach. A sample run was prepared for the EMU with a distance of 2000m between stops and several changes of gradient and track curvature. Adhesion was varied along the run between coefficients of 0.03 and 0.2. Figure 5.3.1 compares the velocity curves for the existing scheme and the GA/NN controller. The existing controller detects poor adhesion conditions, but loses time finding a satisfactory operating point. The new approach controls the adhesion more effectively <u>without requiring any update of the control law.</u> The training carried out previously on typical track for this application was sufficient for the control law. The graph clearly shows not only a saving in time between stops, but also a reduced use of energy. Five power applications are required compared to the six of the existing scheme.

6 Discussion and Conclusions

This work has demonstrated the potential of applying sophisticated techniques using AI to the problem of adhesion. Journey times could be either reduced, or at least more reliably timetabled and energy consumption could also be reduced. However this technology represents a big step for the industry with regard to safety, overall performance and acceptance, and more investigation is required using models and test beds before use on passenger vehicles. Additionally the benefits of such a scheme assume that adhesion is variable and poor at some point.

7 Acknowledgements

The Authors would like to thank the management of European Gas Turbines and GEC Alsthom Traction for their permission to publish this work.

8 References

1. **Muller B.,** Applications of Modern Power Semiconductors, *IEE Int. Conf. Electric Railway Systems for a New Century,* 1987, IEE Press, 35-37

2. **Whiting J.M.W., Taufiq J.A.** Recent Developments and Future Trends in Traction Propulsion Drives, *Int. Conf. on Electrical Machines,* 1992, UMIST/ICEM Secretariat, 195-199

3. **Leigh, M. J.,** Brakes along the line, *Proc. IMechE,* F, **206**, 1992, 79-91

4. **Wilkinson, D. T.,** Electric braking performance of multiple unit trains, *Proc. IMechE,* D, **199**, 1985, 309-316

5. Adhesion Characteristics, Report of ORE Committee B34, 1992

6. **Geering D.R.,** Traction System Combined Testing, *R.I.A. Motive Power Course,* 1994 and GEC Alsthom Traction Ltd.

7. **Warick, K., Irwin, G. W., Hunt, K. J.** (Eds) Neural networks for control and systems, 1992, Peregrinus

8. **Hecht-Nielsen, R.** Neurocomputing, 1990, Addison-Wesley

9. **Goldberg, D. E.** Genetic Algorithms in search, optimisation and machine learning, 1989, Addison Wesley

C514/009/96

Method for flexual vibration damping of rolling stock carbody

Y SUZUKI MS, MJSME, **K AKUTSU, E MAEBASHI,** and **M SASAKURA**
Railway Technical Research Institute, Japan
S CHONAN PhD, MJSME
Tohoku University, Japan

A new method for enhancing the riding comfort by reducing vertical flexural vibration of a carbody is studied in this paper. In order to dissipate energy of flexural vibration while minimizing the weight to be added, viscoelastic layers and constraint layers are pasted partially on the outside sheathing of the carbody. Riding comfort regarding the vertical track irregularities is analyzed theoretically by assuming that the carbody is a partially multi-layered beam. As results of the analysis and the field tests on Shinkansen test car, it is concluded that this method can be a useful measure for enhancing the riding comfort of a lightweight carbody.

NOTATION

b	width of damping layer
h_v	thickness of viscoelastic layer
G_v	shear modulus of viscoelastic layer
M	bending moment of base beam
Q	shearing force of base beam
T	transfer matrix
U	maximum strain energy
u_c	displacement of constraint layer in x-direction
W	displacement of base beam in y-direction
X	state vector of base beam
η	loss factor
Φ	rotation angle of viscoelastic layer
Ψ	rotation angle of base beam

Subscript

c	constraint layer
e	longitudinal strain of constraint layer
f	bending of constraint layer
F	bending of base beam
F_j $(j=0,..,5)$	transfer matrix for each part of base beam
P_j $(j=1,2)$	transfer matrix for each support point of air spring

s shear deformation
v viscoelastic layer

1 INTRODUCTION

Speed-up of railway system needs light-weight vehicles. Because of lower rigidity of the carbody shell caused by its weight reduction, flexural vibration of carbody tends to occur, and it leads to poor riding comfort. The flexural vibration of carbody is considered to be a bending vibration of a beam and its first bending mode is the most influential to the riding comfort [1]. In order to suppress the first mode some ideas have been proposed, i.e., to utilize an underfloor equipment as a dynamic damper [2] ; to control the suspension actively [3] ; to choose its flexural rigidity such that the natural frequency coincides with the frequency on which two trucks are excited by vertical track irregularities out of phase each other [4]. But as a countermeasure for improving the riding comfort, less additional weight is needed. Therefore, a method for increasing the damping characteristics of the carbody would be effective.
    In this paper, a method for reducing the vertical flexural vibration of a carbody and increasing its damping characteristics is studied. Viscoelastic layers and constraint layers (these layers are hereinafter called "damping layers") are pasted partially on the outside sheathing of the carbody. Their effectiveness investigated theoretically. The results on the field test are also presented.
  From these results, it is concluded that this method can be a useful measure to get better riding comfort on a lightweight carbody.

2 METHOD FOR FLEXURAL VIBRATION DAMPING

Pasting a high modulus constraint layer on a viscoelastic layer is a general way of gaining large energy dissipation caused by shear deformation of viscoelastic layer. This concept is applied to an actual carbody as shown in Fig.1. Damping layers are pasted at places far from the neutral axis of bending such as side sills and cant rails. In application of this method to an actual carbody, it is important to determine the optimum length and choose the thickness, length and material properties of the damping layer so that the layers produce the maximum damping.
    In this paper, a method for studying the bending vibration of a partially multi-layered beam is presented first. Then, by using this analytical method, the flexural vibration of a carbody excited by vertical track irregularities is analyzed and the relationships between the material characteristics, the dimensions of damping layers and the riding comfort are investigated. Finally, results on application of this method are presented.

3 FORMULATION OF PROBLEM

3.1 Geometry of problem

Fig.2 shows a carbody whose center part has damping layers. The center part is composed of 5 layers; two viscoelastic layers, two constraint layers and one base beam (the carbody). The base beam is represented by 5 parts which are connected together at 4 joining points. The base beam is supported by two air springs that are mounted on the front and rear truck frames.

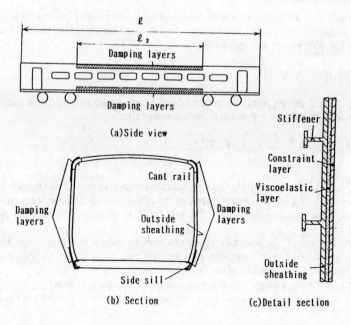

Fig. 1 Application to carbody of method for reduction of flexural vibration

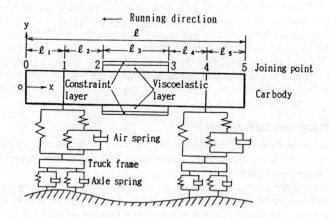

Fig. 2 Geometry of problem and co-ordinates

3.2 Governing equation

The Bernoulli-Euler beam theory is introduced and the transfer matrix method [5] is applied to study the dynamics of the elastically supported beam based on the assumption of steady state vibration. The state vector X is defined as

$$X = \{W \ \Psi \ M \ Q \ 1\}^{\mathrm{T}}. \tag{1}$$

As shown in Fig.2, joining points are numbered 1 to 5 from the left end. The state vectors of the base beam at both ends are connected in the following form,

$$
\begin{aligned}
X_5^{\mathrm{L}} &= T_{F5} \, T_{P2} \, T_{F4} \, T_{F3} \, T_{F2} \, T_{P1} \, T_{F1} X_0^{\mathrm{R}} \\
&= T \, X_0^{\mathrm{R}},
\end{aligned} \tag{2}
$$

where X_0^{R} and X_5^{L} are respectively the right side state vector at point 0 and the left side state vector at point 5. T_{F3} is the transfer matrix of the center part of the base beam which has a higher loss factor than the other parts. The loss factor is determined through the procedure given in 3.3.

With the use of the boundary conditions at both ends of the base beam $M=Q= 0$, equation (2) is solved and the state vectors at the both ends of the base beam are obtained, for example, the vibrational response of the carbody.

In this study, the riding comfort level is calculated in order to evaluate the effectiveness of this new method. The procedure for calculation is as shown in Fig.3.

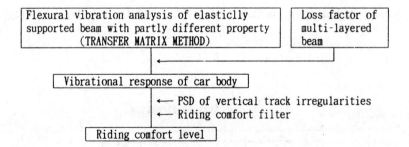

Fig. 3   Procedure of analysis

3.2 Loss factor on multi-layered part

The loss factor on the multi-layered part is obtained based on the following assumptions :
1) Only the first bending mode of base beam is considered and its mode shape is approximated by a sinusoidal function.
2) The mode shape is not affected by the material characteristics of the damping layers.
3) Deflections in each layer are the same and there are no slips between these layers.
4) Since Young's modulus of the viscoelastic layer is very low compared with that of the base beam, the effects of the layer's axial force and bending stress are neglected.

Multi-layered part is assumed to be symmetric with respect to the center of the base beam.

   © IMechE 1996 C514/009

The boundary conditions on the constraint layers are

$$u_c = 0 \qquad \text{at the center and} \tag{3}$$

$$du_c/dx = 0 \qquad \text{at the end.} \tag{4}$$

The shear strain of the viscoelastic layer can be obtained by considering the equilibrium of forces in the constraint layers in x-direction and the boundary conditions given above. The shear strain energy of the viscoelastic layer is

$$U_s = 4 G_v h_v b \int_0^{b/2} \Phi^2 dx, \tag{5}$$

where the origin of co-ordinate frame is set at the center of the base beam. The loss factor of multi-layered part η_3 is given by the ratio of the dissipative energy to the maximum total strain energy in one vibration cycle, which is as shown below.

$$\eta_3 = \frac{\eta_v U_s + \eta_c (U_e + U_f) + \eta_b U_F}{U_s + U_e + U_f + U_F}. \tag{6}$$

The loss factor η_3 is introduced in T_{F3} of equation (2).

4   NUMERICAL RESULTS

The following numerical results are for the Shinkansen electric car. The physical parameters used are listed in Table 1. The properties of suspension system and the power spectral density of the track irregularities are as the ones presented in Reference [1].

For convenience, it is assumed that the riding quality of a vehicle can be evaluated by the riding comfort level at the center of the carbody. Figs.4 (a) and (b) show the riding comfort level at the center of the carbody running at the speeds of 220km/h and 350km/h, respectively. It is noted that the mass of carbody and truck, the rigidity of carbody and the properties of suspension used in the case of 350km/h are half those of ones in the case of 220km/h because the higher speed railway system needs the lighter vehicle. It is observed that there exists an optimum length of damping layers which produces the maximum improvement of riding quality. The optimum length depends on the characteristics of the layers. It is noted that this method can improve the riding comfort level by 5 dB maximum for both 220km/h and 350km/h. The weight of the damping layers having 40% length of carbody is less than 100kg because of the constraint layers made up of CFRP.

5   APPLICATION TO ACTUAL CAR

5.1   Test procedure

In order to confirm the effectiveness of the present method for reducing the vertical flexural vibration of the carbody, it was applied to the test on an actual Shinkansen test car. Following

Table 1 Physical parameters

Carbody		
Length	: ℓ (m)	24.5
	: ℓ_1(m)	3.5
	: ℓ_s(m)	3.5
Height	: (m)	2.5
Loss factor	: η_b	0.022
Mass	(kg)	3.6×10^4
Flexural rigidity	(GNm^2)	1.71
Truck mass	(kg)	5.56×10^3
Viscoelastic layer		
Width	: b (m)	0.6
Thickness	: 2h$_v$(m)	2.0×10^{-3}
Shear modulus	: G$_v$ (N/m^2)	5.0×10^5-1.0×10^7
Loss factor	: η_v	0.4
Constraint layer		
Width	: b (m)	0.6
Thickness	: 2h$_c$(m)	4.0×10^{-3}
Young's modulus	: E$_c$ (N/m^2)	3.5×10^{11}(CFRP)
Loss factor	: η_c	0.002

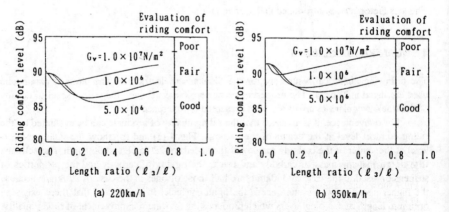

(a) 220km/h

(b) 350km/h

Fig. 4 Riding comfort level at carbody center (calculated)

tests were carried out.

(1) Stationary vibration test : The carbody was excited at the floor center using a rotative-imbalance-mass type vibrator.

  (a) Measurement of the stress in carbody shell under flexural vibration to determine the places on which the damping layers should be pasted.

  (b) Measurement of the loss factor of carbody without and with damping layers under flexural vibration.

(2) Running test : Measurement of the riding comfort level.

5.2 Physical parameters of damping layer

Damping rubber sheet of 2 mm thickness was used as the viscoelastic layer. The constraint layer was 5mm thick CFRP whose Young's modulus is 5.0 $\times 10^{11}$ N/m2. The width of the damping layer was 1.02m in total, which was divided into 8 parts as shown in Fig.5. The length was 2.5m, which is about 10% of the carbody length. The length was determined based on the results of calculation for the temperature and the damping layer property at test. The damping layers were pasted on the inside surface of the roof because the maximum longitudinal strain appears at the roof as shown in Fig.6.

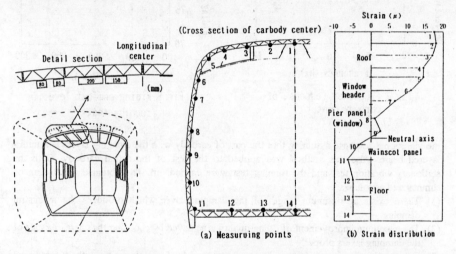

(a) Measuruing points   (b) Strain distribution

Fig. 5 Location of damping layers

Fig. 6 Longitudinal strain distribution of outside sheathing

5.3 Test results

Fig.7 shows the frequency response of the carbody center without and with the damping layers in the stationary vibration test. It is seen that the loss factor of carbody was increased 34% from 0.064 to 0.086 by pasting the damping layers. Fig.8 shows the riding comfort level at the floor center obtained for the running test. The riding comfort level at the same track position is compared for both cases of without and with the damping layers. It is observed that at 300km/h the riding comfort level was improved by 2-3dB by the damping layers.

6 CONCLUSIONS

A method for reducing the flexural vibration has been studied in this paper. In the method, in order to dissipate the energy of flexural vibration while minimizing the weight to be added, the viscoelastic layers and the constraint layers are pasted partially on the outside sheathing of the carbody. The effectiveness of this method was investigated theoretically through the analysis of

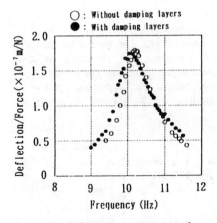

Fig. 7 Frequency response of
carbody center

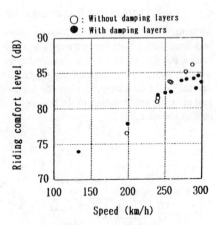

Fig. 8 Riding comfort level in
running test

the vibrational response assuming that the part of carbody with the damping layers is a multi-layered beam. Also this method was applied to the test of the actual carbody. Both the stationary vibration test and the running test were carried out. The obtained results can be summarized as follows.

(1) There exists an optimum length of the damping layer which produces the maximum damping.
(2) The maximum improvement of riding quality is achieved by choosing the characteristics of the damping layers properly .
(3) This method can improve the riding comfort level by 5 dB maximum with an additional weight of less than 100kg.
(4) The riding comfort level was improved by 2-3dB at 300km/h for the actual Shinkansen test car.

Those results suggest the possibility of improving the riding comfort of a light-weight railway vehicle by the present method proposed.

ACKNOWLEDGMENTS

The authors would like to thank EAST JAPAN RAILWAY COMPANY, which contributed to the field tests. Sincere thanks are also to Mitsubishi Chemical Corporation and THE YOKOHAMA RUBBER CO., LTD. for their supporting this study with technical information on the damping layers.

REFERENCES

(1) Y.Suzuki, K.Akutsu. Theoretical Analysis of Flexural Vibration of Car Body, Quarterly Report of Railway Technical Research Institute, 1990, Vol.31, No.1, 42.
(2) R.Ishikawa, Y.Sato. Decrease of Vehicle Body Bending Vibration by Dynamic Damper,

Proceedings of the Japan Society of Mechanical Engineers, 1991, No.910-17 Vol.c, 531.

(3) M.Nagai, Y.Sawada. Active Suspension Control for the Flexible Structure of an Elastic Vehicle Body, Transaction of the Japan Society of Mechanical Engineers(C), 1987, Vol.53, No.492, 1750.

(4) K.Tanifuji, K.Nagai, K.Nagaya. Vehicle Body-Bending Vibration of a Bogie Car Running at High Speed, Transaction of the Japan Society of Mechanical Engineers (C), 1990, Vol.56, No.529, 2327.

(5) E.C.Pestel, F.A.Leckie. Matrix Method in Elastomechanics, 1963, McGraw-Hill.

C514/019/96

Improvement of curving performance of steerable bogie in urban railway

A MATSUMOTO MJSME, MIEEJ, MJREA and **Y SATO** MJSCE
Ministry of Transport, Tokyo, Japan
M TANIMOTO and **Y OKA**
Sumitomo Metal Technology Inc., Japan

To consider the curving performance of trains, it is very important to grasp the practical attitudes of bogies against track. By using new methods, we successfully measured the relative displacement between rail and wheel of running trains as well as contact forces and vibrations. According to the measured results we found that attack angle of leading wheelset and shortage of lateral displacement of trailing wheelset are harmful. Methods to improve those problems were taken, and the improved curving performance in commercial line run is also shown in this paper.

1. INTRODUCTION

To improve the curving performance of bogies, it is necessary to reduce attack angle and undesirable creepage between wheel and rail. The reduction of them leads not only high velocity in curved section but also prevention of abnormal wear such as rail corrugation.

To consider the curving performance, it is very important to grasp the practical attitudes of bogies against track, but very few papers have made them clear precisely because of the difficulty of measurement(1). By using new methods, we successfully measured attack angles and wheel lateral displacement of running trains against rails as well as contact forces and vibrations. In this paper we show the measured results of curving characteristics of bogies, for example, attack angle, lateral displacement, contact force, etc. These data will give useful information for bogie and track design.

According to these results, we found that attack angle of leading wheelset and shortage of lateral displacement of trailing wheelset are harmful. After numerical analysis we concluded that sufficient difference of wheel rolling radius between inside and outside is most important for improving the curving performance of bogie. In this paper, the results of analysis are shown about the relationship between profiles of wheel-tread/ rail-head and getable value of wheel rolling radius difference. After these analysis we gained the methods of improvement, for example, asymmetrical grinding of inside rail head, lubrication of the contact surface, etc. The measured results of improved curving performance in commercial line experiments after the countermeasures were carried out are also shown at the end of this paper.

2. MEASUREMENT OF BOGIE'S BEHAVIOR IN CURVING

2.1 Methods for Measuring the Interface Characteristics between Rail and Wheel

2.1.1 Relative Displacement
In order to fix the curving characteristics of the bogie in commercial line run, the relative

displacement and angle among rail, wheelset, bogie frame and car body are measured. Fig 1 shows the measuring principles and sensor positions(2). The method used in measuring the relative displacement between rail and wheel is newly developed by using remote sensors mounted on axle box. By using the new method, the relative displacement between rail and wheel on commercial line run is successfully measured. From the output of remote sensors, the lateral displacement y and attack angle ϕ of wheel can be calculated by using following equations:

$$y = (s_1 + s_2) / 2$$
$$\phi = (s_1 - s_2) / L$$

Here, s_1 and s_2 are the out put of remote sensors mounted on the same axle box, L is the span of the sensors.

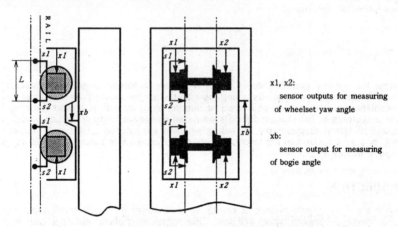

x1, x2:
   sensor outputs for measuring
   of wheelset yaw angle

xb:
   sensor output for measuring
   of bogie angle

Fig 1 Measuring principle of relative displacement

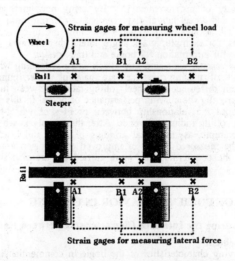

Fig 2 Measuring principle of contact forces

2.1.2 Contact Forces
If wheel load and lateral force of each wheelset can be measured in a continuous section on wayside, it will be very useful in grasping the characteristics of running train. This is realized by improving the general method used in detecting the rail stress of one section of track. The principle of this method is shown in Fig 2.

In the general method, the strain gage pair for measuring the sheering strain of rail usually has a span of approximately 200mm. By extending the span of gage pair as far as possible to the interval of sleepers (A1-A2), the wheel load and lateral force within the gage span can be measured without concerning the position of wheel along the rail. The wheel load or the lateral force will be detected as a change of strain proportional to them. Further more, we use another pair of strain gages over a sleeper (B1-B2) to measure the forces as wheel passing over the sleeper. The outputs of strain gages 'between' and 'over' the sleeper have a cross over section. The measured waveform from both pair of gages are connected into a continuous waveform by using a digital computer based on load calibration data at each pair of gages. Theoretically, if strain gage pairs were applied along the rail one after another, wheel load and lateral force of all wheels passing the measuring section would be measured continuously even if the length of the section was long.

Because the sensitivity of the strain gage pair applied between sleepers is different from that of the gage pair over the sleeper, both sensitivities are arranged by multiplying the calibration coefficient to the original waveform each other before the waveform are connected. As for lateral force, since the sensitivity along the rail changes, the output is multiplied by calibrating function before the waveforms are connected (3).

2.2 Measured Curving Performance of Bogie

2.2.1 Relative Displacement Between Rail and Wheel
From the measured relative displacement between rail and wheel by remote sensors, the relative lateral displacement and attack angle of wheelset are calculated. Fig 3 shows the calculated results. In the figure, the plus direction of horizontal axis means the car running forward (the measuring bogie located as leading bogie in the car, and the curve turns to the right direction), the minus direction means the car running backward.

From the calculated results shown in Fig 3(a) it can be found that although the wheelset moves toward the outside of track in all cases, the displacement from track center changes greatly according to the wheelset location in the bogie. That is, compared with the small displacement of trailing wheelset, the displacement of leading wheelset is very large. This tendency is more remarkable in leading bogie than in the trailing bogie.

In the leading bogie, the maximum lateral displacement of leading wheelset is about 17.5mm. From the simulation result mentioning bellow, it can be seen that the relative lateral displacement in this case is the limit for flange contact. But in the same bogie, the trailing wheelset locates nearly in neutral position with no lateral displacement absolutely.

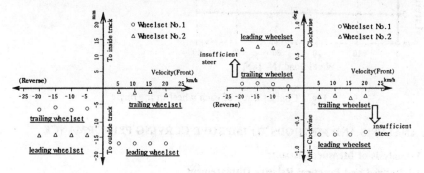

(a) Measured wheelset lateral displacement   (b) Measured wheelset attack angle

Fig 3   Measured bogie attitudes on commercial line

This means the trailing wheelset locates in the state of 'insufficiency of wheel rolling radius difference'. The train speed has almost no influence on the lateral displacement of leading wheelset for the reason of flange to contact the rail. But the lateral displacement of other axles changes with car speed slightly. This is considered for the influence of centrifugal force.

From the calculated results of attack angle shown in Fig 3(b) it can be found, the attack angles of trailing wheelset of both leading and trailing bogie are nearly to zero. This means the trailing wheelset yaws almost in radial direction. But the leading wheelset has a great attack angle against rail. This means the leading wheelset locating in insufficient steer state. In leading bogie, the attack angle of leading wheelset is about 0.75 degree. This is slightly larger than that of wheelset in trailing bogie.

Apart of relative displacement between rail and wheelset, the bogie angle between bogie frame and car body is also measured directly. According to the measured results the bogie angles of leading and trailing bogie are 3.0 and 4.5 degree respectively, and did not change with train speed and contact conditions such as rail lubrication etc (2). The ideal bogie angle at the test curve is 3.75 degree. This means the bogies have a insufficient steer angle of 0.75 degree at either direction.

2.2.2 Rail/Wheel Contact Forces
The measured results of wheel load and lateral force at four wheels of a bogie passing through the curve are shown in Fig 4. As for lateral force, plus shows the direction of force expanding the gauge. From the figure, we found following facts: As for wheel load, the largest value occurs at the outside wheel of leading wheelset, the second largest value appears at the inside wheel of trailing wheelset. These two wheels locate diagonal position in the bogie. This is considered for the reason of wheelset rolling caused by the lateral displacement of leading wheelset. As for lateral force, it is the greatest at the outside wheel of leading wheelset, and the second greatest at the inside wheel of leading wheelset. Since the flange of outside wheel keeps in contact with the rail, lateral force of outside wheel is considered due to the reaction force from the flange, and the lateral force of the inside wheel due to lateral creep force caused by large attack angle. In regard to the outside wheel of trailing wheelset, lateral force becomes almost zero or minus. This is considered for the cancelling of the lateral creep force and the component of gravity force by tread slope.

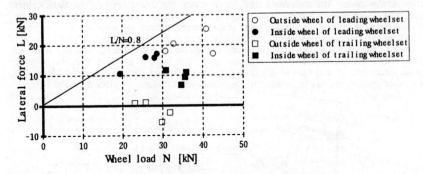

Fig 4 Contact force of each wheel in the bogie

3. ANALYSIS AND METHODS TO IMPROVE CURVING PERFORMANCE

3.1 Analysis of Measured Results

3.1.1 Desired and Practical Relative Displacement
In curving, it is ideal if the car takes a attitude as following. The geometrical center line of car body and bogie frame are tangential to the track center line, the wheelsets are radial to the curve and have a proper relative lateral displacement to the track for the wheelset to

negotiates the curve without insufficiency of rolling radius difference between inside and outside wheels.

Table 1 shows the desired and practically measured attitude of the car. Fig 5 shows the attitude of bogie and contact forces between rail and wheel according to the measured data. The desired values of attitude are calculated under the condition as the experimental car running through the test curve of 84m radii. The estimated wheel radius difference is calculated through contact simulation according to the measured relative lateral displacement between rail and wheel. The lateral displacement of 17.5mm is the limit value for wheel flange to contact the rail. In this state, the difference of wheel radius is about 3.6mm according the contact simulation.

If the rail gauge is G and wheel rolling radii is r, the required wheel radius difference $\varDelta r$ for the wheelset to negotiate the curve of radii R without longitudinal creep can be calculated from the equation below:

$$\varDelta r = r \times G / R$$

The ideal relative angles among wheelset, bogie frame and bogie can also be calculated from the structure parameters of the car and curve radii. By comparing the measured value with the desired value, the performance and the problem in curving can be discussed.

In the case of leading bogie, although the lateral displacement of the leading wheelset can reach to the range of flange contact, the shortage of wheel radius difference of 1.8mm still remain. Let one side of wheel rolls perfectly without creep and another side of wheel creeps to absorb the whole shortage of wheel radius difference, this shortage will give a longitudinal creep rate of 6 ‰ at most. But from the measured attack angle it can be calculated that the lateral creep rate of leading wheelset can reach to 13 ‰. Although the attack angle of tailing wheelset is very small, the lateral displacement of absolutely zero can lead to a longitudinal creep rate of 18 ‰ at most. This means the longitudinal creep force of trailing wheelset can reach to a considerable high level. As this creep force is counteracting to the curving force, abnormal flange force of leading wheelset can be raised for the curving force can only be supplied by the flange force of the leading wheelset in the same bogie.

In the case of trailing bogie, the lateral displacement and attack angle of leading wheelset are about 15mm and 0.6°, those lead to creep rate of about 10 ‰ in longitudinal and lateral directions. Although the lateral displacement of trailing wheelset is slightly larger in comparing with that of trailing wheelset in leading bogie, the longitudinal creep rate can still reach to 17 ‰, but 0° of attack angle is near to the desired attitude.

From the comparison between desired and measured attitude, it can be found that it is necessary to reduce the creep force caused by the abnormal creep rates in order to improve

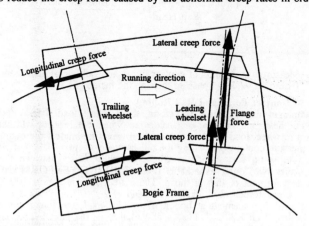

Fig 5 Bogie attitude and contact force on wheel in curving

curving performance of the car by improving the following points.
a) to reduce the attack angle of leading wheelset.
b) to reduce the insufficiency of wheel rolling radius difference of trailing wheelset.

Table 1 Comparison of bogie attitude between desired and practical condition

Parameter of curving performance	Desired attitude	Measured attitude			
		Leading bogie		Trailing bogie	
		Leading wheelset	Trailing wheelset	Leading wheelset	Trailing wheelset
Wheelset lateral displacement (mm)	17.5	17.5	1~4	15	6~7
Estimated wheel radius difference	(5.4)	3.6	0.01~0.02	2.2	0.3
Required wheel radius difference	5.4	5.4	5.4	5.4	5.4
Shortage of radius difference	(0)	1.8	5.4	3.2	5.1
Estimated longitud. creep rate(‰)	(0)	6	18	10	17
Wheelset attack angle vs. rail (deg)	0	▽0.75	▽0.2	▽0.6~0.7	0
Estimated lateral creep rate(‰)	0	13	3	10	0
Wheelset yaw angle vs. bogie (deg)	0.65	0	0.05~0.1	0	0.06~0.07
Bogie yaw angle vs. car body (deg)	3.75	3		4.5	
Difference from desired value	-	▽ 0.75		▽ 0.75	

▽ indicates "insufficient steer". ( ) may be realized by using improved profile.

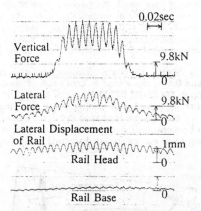

0.02sec

Vertical Force   9.8kN
                      0

Lateral Force   9.8kN
                      0

Lateral Displacement of Rail
Rail Head   1mm
                      0

Rail Base   0

Fig 6   Measured waveforms of contact forces and rail lateral displacement
(Leading wheelset passing through the inside rail)

3.1.2 Oscillation of Contact Force and Rail Rolling

Fig 6 is an example of the measured waveform about wheel load, lateral force, lateral displacement of rail head and rail flange (Inside rail when the leading wheelset passing through). In regard to wheel load and lateral force, the waveform shows the output of strain gages applied to measure continuously along the rail between sleepers. The part where wave of the wheel load turns up corresponds to when the wheel exists between the sleepers. Observing these waves, we found remarkable oscillating variation in the wheel load, the lateral force and displacement of the rail head. Those frequency is almost 100 Hz (Wheel load:30 ± 10kN, Lateral force: 16 ± 4kN, Rail head: 1.2 ± 0.3mm). As for rail lateral displacement, because displacement at bottom flange is small, we found that the rail is oscillating as so called "Rail Rolling". Phases of variation of the lateral force and the rail rolling are slightly behind that of the wheel load. The rail rolling oscillation appears far ahead of the wheel passing through, and we found that oscillation caused at the previous part of rail is conveyed to this part along the rail (4).

Though it is not shown in the figure, as for lateral force and rail rolling at outside rail, amplitude of such oscillation is smaller. The lateral force as well as oscillation of wheel load are very small at the trailing wheelset, and the rail rolling is tend to be rather suppressed by the trailing wheelset passing.

Considering these results and the measured results concerning to attitudes of bogies in curve section, we found that: 1) as for leading wheelset, attack angle is large, and oscillation of wheel load, lateral force and rail rolling occur at inside track where wheel and rail are free from restriction in lateral direction, compared with outside track where the flange of wheel contacts to rail by lateral displacement of wheelset; 2) at trailing wheelset, as attack angle is almost zero, there is no lateral creep and no occurrence of lateral force, then rail rolling would not happen.

3.2 Methods for Improvement of Curving Performance

In order to solve the problem in curving performance, it is important to reduce the longitudinal creepage of trailing wheelset by dismissing the insufficiency of rolling radius difference between inside and outside wheel. Following improvements are suggested for decreasing the creepage.

a. To expand the slack

As freedom of wheelset lateral displacement becomes larger, it is expectant to get more difference of wheel rolling radius in curve. But sufficient effect can not be expected for the geometrical limit of the slack expansion.

b. To grind the rail head asymmetrically

By shifting the position of contact point it is possible to get the desired difference of wheel rolling radius (more details are discussed later).

c. To use high conical tread profile of wheel

It is expectant to get more difference of wheel rolling radius. But the use of high conical tread profile may decrease the hunting stability.

d. To lubricate the contact surface

Lubrication by using solid lubricant or oil can reduce the tangential contact force. This may lead to the reducing of stick-slip vibration and abnormal wear (5). But oil lubrication may lead to problem of breaking performance.

Among the improvements, the effects of **a** and **c** can only be obtained in the case of the trailing wheelset laterally displacing toward the outside of track. For these reasons, to grind rail head asymmetrically is proposed as most effective method. From numerical simulation and stand test the effectiveness of using asymmetric rail head was confirmed.

Fig 7 shows the change of contact position with wheelset lateral displacement in the case to grind the head of inside rail asymmetrically. The head of asymmetrically ground rail declines to inside of track with a slope of 1/13. Table 2 shows the getable wheel rolling radius difference on asymmetrically groun rail. The table shows the sufficient wheel rolling radious can be got on asymmetrically ground rail if wheelset moves up to flange contact.

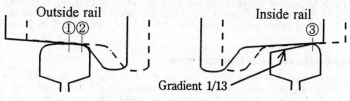

Contact point ①, ③ : wheelset at neutral position   (broken line)
Contact point ②, ③ : wheelset lateral moving toward
                   outside wheel flange contact   (solid line)

Fig 7 Lateral displacement of wheelset and change of contact position
(Inside rail with asymmetrical head profile)

Table 2 Rail head shape and getable wheel rolling radius difference

Rail head shape		Slack=16mm		Slack=20mm	
		Wheelset lateral displacement		Wheelset lateral displacement	
Inside rail	Outside rail	Neutral position	Flange corner	Neutral position	Flange corner
50kgN	50kgN	*①, ② 0mm (17.7 ‰)	3.42mm (6.5 ‰)	*③ 0mm (17.7 ‰)	3.83mm (5.1 ‰)
Asymmetry ground depth=1.0mm	50kgN	0.72 (13.5)	5.28 (0)	*④ 0.75 (15.2)	5.68 (0)
Asymmetry ground depth=1.6mm	50kgN	1.06 (14.2)	5.83 (0)	*⑤, ⑥ 1.23 (14.0)	6.46 (0)

The value inside ( ) is an approximate value of longitudinal creepage at 84m radii curve, and in this curve the sufficient wheel rolling radius is 5.4mm.
* ○ shows the test condition number.

4. EXPERIMENTS FOR PROVING THE METHODS

4.1 Decreasing the Average Contact Forces

As for the tests on improvement of curving performance, following 5 test conditions were selected.
① Present condition with slack of 16mm
② Lubricated wheel tread with solid lubricant
③ Expanding slack of 20mm (the following test was performed on the same slack)
④ Inside rail head was ground asymmetrically (depth was 1.0mm)
⑤ Inside rail head was ground asymmetrically (depth was 1.6mm)
⑥ Asymmetric rail head of the ⑤ and lubricated with oil.

To evaluate the curving performance of the bogie, average forces acting between rail and wheel were measured as shown in Fig 8(a), lateral forces acting on inside and outside wheel of leading wheelset were reduced by several improvements. The reduction of lateral force by using asymmetric rail head is caused by the reduction of longitudinal contact force of the trailing wheelset and the reduction of attack angle of leading wheelset. This is because the reduction of longitudinal contact force in tailing wheelset as shown in Fig 5 leads to the reduction of counteracting curving force directly. And the sufficient rolling radius difference of leading wheelset did not need the wheelset to lateral displacing farther in order to get rolling radius difference, this leads the reduction of attack angle and lateral creep force. In the oil lubricated rail condition lateral force was reduced by lower friction coefficient.

4.2 Reducing the Oscillation of Contact Forces

Fig 8(b),(c) shows the RMS value of the dynamical wheel load and lateral force on inside and outside wheel. By expanding the slack and grinding the rail head asymmetrically, the oscillation of wheel load and lateral force were decreased. But only in the case of using solid lubrication the oscillation were not decreased so much. Lateral force oscillation of inside wheel, which is thought to be most harmful for corrugation formation, was reduced up to half level of present condition by grinding the rail head asymmetrically. Moreover, to grind the head of rail asymmetrically also lead to the reduction of noise level of running train compared with that of the present.

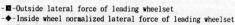

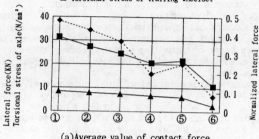

(a)Average value of contact force

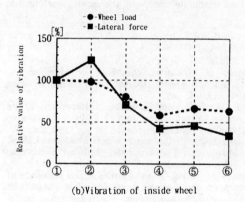

(b)Vibration of inside wheel

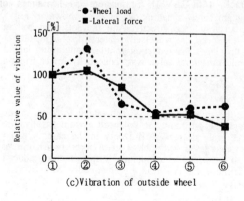

(c)Vibration of outside wheel

Test conditions

① Present condition
  (normal rail head profile. Slack=16mm)
② ①-Lubricated by solid lubricant
③ ①-Expanding slack(16→20mm)
④ Asymmetrical rail head
  (ground depth=1.0mm)
⑤ Asymmetrical rail head
  (ground depth=1.6mm)
⑥ ⑤-Oil lubrication

Fig 8 Effects of improvement methods

5. CONCLUSION

In order to catch the curving performance of bogie, new methods for measuring curving bogie characteristics were developed, such as 1) wheel lateral displacement and attack angle against rail and 2) contact forces between rail and wheel continuously from wayside. By using these methods, we successfully measured the practical curving performance of bogie on commercial line run, and grasped the characteristics of bogie, which are different distinctively between inside and outside wheel, and between leading and trailing wheelset.

Large attack angles of leading wheelset and shortage of lateral displacement of trailing wheelset are observed in every bogie of trains. These values produce extensively large lateral and longitudinal creepage respectively, and large creepage leads undesirable phenomena, such as, noise and abnormal wear.

Considering these measured results, it is thought that sufficient difference of wheel rolling radius between inside and outside is most important for improving curving performance of bogie. According to numerical analysis, such as contact simulation which can find the relationship between profiles of wheel-tread / rail-head and the getable value of wheel rolling radius difference, we gained the methods of improvement, for example, asymmetrical grinding of rail head, oil or solid lubrication of the contact surface.

After we carried out the experiment for validation of these countermeasures in commercial line, we found that asymmetrical grinding of rail head, oil lubrication of rail head and solid lubrication of wheel tread are useful for improving the curving performance of bogies.

References

(1) Elkins,J.A. and Eickhoff B.M. Advances in Nonlinear Wheel/Rail Force Prediction Methods and Their Validation, Transaction of ASME, Journal of Dynamic Systems, Measurement, and Conrol, 1982, Vol.104, 133-142
(2) Matsumoto,A., Sato,Y., Tanimoto,M., Kang Q. and Furuta,M. Observation of the Attitude of Bogie and Wheelsets in Sharply Curved Track, Jointed Railway Technology Symposium J-Rail '94(JSME), 1994, 97-100 (in Japanese)
(3) Sato,Y., Matsumoto,A., Tanimoto,M. and Kang Q. Continuos Wheel Load and Lateral Force by Expanded Wayside Load Measurement, 2nd Jointed Railway Technology Symposium J-Rail '95(IEEJ), 1995, 97-100 (in Japanese)
(4) Matsumoto,A., Sato,Y., Fuji,M., Tanimoto,M., Kang Q. and Furuta,M. The Influence of Bogie and Track Dynamical Behaviours on the Formation of Rail Corrugation (2nd Report), J-Rail '95 (IEEJ), 1995, 149-152, (in Japanese)
(5) Matsumoto,A., Sato,Y., Tanimoto,M. and Kang Q. Study on the Formation Mechanism of Rail Corrugation on Curved Track, 14th IAVSD Symposium on Dynamics of Vehicles, Michigan, 1995, 87-89

C514/039/96

Solving aerodynamic environmental problems arising from train speed-up

Y NOGUCHI MSc, **Y OKAMURA, K UCHIDA, T ISHIHARA** MEng, **S MASHIMO** MEng, and **M KAGEYAMA**
West Japan Railway Company, Japan

SYNOPSIS

This paper describes aerodynamic environmental problems arising from train speedup and about the R&D for solving such problems by West Japan Railway Company (JR-West).

JR-West conducted experimental runs of WIN350, wind tunnel measurements, and numerical simulations to solve the problems of the prospective 300 km/h operation.

1 INTRODUCTION

1.1 Speedup of Shinkansen

JR-West operates Sanyo-Shinknasen Line which is 554 km long between Shin-Osaka and Hakata. As the line is the major source of income for JR-West, it is substantial to maintain the competitiveness of Sanyo-Shinknasen Line against other means of high speed transportation such as airlines and expressways. Speed is a major factor for the competitiveness.

The 300N-series Shinkansen 'Nozomi' is the fastest train on Sanyo-Shinkansen Line since 1993; its current top speed is 270 km/h. JR-West is aiming at 300 km/h commercial operation.

1.2 Aerodynamic environmental problems

Aerodynamic environmental problems must be solved for the speed increase. In 1992, JR-West started a project for developing faster commercial trains and, for that purpose, made an experimental train named 'WIN350' to solve the problems (see Fig. 1).

There are several kinds of aerodynamic environmental problems for high speed trains: wayside noise, pressure fluctuation during train passing by, and micro-pressure waves at tunnel exits.

1.2.1 Wayside noise

Wayside noise emitted from Shinkansen must meet the Japanese regulation set by Environment Agency. In populated regions, slow peak levels of noise (Lmax) should be no more than 75 dB(A) at 25 meters from a track center. Therefore, effective measures have to be taken against wayside noise to operate trains at 300 km/h.

Sound barriers are fundamental measures to reduce wayside noise, but they are costly and their effect is limited because high barriers obstruct views from passengers. Noise emission from sound sources on a train must be reduced.

1.2.2 Pressure fluctuation during train passing by

Pressure fluctuation can be a new problem at open track sections. The pressure fluctuation rattles windows on some houses built very close to a track when a high speed train is passing by.

As shown in Fig. 3, pressure distribution around a running train moves with the train and is observed as pressure fluctuation on the ground near the track [1].

A high wall which is approximately 4 meters in height can shield such pressure fluctuation but it is ineffective and too costly. Reducing the pressure fluctuation by optimizing a train shape is necessary.

1.2.3 Micro-pressure wave at tunnel exits

It is very important for JR-West to solve environmental problems concerning tunnels because the summed length of the all tunnels on Sanyo-Shinkansen Line is approximately as long as half of the entire length of the line.   As depicted in Fig. 4, a micro-pressure wave is made at a tunnel exit by a high speed train entering the tunnel entrance and accompanies impulsive noise radiation at the exit [2].

Tunnel entrance hoods are the most effective measure to reduce the micro-pressure waves but such hoods are costly. Reduction of the pressure level by optimizing a train shape is necessary for higher speed operation.

2   DEVELOPMENT OF CURRENT COLLECTION SYSTEM

JR-West developed a low noise current collector and its relating technologies on current collection systems to reduce aerodynamic noise from the current collectors and solve other environmental problems.

A noise wave form, measured with a microphone array, of a 300N-series Shinkansen train is shown in Fig. 2. The train has several large noise sources. Their positions are indicated by the peaks of the wave form at the car nose, the car coupling areas, the current collectors, and the train tail. The noise emission from the current collectors is dominant among those noise sources.

The noise from a current collector consists of spark noise, aerodynamic noise, and friction noise. A high voltage bus cable which electrically connects two pantographs reduces sparks. The friction noise from wearing strips is negligibly low compared with the other noise. The aerodynamic noise from current collectors becomes largest at high speed.

2.1   Noise reduction by pantograph covers

A conventional pantograph is composed of wearing strips, bows, horns, and frame pipes as shown in Fig. 5a. The rectangular bows have stable aerodynamic lift but emit large aerodynamic noise. The frame also emits significant noise.

To reduce the aerodynamic noise from a conventional pantograph, a pantograph cover has been used. The front bulkheads of the pantograph cover decrease airflow velocity around the pantograph and reduce aerodynamic noise emission from the pantograph. The side bulkheads shield noise from the pantograph.

Fig. 5b shows a long-slope type pantograph cover developed by JR-West. WIN350 equipped with a conventional pantograph and the long-slope type pantograph cover cleared the regulation of 75 dB(A) at 300 km/h. It is the lowest record of wayside noise of WIN350 using a conventional pantograph [3].

The authors consider the record will not be broken if a conventional pantograph is used. Moreover, the conventional pantograph requires a pantograph cover but the pantograph cover is heavy and has large aerodynamic drag. Therefore JR-West developed a low noise wing-shaped current collector.

2.2 Wing-shaped current collector

The wing-shaped current collector has a wing-shaped collector head supported by a column as shown in Fig. 6a.

On the early stage of its development, the wing-shaped current collector had large and unstable aerodynamic lift. Large lift force of a collector head would make a contact wire move up beyond the upper limit and damage the catenary. Therefore, the aerodynamic lift force should be kept within appropriate levels.

It was discovered that the supporting column effected the airflow around the collector head and increased the lift. Making the aerodynamic lift insensitive to the attack angle of the flow to the collector head also decrease the aerodynamic lift. After repeated wind tunnel tests and field measurement by JR-West, the optimal shape of the collector head which has moderate aerodynamic lift was developed [4].

The wing-shaped current collector radically reduces noise emission. The peak levels of the wing-shaped current collector and the conventional pantograph measured with a microphone array are compared in Fig. 7. The wing-shaped current collector is used without the front bulkheads of the pantograph cover. The wing-shaped current collector emits lower noise by approximately 5 dB(A).

2.3 Insulator cover

The low noise wing-shaped current collector changes the design concepts of pantograph covers. The new concepts are as follows.

First, the pantograph cover does not need to decrease the velocity of the airflow around the wing-shaped current collector because the wing-shaped current collector is quiet by itself. The height of the front bulk head is lowered to the top of the insulators which support the collectors. So the new pantograph cover is called an insulator cover. The lower front bulkhead of the insulator cover does not disturb the air flow around the wing-shaped current collector.

Second, the noise emission from the insulator cover itself must be reduced. The insulator cover is streamlined and smooth-faced.

Third, the cross sectional area of its projection should be decreased to reduce the pressure fluctuation around the train.

Finally, the insulator cover should be as light as possible to reduce ground vibration caused by the train.

Following the above concepts, 1/20 models of streamlined insulator covers were designed and noise levels emitted from the models were measured in a low-noise wind tunnel to find the optimal shape. The measured noise levels are almost equally low among the models if the gradient of the front bulkheads is small enough.

The actual insulator covers were installed on WIN350 as shown in Fig. 6b and 6c. They are 1.44 square meters in cross-sectional area and 11.7 meters long. The cross-sectional area is smaller by 59% and the length is shorter by 48% than those of the long-slope pantograph cover.

The pressure wave form of running WIN350, measured at the point 7 meters away from the train, is shown in Fig. 8. The pressure fluctuation of the insulator cover is lower than that of a long-slope pantograph cover.

2.4   Further noise reduction of the wing-shaped current collector

The side bulkheads of the insulator covers on WIN350 is as high as the supporting insulators and does not shield noise from the wing-shaped current collector. There are two kinds of measures for further noise reduction of the current collection system.

The first measure is to reduce the noise emission from the wing-shaped current collector. The shapes of the components of the wing-shaped current collector need to be improved.

The second measure is to increase the shielding effect of the insulator cover.

This paper describes the first one.

2.4.1   Detecting noise sources of the wing-shaped current collector

Noise from the actual wing-shaped current collector was measured at the flow velocity of 300 km/h in a low-noise wind tunnel to detect noise sources of the collector.

The elliptic vertical column emitted the largest noise with peak frequency at 500 Hz. The horn was also a large noise source which had peaks at 800 Hz and 2 kHz.

Noise reduction measures were also tested in the wind tunnel [5].

2.4.2   Reduction of the noise from the column

Trip-wires and vortex generators were attached on the surface of the supporting column to eliminate vortices and such measures were tested to evaluate their effectiveness.

Attaching vortex generators in two rows as in Fig. 9 was the most effective way to suppress the peak at 500 Hz and reduce the overall noise level.

2.4.3   Reduction of the noise from the horns

Two shapes were compared. One shape had a circle cross section, and the other had an elliptic cross section in the ratio of 1:2. To suppress the peaks at 800 Hz and 2 kHz, 5 mm holes in diameter were made through the horns every 15 mm as shown in Fig. 10.

The elliptic horn with holes was selected based on the measurement.

2.4.4   The field measurement

The wing-shaped current collector with the improved column and horns were set on WIN350. The insulator covers were also used together. Wayside noise was measured with a microphone array.

As shown in Fig. 11, the peak levels at the current collector on WIN350 were decreased by 1.7 dB(A) at 300 km/h after the improvement.

The interruption rate of contact between the current collector and contact wires was satisfactory, too.

3   OPTIMIZING THE SHAPE OF THE TRAIN NOSE

To develop an optimized shape of a high speed train, many problems have to be taken into consideration, for example, wayside noise, pressure fluctuation, micro-pressure waves, aerodynamic drag, riding comfort, and pressure to other trains passing by. Moreover, space for train operators and the front bogie should be considered for practical application.

3.1   Micro-pressure wave at tunnel exits

The problem of micro-pressure wave at tunnel exits is the first thing to be solved for the operation of 500-series Shinkansen at 300km/h.

To reduce the micro-pressure wave, the followings are required.[2]
(1) Reducing the cross-sectional area.
(2) Equalizing the change of cross-sectional area along the train nose.
(3) Extending the train nose and reducing the change of cross-sectional area along the train nose.

The cross-section of 500-series Shinkansen is down-sized as possible, securing the transport capacity and the space for passengers. The cross-sectional area is 10.2(m2), where that of 300N-series Shinkansen is 11.4(m2).

The change of cross sectional area of 500-series Shinkansen is also equalized as possible. The space for train operators and the front bogie cannot be avoided.

Under the condition mentioned above, the authors need to extend the nose length of 500-series Shinkansen to 15 meters, more than twice as much as that of 300N-series.The nose of 500-series Shinkansen and the comparisson of cross-sectional area are shown in Fig.12.

Micro-pressure waves made by the 500-series Shinkansen were estimated by Railway Technical Research Institute for JR-West. The proposed nose shape is sufficiently able to suppress the increase of micro-pressure waves.

3.2   Pressure fluctuation during train passing by

The maximum peak level of the pressure fluctuation is made by the nose of a train. The pressure levels depend on the train speed, the nose shape, and the cross-sectional area of the train.

As shown in Fig. 13, the pressure peak levels at the nose of running trains on an elevated track were measured from a point horizontally 7 meters away from the train centers. The levels increase proportionally to the square of the train speed if the nose shape is identical.

To determine the nose shape of Shinkansen 500 series, 1/20 scaled models were made for wind tunnel measurement. A wind tunnel at Osaka University was used. The wind tunnel has a test section 1800 mm in width, 1800 mm in height, and 9500 mm in length. The maximum flow velocity is 20 m/s.

The 15 meters long 2-dimensional and 3-dimensional nose shapes are compared in this test. The nose of 300N-series Shinkansen is also tested.

The measured pressure coefficients Cp are shown in Fig. 14. As a result of the wind tunnel measurement, the level of pressure fluctuation made by the 500 series at 300 km/h is supposed to be decreased less than the level of the 300 series at 270 km/h.

The authors also comfirm that there is no differnce between the the pressure fluctuation of 2-dimentional nose shape and that of 3-dimentional nose shape, as far as the nose length is 15 meters .

4 CONCLUSION

As a conclusion, it is possible to solve the aerodynamic environmental problems at train speed up to 300 km/h.

The 500-series Shinkansen was made in January 1996. Based on the measurement by using WIN350 and the wind tunnels, it is estimated that the wayside noise, the pressure fluctuation, and the micro-pressure waves of the 500-series running at 300 km/h remain the same level of 300-series running at 270 km/h. This conclusion is confirmed by the actual test run of the 500-series.

5 ACKNOWLEDGMENT

The authors are grateful for the researchers at Railway Technical Research Institute, Prof. Miyake and Mr. Igarashi at Osaka University, Mr. Miyamura and Mr. Yajima, engineers at Showa Corporation and many others who cooperated in the speedup project.

6 REFERENCES

[1] KIKUCHI, K., et al., Numerical Analysis of Pressure Variation under Train Passage using Boundary Element Method (in Japanese) , The 4th Transportation and Logistics Conference, 1995, pp357-358 (The Japan Society of Mechanical Engineers).

[2] MAEDA, T., et al., Effect of Train Nose on Compressive Wave Generated by Train Entering Tunnel, STECH'93, 1993, Vol.2, pp.315-319.

[3] HIGASHI, A., et al., Aerodynamic Noise from Car Bodies and Pantograph of WIN350, STECH'93, 1993, Vol.2, pp59-64.

[4] FUJITA,Y., et al., The Reduction of Excessive Aerodynamic Lifting Force and the Performance on Current Collection at 300km/h with Low Noise Current Collectors, SCTECH'96, 1996, Submitted Papers.

[5] ASO,T., et al., Reduction of Current Collecting Noise by Wing-shaped Current Collector (in Japanese) , The 4th Transportation and Logistics Conference, 1995 ,pp.342-344 (The Japan Society of Mechanical Engineers).

Fig.1 JR-WEST's High Speed Experimental Train "WIN350 (6 cars)"

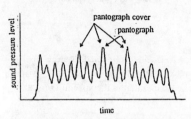

Fig.2 Wayside Noise of 300N Series Shinkansen (measured by ultra directional microphone array)

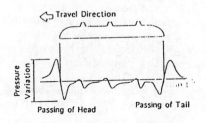

Fig.3 Pressure Fluctuation Induced By Trains [1]

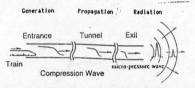

Fig.4 Compression Wave and Micro-pressure Wave [2]

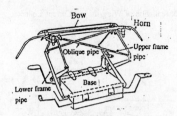

(a) Configuration of Conventional Pantograph

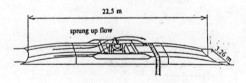

(b) Long-slope+H Type Pantograph Cover

Fig.5 Conventional Pantograph and Long-slope Pantograph Cover

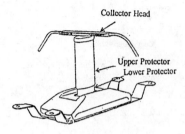

(a) Configulation of Wing-shaped Current Collector

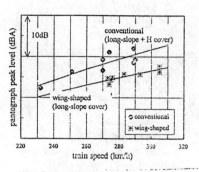

Fig.7 Comparison of Pantograph Noises (loaded in long-slope pantograph cover)

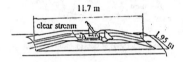

(b) Small Slope type Pantograph Cover (Insulator Cover)

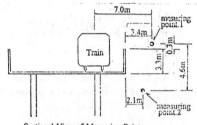

Sectional View of Measuring Point

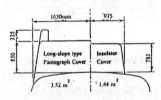

(c) Comparison of Pantograph Covers' Front View (loaded on WIN350)

Fig.6 Comparison of Pantograph Covers

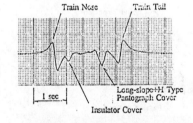

Fig.8 Wave Form of Pressure Fluctuation Measured at the Point 7meters from the Rail Center

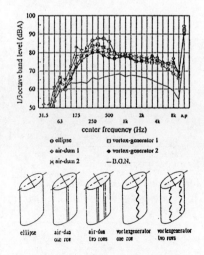

- ⊖ ellipse
- ◇ air-dum 1
- ✕ air-dum 2
- ⊟ vortex-generator 1
- ◆ vortex-generator 2
- — B.G.N.

ellipse | air-dum one row | air-dum two rows | vortexgenerator one row | vortexgenerator two rows

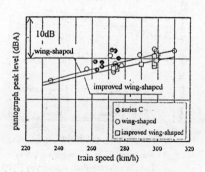

Fig.11 Effect of Noise Reduction by the Improvement of Wing-shaped Current Collector(measured by ultra directional microphone array)

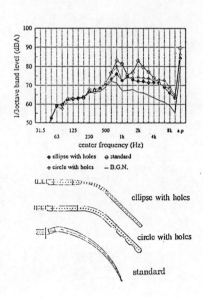

- ◇ ellipse with holes
- ◇ circle with holes
- ⊖ standard
- — B.G.N.

ellipse with holes

circle with holes

standard

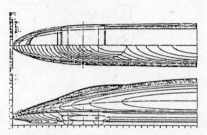

The nose of 500-series Shinkansen

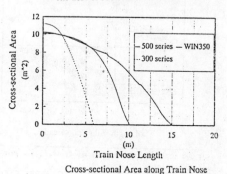

Train Nose Length

Cross-sectional Area along Train Nose

Fig.12 The nose of 500-series Shinkansen and Comparisson of cross-sectional area along Train nose.

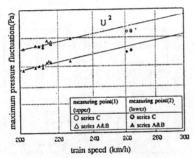

Fig.13 Maximum Pressure Fluctuation versus
Train Speed(Peak-Peak at the Train Nose)
(measured at the point 7 meters from
the Rail Center)

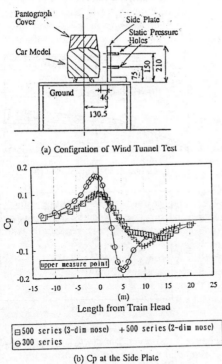

(a) Configration of Wind Tunnel Test

(b) Cp at the Side Plate

Fig.14 Pressure Coeficient Cp measured in
Wind Tunnel Test

C514/057/96

Problems of wear in railway track – identification and evaluation

D M LILLEY BSc, PhD, MICE, MIStructE and **J ROBINSON** BEng
Department of Civil Engineering, University of Newcastle, Newcastle-Upon-Tyne, UK

SYNOPSIS

Maintaining railway track in good condition is an essential feature of ensuring better journey times. Contact between wheels of railway vehicles and steel rails inevitably creates sound and vibration, and can be a nuisance in sensitive locations.

Sound and vibration generally increase with increasing train speed, and increase dramatically if track or wheels become worn or corrugated. Interdependent damage or wear is produced if either wheels or track are in poor condition.

This paper describes and demonstrates a low-cost method for identifying sections of track where remedial maintenance work is likely to be most cost-effective.

1 INTRODUCTION

A 'better journey time' may be the result of improvements to a railway system leading to either a reduction in the time taken to travel within the system, and/or a more comfortable journey for those in transit. In either case, the condition of the railway track is important.

Trains travelling over track identified as being in poor condition will be subject to speed restrictions, inevitably resulting in longer journey times. Increased levels of noise and jolting movement within carriages reduces passenger comfort, potentially leading to complaints and reduced numbers of passengers.

Rolling steel-on-steel contact between the wheels of a train and the rails allows frictional wear and 'impact' damage to occur on both sides of the contact. The rate of wear within each component is related to the level of exposure, the relative material properties and the smoothness of the contact surfaces.

Damage or wear within the running surface of a wheel or a rail can be expected to increase if the other contact surface is in poor condition. Mutual benefit is obtained by keeping both wheels and rails in good repair. If, however, the contact surfaces of the wheels are allowed to deteriorate without remedial measures at appropriate times, the condition of the running surfaces of the track can be expected to degrade. Similarly, the passing of trains over track in poor condition will lead to increased wear on the wheels and increased variation from their design profile.

The problem is exacerbated if rolling stock operated and maintained by different commercial organisations is allowed access to a common length of track. The interdependence of wear of surfaces in contact is shown in simplistic form in Figure 1. If one company operates rail traffic with wheels in poor condition over a period of time, the effect will gradually spread to the running surfaces of the rails, and then to the wheels of rolling stock operated by other companies.

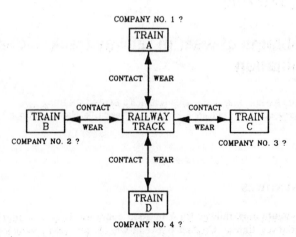

Figure 1 Inter-dependency of contact wear in track and rolling stock

Remedial work on wheels can be undertaken on a regular basis by machining worn surfaces and returning the profiles of the wheels to the intended shape. This can be achieved using modern equipment without removing the wheels from the train. The cost of this work increases with the amount and rate of wear on the wheels.

Responsibility for the condition of wheels of rolling stock and running surfaces of rail track is likely to become a contentious issue (with legal implications), particularly in situations where rolling stock and railway track are the responsibilities of different organisations.

2 TRACK MAINTENANCE

Track deterioration can be the result of several factors, eg. inadequacy of the original design of the track, breakdown of supporting ground strata beneath the track, formation of vertical corrugations along the running surfaces of the rails, etc., etc..

Funding for planned maintenance of railway track is likely to be allocated from limited budgets. Decisions regarding track priorities for expensive remedial work are sometimes made by engineers making track-side observations without sophisticated measurement techniques or equipment.

In an extensive railway system some sections of track can be expected to deteriorate at rates faster than others. This deterioration can take many forms, some of which quickly become apparent. For example, 'pumping' of mud and water onto the track by a passing train indicates local failure of the foundation below sleepers; modern signalling incorporating train detection systems can rapidly detect a cracked rail.

Less obvious are small deviations from the intended line and level of the track, but these may lead to unusual or excessive wear on the rails and the wheels of the rolling stock. Vertical undulations or corrugations in the running surface along the length of a rail occur in many railway systems. In general, the horizontal distance between two adjacent peaks (or troughs) in a corrugated rail varies from about 50 mm to 125 mm; in extreme cases the vertical height of the rail may vary as much as 1 mm between peaks. Research has shown that some of the causes of corrugations have been recognised (1); others remain to be identified.

In urban environments corrugated rails present a likely source of noise nuisance rather than jeopardising the safety or efficiency of the transport system. Remedial work in the form of grinding the running surface of the rails is usually undertaken. This removes the 'peaks' and attempts to return the profile of the top of the rail to its original shape. As rail grinding is an expensive process (costing typically about £6000-£7000 per kilometre of conventional two-rail track) it is important to identify sections of track with the worst corrugations as priorities for this form of track maintenance.

Assessment of corrugations is usually made visually or sometimes using a hand-held device to produce a visual record of the surface profile of the rail. Very short sections of track (up to 1 m) can be examined using this method of measurement, but it is time-consuming and is too expensive for determining the condition of the rail running surfaces throughout a substantial railway system.

3 PREVIOUS SOUND MEASUREMENTS

Sound levels generated by railway or other forms of transport have previously been measured using wayside microphones. Such measurements have been made adjacent to the track of Tyne and Wear Metro by the authors, in addition to those taken previously by others.

Tyne and Wear Metro is a light railway system forming an important part of the public transport infrastructure of the city and suburbs of Newcastle upon Tyne. The current system incorporates approximately 56 km of twin track, 43 km of which were previously operated by British Rail, approximately 8 km is new but similar construction, and a further 5 km are within new tunnel sections under the city centres of Newcastle and Gateshead (2). In the 1970s, two major bridges (Byker Viaduct and the Queen Elizabeth II Bridge) were constructed and Howdon Viaduct (an existing major structure) was strengthened. In addition, new stations were built and numerous others were refurbished. The system is passenger-orientated, and uses light trains powered by overhead electrification.

Sound levels close to a special test track were measured (3,4) during prototype testing and used to predict train-generated sound levels during normal Metro operation. After the system had been brought into operation, sound exposure levels at a particular site were measured (5) in order to examine the effect of different types of track construction. Further measurements were taken (6) as part of a social study into the impact of train-generated sound from the Metro. In each case, the instrumentation and test conditions were similar, comprising a microphone connected to a sound level meter taking measurements at trackside locations as trains went past. In some cases the microphone was connected on site to a high-quality tape recorder and the recorded signal analysed later to determine its frequency content.

Measurements have been made at trackside locations where sound levels have been perceived to be unusually high to provide information relating to the instantaneous and time-averaged (L_{EQ}) values of sound pressure level. This information is additional to the frequency content and site-specific attenuation of sound with increasing distance from the track. However, use of this technique provides no information about sound levels experienced by passengers, and considerable time and effort is usually required to determine sound levels at a single site.

4 CURRENT RESEARCH

An on-going programme of collaborative research is currently being undertaken by Tyne and Wear Metro and the University of Newcastle to find acceptable methods of reducing train-generated sound and vibration. One result from this work is a method of measuring sound levels within a train travelling throughout the entire railway system. Measurements are taken over a few hours within a single train, which then returns to normal operation at the end of the test.

A similar technique has been described previously (7) in which external microphones were fixed to a prototype vehicle so as to measure sound levels generated at the rail-wheel interface. The current programme of research with Tyne and Wear Metro is, however, is believed to be the first time that a fully operational system has been used in this form of test.

Sound levels at a particular point within the train environment are monitored and a means of assessing track quality is also provided. This is possible because train-generated noise and vibration are often present as a result of wear or component failure. Monitoring of sound levels within a railway system on a regular basis can be expected to indicate sections of track which wear at faster rates than other sections. Track in greatest need of maintenance is identified, providing the potential for maximum benefit from available finance.

The original sections of the Metro system have now been operating for approximately 17 years, making the development of this technique particularly timely.

5 IN-TRAIN MEASUREMENTS

Figure 2 illustrates a simple schematic diagram of the instrumentation for measuring and recording sound pressure levels, train speed, and synchronised video recording of the front and rear views from the train.

The equipment comprised a computer-controlled data logger, an existing system within the train that generated an electrical voltage proportional to train speed, a sound level meter, and two video cameras. The aim was to record auto-

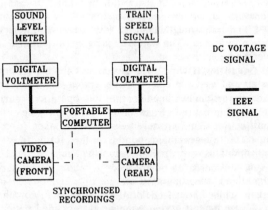

Figure 2 Schematic diagram of instrumentation

matically at one-second intervals values of train speed and interior sound pressure levels while the train travelled normally around the system. These values could then be related to precise locations along the track.

Values of A-weighted sound pressure level (SPL) were measured at 1-second intervals at a pre-determined position within a Metro train using a high-quality sound level meter. These values were recorded electronically using a d.c. voltage output signal, the magnitude of which varied in direct proportion to the value of SPL recorded by the meter.

Video cameras positioned inside the train gave a clear view of the track from the front and rear windows. The Tyne and Wear Metro system does not allow trains to follow a continuous route. Thus the use of a camera at each end of the train minimised delays when the direction of travel changed at each terminus. The date and time (to an accuracy of one second) in conventional numerical format were superimposed on the video images. Time signals from each camera were synchronised with that of the computer at the start of the test, thus allowing values of SPL and train speed to be related to positions along the track.

The influence of extraneous effects other than rail-generated sound was minimised by closing all opening windows on the train. The dedicated train was unladen except for the train driver and personnel involved in monitoring and controlling the recording equipment.

An increase in sound level occurred when another train passed on the adjacent track; subsequent checking of the video recordings enabled these events to be removed from the measured data before detailed analysis commenced. Measurements were taken on a Sunday when trains operate less frequently than on other days of the week, thus keeping the number of false measurements resulting from passing trains to a minimum.

The continuous operation of the Metro system while measurements were taken required the train to maintain its position relative to other trains on the system. This resulted in measurements being repeated along substantial sections of track, providing an opportunity to examine the repeatability of the measurement process.

The test produced results in two forms; a computer data file containing values of voltages relating to SPL and train speed at known instants of time, and video recordings with superimposed details of date and time. Numerical data were processed using a computer taking into account pre-determined calibration and offset factors. Values of SPL and train speed were obtained in units of dB(A) and km/h, respectively.

6 CALIBRATION

An investigation was made of the relationship between train speed and interior sound pressure level by taking measurements as the train travelled on relatively new track considered to be in good condition. Measurements were taken while the speed of the train was held at 20 km/h for a period of approximately 30 seconds. The train speed was then increased to 30 km/h and a further set of measurements was

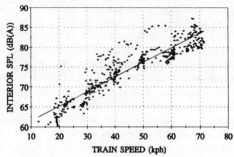

Figure 3 Relationship between train speed and interior sound pressure level

recorded automatically for a further period of about 30 seconds. This process was repeated several times as the train speed was increased in increments of 10 km/h up to 80 km/h (maximum speed within the Metro system).

Figure 3 illustrates values of train speed and interior sound pressure level obtained when the train was travelling at speeds between 20-80 km/h. A 'least-squares' fit of an exponential curve to this data produced a near-straight line indicating that, for practical purposes, the relationship between speed and SPL can be assumed to be linear. The results showed that, in general, an increase in train speed of 10 km/h in the range 20-80 km/h caused an increase in SPL of approximately 3.7 dB(A). Using the relationship provided by this data it became possible to predict a value of sound pressure level within the train when travelling at a given speed over track in good condition.

These values relate to a Metro train travelling on surface track in good condition constructed predominantly from continuously-welded flat-bottomed rail supported on timber sleepers and ballasted track. Earlier (unpublished) studies have shown that, at train speeds less than 20 km/h, external machinery on the train (such as compressors providing air to the braking system) had undue influence on the values of SPL within the trains. As a result any measurements obtained at speeds below 20 km/h were rejected from the analysis.

Although the track over which these results were obtained is relatively new, some natural variation in its quality was expected. This is variation is illustrated in Figure 3 as scatter within the results.

7 SURVEY OF THE METRO SYSTEM

After synchronising and calibrating the equipment, the train began its journey around the entire system as if it was operating normally, ie. briefly stopping at each station and obeying signals and track speed restrictions.

Predicted values of interior SPL were calculated within a computer spread-sheet using measured values of train speed recorded during the test. A comparison was made between predicted SPL values with those measured in-situ for each second of the train journey. Positions along the track where the train generated values of SPL greater than the predicted value for its speed suggested possible sites of excessive wear and deterioration.

The testing of the entire Metro system produced a large amount of computer-based data in a relatively short period of time. At the end of the test period an analysis was made of the sections of track between pairs of adjacent stations, and the results prepared in graphical format. Figure 4 illustrates the results between Hadrian Road Station and Howdon Station, and represents a typical data set obtained from the measurements. The results presented in Figure 4 relate to the time when the measurements were taken, the value of sound pressure level inside the train at that time, the train speed and, of most significance, the algebraic difference between measured and predicted values of sound pressure level. In all more than 90 individual graphs were produced, each relating to the track between two adjacent Metro stations. Variations in the values of SPL were clearly apparent as the train travelled throughout the system, and numerous structural features (such as bridges and tunnels) were readily identified.

No reliable information was obtained when the train was travelling at less than 20 km/h. At these points a nominal zero value was assigned to the difference between measured and predicted values of sound pressure level. These values are shown in

Figure 4 using a different style of line in order to distinguish them from places where predicted SPL values truly coincided with measured values.

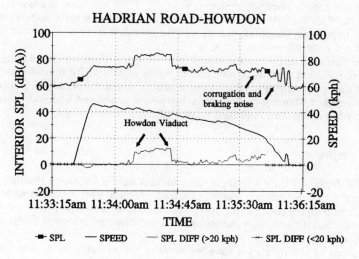

Figure 4 Speed and SPL measurements

Compilation and editing of the video recordings using an audio-visual editing facility removed unwanted material, such as time spent waiting at stations. Uninterrupted recordings of the views from the front and rear of the train were produced as the train travelled the full extent of the system.

Figure 5 Enhanced video recording

Numerical values of SPL, train speed, and differences between measured and predicted values of SPL were visually implanted into the video recordings at one-second intervals using synchronised results and advanced computer technology. Retention of the original audio signals recorded by the video cameras allowed the path of the train to be seen and sounds heard whilst numerical values of SPL etc. were visible within the video image. This sophisticated editing of video tape allowed specific features within the system to be easily identified, and allowed the occasions when trains passed on the adjacent track to be eliminated.

Figure 5 shows a single frame obtained from the video recording after editing. On completion, sufficient records of train speed, values of SPL and differences between measured and predicted SPL values were available within the video recording and graphs to enable features such as bridges, fish-plated joints, and significant levels of corrugation to be identified, and their relative effects to be evaluated.

8 CASE STUDY

The results shown in Figure 4 relate to the train as it travelled between Hadrian Road Station and Howdon Station. Values of sound pressure level recorded within the interior of the train are shown as the upper trace. Using the left-hand axis of the graph, it can be seen that the maximum and minimum values were approximately 84 db(A) and 58 dB(A), respectively.

The speed of the train as it travelled between the two stations is shown in the central trace, and is related to the right-hand axis of the graph. The results indicate that the train gained speed quickly from rest at Hadrian Road Station until it reached a section of track which is subject to a track speed restriction of 45 km/h. It then gradually slowed to approximately 42 km/h over the next 30 seconds during which time the values of sound pressure level inside the train remained sensibly constant. The train then came onto Howdon Viaduct (see Figure 6), producing a sudden increase in values of SPL even though the train was gradually reducing its speed.

Figure 6 Howdon Viaduct

Differences between the measured and predicted values of SPL are illustrated in the lower trace in Figure 4, and relate to the left-hand axis. These results indicate that when the train is travelling along Howdon Viaduct it generates values of SPL which are about 13 dB(A) higher than would be expected from conventionally ballasted track at the same train speed. This represents a very significant localised increase in sound level, and justifies the speed restriction for trains travelling over the bridge.

Howdon Viaduct is an old bridge structure comprising seven arched spans with an overall length of approximately 320 m. Originally built in 1839 (8,9), the bridge has been strengthened and modified several times to allow for increased loading from trains and normal deterioration of the structure as a result of environmental exposure. A brief outline of the changes made to the structure has been published earlier (10).

Local residents recognise Howdon Viaduct as a source of significant levels of train-generated sound. The rails forming the track are jointed using conventional bolted fish-plates which produce impact noise when traversed by train wheels. Vibrations caused by trains are amplified within heavy cast-iron plates forming the deck panels of the Viaduct.

In Figure 4 the lower trace suggests that the difference between the measured and predicted values of SPL became less as the train left the Viaduct, but the measured values were still greater than expected. Thus it can be concluded that the condition of the section of track beyond the Viaduct was worse than that on its approach, and a clear indication is given for the priority of track maintenance in this section of the Metro system. A site inspection of the track after the test revealed that the section of track beyond the Viaduct comprises continuously-welded flat-bottomed rail with supports from concrete sleepers in conventional ballasted construction. In addition, observations of vertical movement of the track with passing trains suggest that improvements may be possible to the track by tamping the ballast.

On further inspection several small clay holes were seen in this section of track; the effect of these can be seen in the lower trace of Figure 4 as localised increases in measured values of SPL relative to the corresponding predicted values.

As the train slowed on its approach to Howdon Station, values of SPL remained higher than expected. This is partly as a result of sound produced by the application of the braking system of the train and partly due to the general condition of the running surface of the track.

9 CLOSING COMMENTS

Maintenance of track quality and performance is an essential feature of modern competitive railway systems, and will lead to better journey times. In commercial terms, funds spent on track maintenance for reasons other than safety may not seem to represent value for money, particularly if priority locations for remedial work cannot be easily identified.

The technique described and demonstrated above provides an accurate and rapid means of monitoring the condition of the running surfaces of the track of an entire railway system. Sections of track are readily identified where maintenance work to reduce train-generated sound levels would be most cost-effective.

The condition of sensitive structures and locations can be examined on a regular basis enabling any significant increases in train-generated sound levels to be identified and investigated. Other work (11) has shown the effects produced by trains when travelling through underground sections of track.

In-train measurements can provide regular detailed information about sound pressure levels within complete railway systems, which would never be obtained economically using conventional trackside measurements. Measuring sound pressure levels within a dedicated train is not intended to replace conventional trackside measurements, the latter being important when considering sound levels at locations which may be particularly sensitive.

Correct interpretation of measurements taken within the train is made much simpler using enhanced video recordings of the front and rear views from the train. Examination of all the information produced within a few hours enables potential problems (such as clay holes or block joints in need of realignment) to be identified. Localised increases in values of SPL caused by such features as bridges and tunnels can be eliminated from the interpretation of results as normal maintenance work would not reduce sound levels in these areas.

Important lessons may be learned from considering sections of track which produce lower values of sound pressure level than expected for a given train speed; it may be possible to identify good practice in terms of track design and construction and effective maintenance. The effectiveness of rail-grinding in terms of reducing train-generated sound can be examined, together with the effect of machining the wheel profiles of the rolling stock used for the test.

Reduction of sound levels may not be possible or feasible throughout a complete railway system. Remedial measures can at least be considered for locations shown to produce sound levels greater than expected by the method described above. Research is ongoing to discover further causes of train-generated noise, vibration and particularly rail corrugation.

Further thoughts are being given to the development of permanent in-train and trackside measurements similar to those described above. In particular, equipment may be fitted to a train to enable measurements to be taken more frequently using a semi-automatic system. This might even be undertaken on a daily basis so that effects such as ambient external temperature and gradual wear of the train wheels can be assessed.

A permanent trackside facility is being discussed so that every train that passes a fixed measurement position will be monitored for the sound levels it produces. The speed of each train will be closely monitored, and trains which regularly produce more sound than expected will be selected for further investigation and possible wheel maintenance.

The changes in the ownership and operation of the railway system in the UK and the existence of other light rail systems similar to Tyne and Wear Metro are likely to increase pressure for 'better journey time'. Organisations with interest in maintaining and/or proving rail quality may wish to consider adopting a technique such as that described above. A long-term result of such an approach will be the creation of much detailed information about track performance and deterioration, and should result in improvements in the design of railway track and wheels of rolling stock.

10 ACKNOWLEDGEMENT

The authors would like to thank Mr. Boyd Lucas and other staff at Tyne and Wear Metro for providing the dedicated train, and for assistance and technical advice enabling train speeds to be recorded electronically.

11 REFERENCES

1. Grassie, S.L., Corrugation: variations on an enigma, Railway Gazette International, 1990, Vol. 146, 531-533.
2. Howard, D.F. and Layfield, P., Tyne and Wear Metro: concept, organization and operation, Proceedings of the Institution of Civil Engineers, 1981, Part 1, Vol. 70, 651-668.
3. Leung, C.N., Level, frequency and incidence of rail noise from a rapid transit service, Dissertation submitted in partial fulfilment of Diploma in Highway and Traffic Engineering, 1977, University of Newcastle upon Tyne.
4. Bell, M.C. and Leung, C.N., A study of noise from the Tyne and Wear Metrocar, Research Report 27, 1978, Transport Operations Research Group, University of Newcastle upon Tyne.
5. Hamill, A. and Hills, P.J., The noise nuisance caused by urban railway operations, Research Report 63, 1986, Transport Operations Research Group, University of Newcastle upon Tyne.
6. Thancanamootoo, S., Impact of noise from urban railway operations, Dissertation submitted for the Degree of Doctor of Philosophy, 1987, University of Newcastle upon Tyne.
7. Remington, P.J., Dixon, N.R. and Wittig, L.E., Control of wheel/rail noise and vibration, Report Number UMTA-MA-06-0099-82-5, 1983, US Department of Transportation, Washington D.C., USA.
8. Green B., The description of the arched timber viaducts erected from the designs of Messrs. J. and B. Green, Proceedings of the Institution of Civil Engineers, 1839, Vol. 1, 88-90.
9. Green B., On the arched timber viaducts of the Newcastle and North Shields Railway erected by Messrs. J. and B. Green of Newcastle-upon-Tyne, Proceedings of the Institution of Civil Engineers, 1846, Vol. 6, 219-227.
10. Lilley D.M. and Robinson, J., Research studies in vibration effects within bridge structures supporting a light railway, ASME Energy and Environmental Expo 1995 Conference, Structural Dynamics and Vibration Symposium, Houston, Texas, 1995, PD-Vol. 70, 37-46.
11. Lilley, D.M. and Robinson, J., Cost-effective maintenance of railway track using sound measurements, ASME Engineering Technology Conference - Energy Week 1996, Houston, Texas, 1996, 224-232.

C514/029/96

Analyses on behaviour of track irregularity with use of TOSMA

T OHTAKE MEng, MJSCE and **Y INOUE** MEng, MJSCE
Shinkansen Operations Division, Central Japan Railway Company (JR Central), Tokyo, Japan
Y SATO MJSCE
Nippon Kikai Hosen KK – NKH (JR Central), Japan Mechanized Works and Maintenance of Way Company Limited, Tokyo, Japan

Synopsis
TOkaido Shinkansen track MAintenance system (TOSMA) has been developed so as to realize the system which makes the plan for track maintenance not only fully using the data held in new 'Permanent Way Management System' but also changing the track management from that controlling the size of irregularity to that controlling the irregularity growth. The TOSMA gives the final converged size of irregularity in each lot of 20 m long, that in the section of several hundreds meters and that in the district of several kilometers. The amelioration of the lot with extreme growth makes the exclusion of manual maintenance possible.

1. INTRODUCTION

Tokaido shinkansen, which opened a new era of high speed railways in the world and made great contribution to Japan's economical success, has been running on the main corridor of Japan between Tokyo and Osaka for more than 30 years. Its maximum speed is 270 km/h. 11 trains are operated in an hour or the average train head is 5.5 minutes.

To maintain such fast, steady and safe operation, track maintenance works supported with computer system called SMIS (Shinkansen Management Information System) have been essential. The recent rapid development of computers has made the innovation of the system necessary. Now, the large scale revision and reconstruction of the system is on the way.

Hereafter, at first, the present permanent way system of 'SMIS' and the framework of the new system which is now being developed is reviewed. Following it, the basic idea of the 'TOSMA'(TOkaido Shinkansen track MAintenance system), which constitutes the main part of the new system, is introduced. Finally, the behavior of track irregularity with use of TOSMA are discussed.

2. PRESENT SYSTEM 'SMIS' AND ITS RECONSTRUCTION

A mainframe computer system called SMIS was provided in 1973. It processes data from the track geometry car which runs every 10 days and holds databases of track state, material condition, the history of maintenance works etc. in the permanent section as shown in Fig. 1.

It has been the brain of track maintenance works, but the increasing demands on better riding comfort, the technological advance and the policy change in maintenance works made large scale revision and reconstruction of the present system necessary.

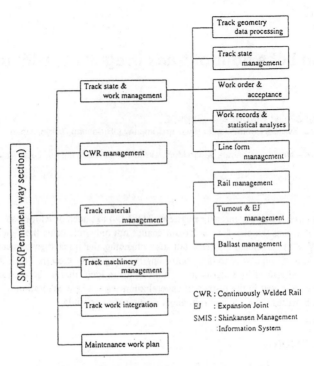

Fig. 1 Composition of SMIS's permanent way section

The composition of the new system is given in Fig.2. The new system consists of 3 parts. Those are a new SMIS, the newly developed track management system and the system for the contractors. In the framework of the new system the role of the permanent way section of SMIS is streamlined and mainly concentrating on processing data from track geometry cars. Instead, other functions are rebuilt and reinforced as a new system depending on the newest theories and computer technologies. These are for constructing an integrated database of track conditions and material conditions, for raising the reliability of the database, for filling the gap between the practices of maintenance work and the computer system, for supporting the optimum decision making processes and for achieving an user-friendly operating interface.

The new system's hardware are composed of UNIX EWS (Engineering Work Station) and Windows' PC etc. as shown in Fig.3.

The new system TOSMA which is dis-cussed hereafter is being developed as a sub-systems of the new system. It is mainly concentrated on the area of forecasting and controlling track irregularity and supporting the decision making processes to be optimum.

3. NEW TRACK CONTROL SYSTEM "TOSMA"

The new system TOSMA has been developed basing on the 'Convergence theory (1)-(4)'. TOSMA constitutes a sub-system of the new 'Permanent Way Management System' and is designed to make the full use of the database provided by the whole system.

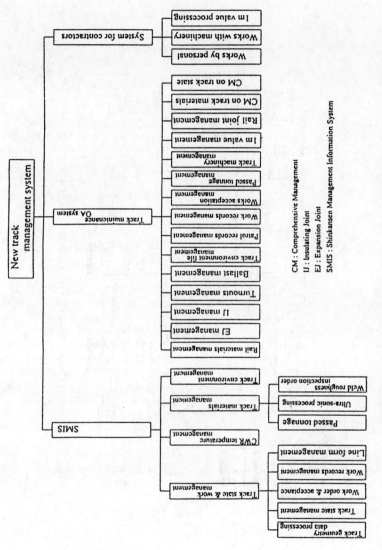

Fig. 2 Composition of new 'Permanent Way Management System'

CM : Comprehensive Management
IJ : Insulating Joint
EJ : Expansion Joint
SMIS : Shinkansen Management Information System

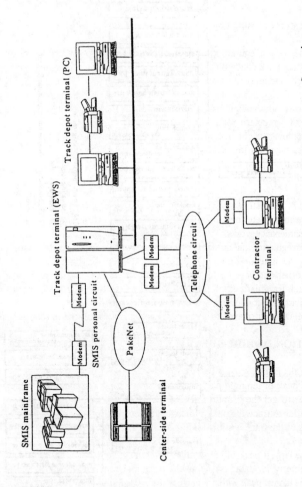

Fig. 3 Composition of equipments for new 'Permanent Way Management System'

3.1 Principle
The main purposes of the system are as follows;

- To give better information on track situation to the decision maker,
- To suppress extreme irregularity growth through the amelioration of track materials or that of roadbed in order to exclude manual tamping as small as possible and
- To assist making the optimum plan for maintenance works including permanent measures to exclude the extreme growth of irregularity (work kind, amount, location and timing of the works)

The basic concepts for constructing the system are as follows;

- Pursuing the track irregularity of longitudinal level so as to calculate its growth ratio,
- Forecasting the track condition in the future basing on irregularity growth rate, rectification ratio of work, and amount of works in a year depending on convergence theory,
- Finding lots with high irregularity growth rate and planning fundamental measures to decrease it through the amelioration of materials or that of the structure so as to decrease the growth of irregularity and
- Recalculating the mean of irregularity size in the section assuming that those fundamental measures are taken for lots and repeating it until the aimed mean is attained.

3.2 Structure of System

The composition of the system is shown in Fig.4. It consist of three parts. The first is composed of 'Action' program to determine the growth rate and the rectification ratio of irregularity, and the amount of maintenance works. The second is for taking a view of the track irregularities, the growth of track irregularity, the numbers of tamping works, and the spots where manual taming is necessary, and the amelioration works to suppress extreme growth. The third is the database working with other database in the whole system by exchanging the data. The data used in TOSMA are track irregularities of 10 m versines, growth of irregularity, work records, track environments, materials conditions, amelioration work records etc.

The fundamental performance is given in (5) and it is ameliorated in followings..

4. CALCULATION OF IRREGULARITY GROWTH IN 'TOSMA'

4.1 Trace of peaks in lot
In TOSMA the peak of longitudinal level irregularity in a lot is pursued on the records of the irregularity so as to get the growth rate. To realize this correctly, some new techniques are introduced in order to trace the peak of irregularity correctly.

The process of the trace is shown in Fig. 5. It is as follows;

- (a) Finding the largest peak of longitudinal level irregularity for each rail of each 20 m lot on track geometry car data just before the track maintenance work is executed,
- (b) Finding the largest peak within 5 m from the peak found in (a) on the track geometry data which is dated just before the one used in (a),
- © Repeating the processes of (b) doing back it to just before the track maintenance work is found from the peak trace to that within 5 m on the former track geometry data,

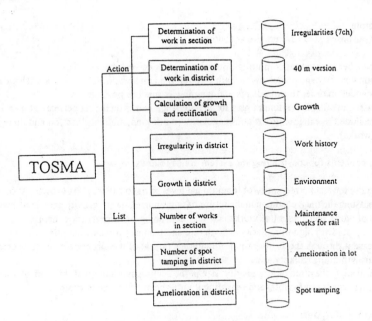

Irregularities (7ch)

40 m version

Growth

Work history

Environment

Maintenance
works for rail

Amelioration in lot

Spot tamping

Fig. 4 Composition of TOSMA

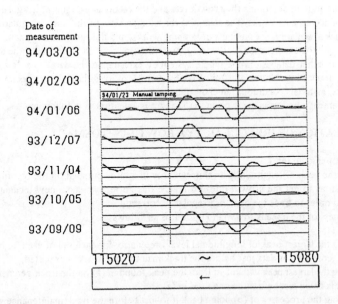

Date of
measurement

94/03/03

94/02/03

94/01/23 Manual tamping

94/01/06

93/12/07

93/11/04

93/10/05

93/09/09

115020　　～　　115080

←

Fig. 5　Process of tracing peaks

(d) Calculating an irregularity growth rate from peaks obtained in (a)-(c) with use of linear regression method for a period between the maintenance works found in (a) and (c) assuming a linear growth rate and
(e) Calculating growth rates for other periods in a year by repeating the process of (a) to (d).

4.2 Characteristics of calculation of irregularity growth
The calculation of irregularity growth by tracing and having the regression of peaks of irregularity in a lot in the above has following characteristics.

(1) The peaks are pursued from the new one to the older one in order to trace the irregularity which finally caused the maintenance work to be executed.
(2) The calculation is reset when another work is found within 5 m of the peak in order to reflect the effect of the work correctly.
(3) The calculation is made for each operation of track geometry car (every 10 days) and for each rail of the track, in order to attain high accuracy for the calculation.

The calculated regression line is given in Fig.6. The value of irregularity growth for the lot is given with the larger one of the means on both rail given in the figure.

5. CALCULATION OF RECTIFICATION RATIO

Getting the regression line for the growth of irregularity, the effect of rectification works is given with the values before and after the work. The value before the work is given by that at the date of work on the regression line of irregularity peaks at A in Fig. 6. The value after the work is

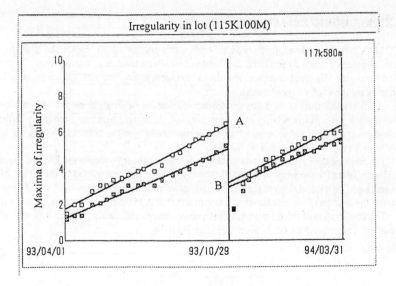

Fig. 6   Application of regression line

given by that at the date of work on the regression lin of irregularity peaks at B in the figure. The value for the irregularity growth before and after the work are given by the slopes of regression lines before and after the work.

Those in a district are given in Fig. 7. The values of irregularity before and after the work is shown at the figure at the top left. They gives the rectification ratio of 0.623 for the regression through the origin. However, it is necessary to take notice on the considerable dispersion of data. The relation on the irregularity growths rate before and after is given at the figure at the bottom left. They show that there is no clear relation between them. The relation between the value of irregularity growth and the irregularity after the work is given at the figure at the bottom right. They show that there is no clear relation between them.

6. DECISION MAKING IN DISTRICT

An example of decision making is given in Fig. 8. The upper left shows the condition and the necessary work volumes for sections. The lower left shows that for the lot. The upper right shows the necessary work volumes and the amount of prices for the work. The lower right shows the bottom for selecting the sections and lots for left figures. This window is moved to the decision making window by clicking the section at which changes are considered and after the change of decision in the section is finished, returned here again.

At the top of upper left the irregularity and the irregularity growth at each lot is given. Under these, the necessary and determined volumes are given for the volume of manual tamping, the assumed number of tamping with multiple tie-tamper, the necessary volume of amelioration works for sections and lots. The amelioration works consists of rail exchange, rail grinding, rail bending-up, exchange of rail pad, increase of tie weight, exchange of ballast, amelioration of roadbed and others.

7. CONCLUDING REMARKS

TOSMA has been developed so as to realize the system which makes the plan for track maintenance not only fully using the data held in new 'Permanent Way Management System' but also changing the track management from that controlling the size of irregularity to that controlling the irregularity growth.

To realize this, at first the automatic calculation of irregularity growth rate and the rectification ratio of irregularity have been pursued. Although there had been many difficulties, now it became possible. As the next step, the decision making in the district including the cost for the works as shown in Fig. 8 is on the way.

The merit of the system depends on the convergence theory because the level of irregularity is directly related to the work volume and the work volume is not so related to the timing of work execution. The TOSMA gives the final converged size of irregularity in each lot of 20 m long, that in the section of several hundreds meters and that in the district of several tens kilometers.

The amelioration of the lot with extreme growth makes the exclusion of manual maintenance possible. These could be easily planned in the TOSMA.

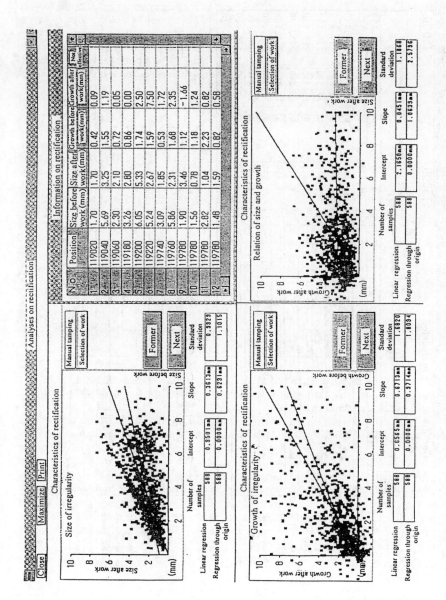

Fig. 7 Anlyzed results on rectification ratio

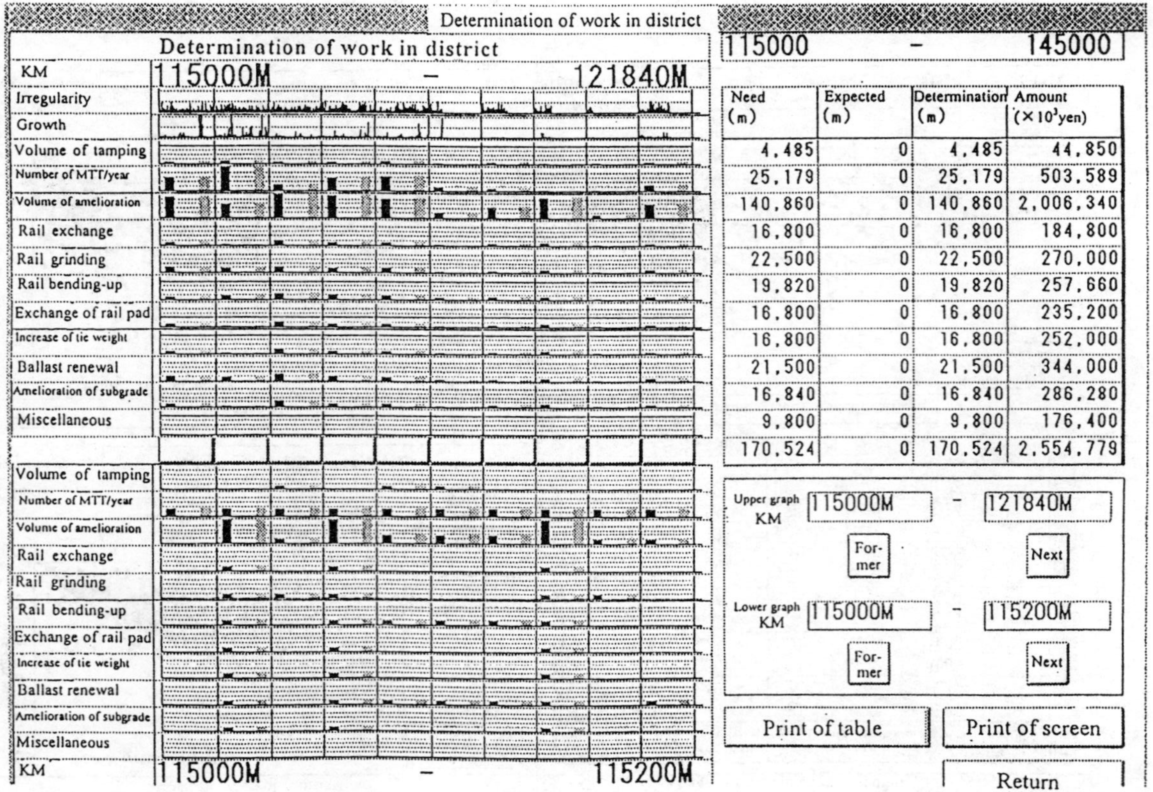

Determination of work in district

| KM 115000 – 145000 | 115000 – | | |

	Need (m)	Expected (m)	Determination (m)	Determination Amount (×10³yen)
Volume of tamping	4,485	0	4,485	44,850
Number of MTT/year	25,179	0	25,179	503,589
Volume of amelioration	140,860	0	140,860	2,006,340
Rail exchange	16,800	0	16,800	184,800
Rail grinding	22,500	0	22,500	270,000
Rail bending-up	19,820	0	19,820	257,660
Exchange of rail pad	16,800	0	16,800	235,200
Increase of tie weight	16,800	0	16,800	252,000
Ballast renewal	21,500	0	21,500	344,000
Amelioration of subgrade	16,840	0	16,840	286,280
Miscellaneous	9,800	0	9,800	176,400
	170,524	0	170,524	2,554,779

Upper graph KM [115000M] Former – [121840M] Next

Lower graph KM [115000M] Former – [115200M] Next

Print of table   Print of screen

Return

Determination of work in district

115000M – 121840M

115000M   121840M

115200M

Irregularity
Growth
Volume of tamping
Number of MTT/year
Volume of amelioration
Rail exchange
Rail grinding
Rail bending-up
Exchange of rail pad
Increase of tie weight
Ballast renewal
Amelioration of subgrade
Miscellaneous
Volume of tamping
Number of MTT/year
Volume of amelioration
Rail exchange
Rail grinding
Rail bending-up
Exchange of rail pad
Increase of tie weight
Ballast renewal
Amelioration of subgrade
Miscellaneous
KM

115000M

Fig. 8   Decision making in district

REFERENCES

(1) SATO, Y. Analysis of system behavior of railway track, 5WCTR(World Conference on Transport science), Yokohama, Japan, 1989, E-09-2.
(2) SATO, Y. Composition of a system for mechanical track maintenance of railway, Proc. of 6WCTR, 1992, SIG5.
(3) SATO, Y. New track maintenance base on convergence theory of track irregularity, Proc. of Fifth International Heavy haul Railway Conference, Beijing, China, 1993.
(4) SATO, Y. Theoretical analysis on converging process of track irregularities in a section, 7WCTR, Sydney, Australia, 1995.
(5) KONDO, K., INOUE, Y. & SATO, Y. Development of new railway track control system (TOSMA) based on convergence theory, 7WCTR, Sydney, Australia, 1995.

C514/066/96

Railway journey times – quality tools for continuous improvement

A N CUTLER BSc, CEng, CPhys, MInstP, AFIMA
Wilmslow, Cheshire, UK
S SCANNALI BSc, CPhys, MInstP
Vectra Technologies Limited, Warrington, Cheshire, UK

SYNOPSIS

Manufacturing industry, especially the automotive and electronics sectors, has radically changed its approach to business over the past decade. In particular, the ideas of Deming, Taguchi and Box, among others, have given manufacturers the tools to enhance quality and reduce costs beyond the scope of previous endeavours.

This paper discusses how techniques such as control charting and ideas arising from statistical process control (SPC) can be imported into a railway business. Furthermore, modern ideas for robust design of products, along with focused mathematical modelling, can be applied to railway services to develop service patterns that possess ruggedness against uncontrollable external factors

1. CAPITAL PROJECTS RELIABILITY

Today's railway service providers are not interested solely in how many passengers they can carry between destinations or even simply in comfort of accommodation and speed. The need to retain and expand market share in a competitive transport industry means that reliability and predictability are every bit as important as getting from A to B. Passengers have growing expectations of services that arrive and depart on time. The service that the passenger sees clearly depends upon the operating company's timetable and upon the reliability of the company's rolling stock, infrastructure and human resources in delivering the advertised service. Timetable and service development through modelling of train performance is older than the electronic computer (1) but still provides opportunities for novel calculation methods and innovative timetable solutions. However, such modelling conventionally relies on using data from either ideal conditions or using pessimistic limit-cases to explore the possible impact of faults and failures.

Railway operators keen to understand and control the variable and failure-prone real world looked to the nuclear power industry where Probabilistic Risk Analysis (PRA) was a well established discipline. PRA was developed during the 1960s as the public's concern about nuclear safety led the industry to find ways of reassuring community fear through a quantified, rational, traceable and auditable presentation of a safety case. PRA's ability to discourse about risk, uncertainty and liability management soon came to the attention of those working in the field of reliability improvement, among others in the railway industry.

Unfortunately, PRA is very good at telling us how unreliable things are but less good at helping us to improve the situation. However, it did enable project managers and procurement professionals to set targets in terms of failures per year for new capital investment. The worst of these targets were arbitrary and based on cost-benefit analyses that considered everyone but the passenger. The best encouraged suppliers to try to improve their performance. The down side was that they focused supplier effort on a paper numbers-game. These contractually defined specifications further fostered a willingness to blame other system suppliers, and their customers, for failure to meet targets. Railways did get more reliable but operators soon found that there were limits beyond which they could not advance.

A key part of this number-driven approach was the Reliability Growth Programme (RGP). The conventional wisdom of the RGP was that, during the early stages of trial operation, design faults in the product would be identified and remedied, as long as it did not cost too much, in which case their would be a confrontation between supplier and customer. Beyond the hope that enough faults could be rectified to demonstrate the reliability target, or at least enough evidence gathered against the rest of the world to cry 'frustrated contract', there was little intention of continuing the improvement beyond the stage payment. Once the target was reached, or discredited, everything had to be accepted as 'good enough'.

As with other aspects of PRA, RGPs had their own rituals for describing how unreliable things were. The Duane plot was a method for processing reliability data from RGPs without giving any clues about what should be done to improve the matter. It is good to encourage people to do things better but without tools to assist them there will be a limit beyond which they cannot go. 'Best endeavours are not enough.'

2. THE QUALITY REVOLUTION

While all this was going on, a quiet revolution was occurring, firstly in the USA and later on the Pacific rim. In the 1920s, Walter Shewhart was trying to find ways of controlling the quality of manufactured parts at the Bell Telephone laboratories. He spent many years developing various techniques and during this time identified two central ideas that were to remain on the fringes of conventional statistics for 40 years but which are now accepted as fundamental concepts in managing uncertainty and variation (2).

The first idea is that most of the statistics that we do in industry is aimed at predicting how processes and systems will perform in the future rather than how they performed when we were collecting the data. This may seem an obvious idea but Shewhart showed that stable and predictable processes (he called them 'in statistical control') are not a state of nature and are only achieved after concerted and continual action on the system.

His second important idea was to draw the distinction between 'special causes' and 'common causes' of variability. Special causes lie outside the system and can, and should, be remedied on a case-by-case basis. Common causes lie in the nature of the system itself. Any attempt to deal with common cause variation on a case-by case-basis only increases the variability of the system.

Shewhart's revolutionary ideas were picked up and exported to Japan in the late 1950s by W Edwards Deming. Deming preached a gospel of continuous improvement of service to the customer (3). He championed the 'control chart', Shewhart's great methodological innovation, for establishing statistical control and reducing variation. He also set these statistical tools in a wider industrial context encouraging workers at all levels to take control of their jobs and, using simple statistical tools, to set about delighting their customers and relearning pride in the job. Deming showed the Japanese that it was poor service and shoddy goods that increased business costs. Quality reduced waste, rework and energy spent handling customer returns.

The early 1960s also brought about the rediscovery, by the Japanese statistician Genuchi Taguchi, of the work of Sir Roland Fisher between the wars. This work described how products, and services, with many factors, could be improved and optimised using highly efficient and effective statistical techniques. A key part of Taguchi's philosophy was to design products and processes that were robust against uncontrollable environmental and final use variability (4), an idea first used at the Guinness brewery in Dublin at the turn of the century.

The revolution wrought by Shewhart, Deming and Taguchi has had a dramatic effect on the economies and industries where it has been practised. There is an opportunity here for the railway industry.

3. A FUSION BETWEEN THE QUALITY AND RELIABILITY PROFESSIONS

Both the PRA and Quality movements seem to have common issues at heart, improving the service that we give to our passengers and keeping us in profitable business. Surely, it is time that reliability professionals started to look hard at quality ideas and to understand what they have to offer. What better place to start doing this than the railway industry.

3.1 Continuous improvement

There are so many opportunities on a railway where we can take a group of workers, encourage them to look at their jobs and the customers, train them in some simple quality tools and give them the job of continuously improving the way that they work. The best people to set to work improving journey times are those responsible for getting trains between destinations: drivers, signalling staff, maintenance engineers, station staff. A group of such people empowered to gather the information that they need, facilitated by a senior manager and encouraged to introduce change into their own work sphere can be trained to use a range of quality tools. Such tools are likely to include:

control charts
cause and effect diagrams
scatter plots
Pareto analysis
tally charts
flow diagrams
histograms.

Control charting is central to the business of improving performance (5). The control chart has been called the *voice of the process*. It tells us what our system is doing, not what we would like it to do. *The voice of the customer* is clear. They want reliable journey times, as fast as comfort and safety will allow. By control charting, we start the process of aligning the voice of the process with the voice of the customer.

For our present exercise, the necessity of control charting journey times is plain. However, it is not the whole story. There are other factors that will affect journey time which we will need to learn about, control and improve. Neither is journey time the whole story. We will not meeting passengers' needs if the train is on time but in need of cleaning. We must aim to improve the whole service. The technique of *Quality Function Deployment* (QFD) (6) is widely used to establish which factors will influence the customer's experience. A preliminary list of variables for control charting might include:

Rolling stock and infrastructure maintenance times
Hazardous incidents on platforms
Signals passed at 'danger'
Power consumption
Frequencies of rolling stock and infrastructure failure
Staff availability and absenteeism

A typical control chart in the early stages of process improvement might look like Fig 1. Nearly all observations should fall between the upper and lower control limits. Any that do not are evidence of a special cause. The root origin of special causes should be hunted down and eradicated by teams of workers active in the job of getting passengers from A to B, relying on management only for vision and support. There are other signs of special causes that can be mastered by any one trained in control charting. Simply chasing every journey time above average (or above some arbitrary target) raises costs, leads to frustrated and demoralised staff and makes the system worse. Continual attention to special causes, using Shewhart's Plan-Do-Study-Act (PDSA) cycle (3), will eventually produce a process that is in control. Until the railway journey times are in a stable state, they cannot be improved as we do not know how they will perform in the future. This will not happen by chance, only by using the right tools. Control charting aspects of the maintenance operation is also central to reducing costs year-on-year, while improving services to customers and society at large.

Managers too need to understand control charting because, once statistical control is achieved, it is their turn to pursue the common causes. These are part of the system and cannot be eliminated simply by tampering with it. Fundamental change is required. The managers will need to think about:

capital investment
redesign of systems of work
procurement of superior supplies
training and skilling
developing the product

or other, possibly radical, changes. The procurement issue is central. Suppliers must become part of the team, committed to improving the goods and services that they offer in just the same way as the railway is committed, in turn, to serving its own customers. The old adversarial environment in which railway and supplier fought each other over contracts, specifications and variation orders must end. Both railway and supplier have a common interest. If we do not produce a service that the public is willing to pay for, in fares or taxes, then we shall all be out of a job. The system of awarding long-term single-supplier contracts on the basis of openness, involvement at all stages of the job, and a commitment to improving quality and reducing costs has been a demonstrable success in the Japanese motor industry.

All managers need to work on thinking of the railway as an entire system without the old barriers between passengers and staff, suppliers and operators, and between functional areas within the organisation. Managers need to lead their staff to similar insights.

Deming warned that there is no 'instant pudding'. These quality methods are only tools that enable railway staff to do their job. They will not bring overnight success. However, without them, any short term benefits from quality initiatives will ultimately be frustrated by lack of understanding of variation and inability to manage real-world chaos.

3.2 Reliability growth

As we observed, only with the commitment of the equipment suppliers will continuous improvement become a reality. Though most suppliers of capital goods talk about RGPs, few have managed to acquire the tools to offer permanent and continuous reliability improvement to their customers. Equipment suppliers are now becoming increasingly aware of their deficiency as they take on the new wave of design, build and maintain contracts. It is cost to society, not our immediate customer, that will ultimately control whether we, as an industry, stay in business.

Any RGP must be based on a Fault Reporting and Corrective Action System (FRACAS). FRACAS development activities are traditionally hampered by the difficulty of defining faults, measuring dependability, and reliably collecting data. This is the most critical part of the whole reliability growth process. The aim of reliability growth is to delight the passenger. We cannot measure passenger experience directly so we need to find things that we can measure that we believe are significant influences. Again, QFD is a useful tool but the important issues for designing a FRACAS are:

concentrate on passenger experience
gather a multifunctional team to put the FRACAS together
involve the staff who will do the reporting
implement focused training in the agreed system
keep it under review
continually improve the system.

Once we have a FRACAS we can get on with control charting. Without control charting we will never be sure that we have permanently improved the process and might be led into costly modifications that only make matters worse. Figure 2 illustrates the plan for running a modern RGP.

Beyond control charting, we might wish to investigate the impact of quite complicated changes to the system on its performance. Typically, there might be several factors capable of adjustment simultaneously. Trying to find out their joint effect by tampering with each factor individually is clearly futile on a system with so many diverse electrical, mechanical and human systems as a modern railway. The strong interactions between various effects would soon swamp any data gathered from adjusting parameters singly. Alternatively, trying all possible combinations would be an impossibly vast task to manage and interpret. Fortunately, a body of knowledge has been developed for designing such experiments effectively and efficiently. The statistician George Box in particular has championed (7) the use of so called Response Surface Methods (RSMs) which enable an engineer, with a little training and a lot of engineering knowledge, to:

find system parameters that improve performance
check if the system is currently near an optimum
locate on optimum
find lower cost system settings that deliver the same performance.

The idea is that various experimental studies can be performed sequentially, as knowledge grows, allowing ever deeper insight. Box introduced the idea of Evolutionary Operation (EVOP) where changes are actually made on a real life operating system and their effects monitored, understood and used to move towards a higher quality arrangement (8, 9).

Learning about the system by adjusting the real thing must be the ultimate goal of all operators but, at the early stages of acquiring knowledge, most will prefer to use mathematical models. Mathematical modelling of railway journey time predates the electronic computer and is, perhaps, the oldest numerical engineering problem solved by mechanical means (1). Mathematical models are now well-developed for power systems, journey times and also system reliability and offer many opportunities for exploring opportunities to improve quality.

All modelling is fruitless unless it offers a valid representation of the real system performance. One fundamental issue is to obtain data for the model that can be used to make predictions. Control charting is again of central importance. Unless that data collected displays statistical control, any mathematical predictions based on it will be unreliable. Furthermore, control charting is important to the effective and efficient validation of mathematical models by measured system data.

RSMs have gained popularity in recent years through their use in exploring the knowledge offered in mathematical models. Effective and efficient studies can be performed that help the engineer adjust system parameters, first to improve, and later to optimise performance. The mathematical model is a complicated function of its inputs. RSMs enable the engineer to study settings near to the nominal or starting conditions and to adjust the inputs towards improved responses. The advantages that RSMs have over conventional numerical optimisation methods is that they can be used with very large models requiring hours or days per run, and that they enable the engineer to use his or her expertise and judgement in guiding the path of the optimisation. These methods have, to date, been largely used for designing complicated mechanical structures to offer specified strength at minimum cost (10). However, there is no reason why they should not be applied to journey time simulations or to power system studies.

Journey times and power consumption are not the only things that can be modelled. We need to make sure that sufficient rolling stock and infrastructure is functioning to deliver the service. Availability models for railways are becoming more realistic and are able to exploit techniques such as markov and semi-markov processes and Petri nets to predict the global effects of equipment reliability, maintenance procedures, recovery plans and timetables. These methods consider the operating railway as a network of operating states. The system moves from one state to another, for example, when a vehicle fails. Figure 3 shows a fragment of such a model. Appropriate reliability information can be included to give an overall description of the system. Of course, it is obvious that reducing recovery and repair times and maximising reliability will improve the system. The aim of the operator should be a continuously improving service. In these cases, sensitivity studies drawing on RSMs will enable managers to target capital investment towards the most rewarding features of the system.

3.3 Robust timetable design

These simple experimental methods enable Taguchi's vision for robust designs to be realised. All engineers know of cases where the most splendid designs can be developed with marvellous performance so long as no dimension departs from nominal and the conditions of use can be monitored and controlled. Unfortunately, when they are exposed to the weather (snow), wear, normal misuse and serial manufacture, their performance often falls below that of a perfect component.

It is common for railway timetables to suffer from similar problems. We want a service that is not only fast but which does not degrade ungracefully when a minor fault occurs in its operation. This idea has already been put into practice for services operating when severe weather is forecast. However, using the right tools enables this aspiration to be taken further. For a design as complicated as a railway timetable or signal layout, the simple methods championed by George Box are unlikely to be sufficient. Fortunately, appropriate methods for studying mathematical models of complicated systems subject to significant variation were developed by Iman and Conover (11). These methods allow the engineer to use a limited number of runs of the model to subject it to the full range of variability that the components of the system will experience. An assessment can then be made of how this affects performance and its variability. RSMs, or perhaps even genetic algorithms (12), can then be applied to these studies to optimise the design for performance and robustness.

All thorough robust design studies require a trade off between mean performance and variability. There is no neat mathematical trick for deciding this, despite the proposals of some statisticians. Just how much performance to trade off against just how much reliability will always be down to the commercial judgment of the manager. Deming frequently claimed that the most important numbers in any business are unknown and unknowable. Just what impact performance and reliability have on passenger experience will never be written down in an equation. There is no substitute for managers knowing their business.

3.4 Railway safety

Hazards do not speed up journey times. The complexity of modern railway systems demands that hazard analysis is thoroughly performed using the PRA methods that we learnt from the nuclear industry. In order to delight our customers, there will be a continual cycle of capital investment in new technologies and practices to reduce journey times. Each new technology must be assessed before it can be allowed to operate in passenger service. By the very nature of this process, we cannot rely on experience with operating the new technology and must rely on relevant data from existing systems and conservative assumptions based on engineering judgment.

Here is another example of using data to make predictions. Historical data is used to assess risks of infrastructure failures in the future. We have already discussed that no prediction can be made without a stable and predictable failure process, one that is in statistical control. Furthermore, such a state is unlikely to exist unless we have been working on it. The reliability engineer's dream that all failures occur as a homogeneous Poisson process has little foundation in real life engineering systems. We can have little confidence in any source of data from a process that is not being control charted and controlled.

This may worry some engineers very little. We usually look for safety cases that are robust against any single one of their assumptions. If some assumptions turn out in practice to be ill founded, there will be a limited exposure to risk of a short duration while additional safety measures are introduced. But again, the importance of control charting is highlighted. How do we know if the assumptions in our safety cases hold into the distant future. We want the earliest possible warning that there is some deterioration and that failures are occurring more frequently. However, we cannot afford to over react the first time we get an early failure. We need to distinguish signal from noise, special from common causes.

4. REFERENCES

(1)   Hartree, D.R. and Ingham, J. Note on the application of the differential analyser to the calculation of train running times, Memoirs and Proceedings of the Manchester Literary and Philosophical Society, 1938-9, 83, 1-15
(2)   Shewhart, W. Economic Control of Quality of Manufactured Product, 1936, Dover
(3)   Deming, W.E. Out of the Crisis, 1986, Cambridge University Press
(4)   Taguchi, G. and Wu, Y. Introduction to Off-Line Quality Control, 1980, Japan

Quality Control Organisation

(5)   Wheeler, D Understanding Variation: The Key to Managing Chaos, 1993, SPC Press

(6)   Moen, Nolan and Provost Improving Quality Through Planned Experimentation, 1991, McGraw-Hill

(7)   Box, G.E.P. *et al.* An explanation and critique of Taguchi's contribution to quality engineering, Quality and Reliability Engineering International, 4, 362-368

(8)   Box, G.E.P. and Draper, N.R. Evolutionary Operation, 1969, Wiley

(9)   Benski, C. and Cabau, E. Unreplicated experimental designs in reliability growth programs, IEEE Transactions on Reliability, 1995, 44, 199-205

(10)   Schoofs, A.J.G. *et al.* Approximation of structural optimization probles by means of designed numerical experiments, Structural Optimization, 1992, 4, 206-212

(11)   Iman, R.L. and Conover, W.J. Small sample sensitivity analysis techniques for computer models, with an application to risk assessment, Communications in Statistics - Theory and Methods, 1980, 9, 1749-1842

(12)   Painton, L. and Campbell, J. Genetic algorithms in optimization of system reliability, IEEE Transactions on Reliability, 1995, 44, 172-178

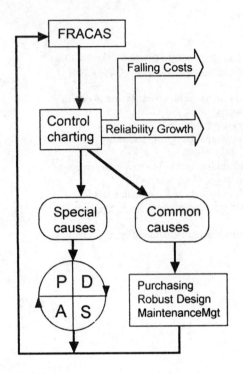

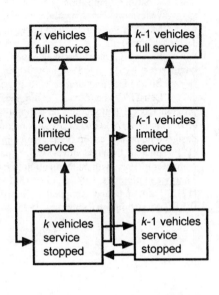

Fig 2: Quality tools within a reliability growth programme

Fig 3: Fragment of state diagram for railway with fleet of vehicles

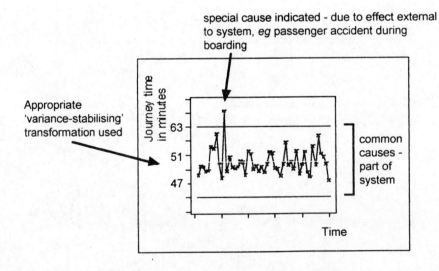

Fig 1: Specimen control chart for railway journey times

© IMechE 1996 C514/066

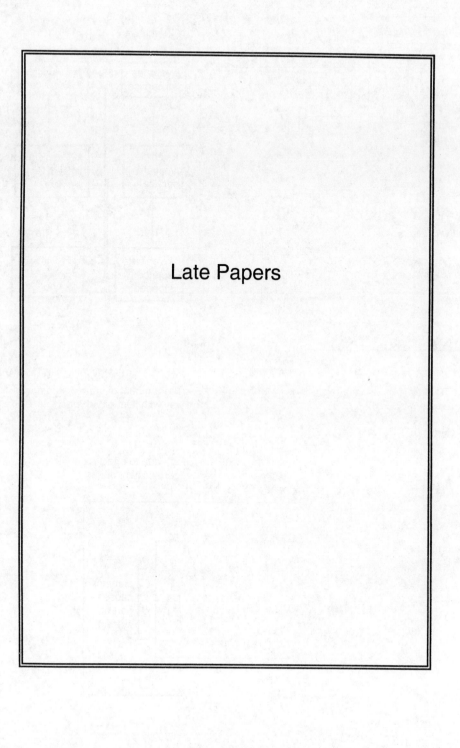

Late Papers

C514/068/96

Every second counts and people matter

C I HORSEY BSc, CEng, MIMechE, MIEE, MCIT
Victoria Line, London Underground Limited, London, UK

SYNOPSIS
The Victoria Line is one of the busiest routes in the London Underground network, carrying 140 million people every year. The Underground's customers expect safe, regular and reliable train services with minimum waiting and journey times, and the paper describes the Victoria Line's approach to delivering the best possible performance. Examples of key improvement areas, such as station and train design, signalling, system reliability and operational processes illustrate this. The paper also discusses the overall systems approach, the modelling tools used and the structured processes used by the line to manage its own evolution.

1   THE IMPORTANCE OF JOURNEY SPEED

In the long history of London's transport, the need for improved speed of transit has always been a strong spur to development. The planners of the first Underground railways in the 19th Century had a vision of a service that would relieve heavily congested inner streets of local traffic and provide new opportunities for rapid journeys to the City from outlying suburbs.

Journey speed is still an important factor for all mass transit systems today. This paper describes how the journey speed and business performance of an underground railway can be linked to the characteristics and performance of its component parts, operated together as a system. Practical examples illustrate how the Victoria Line is working to improve journey times for its customers by enabling all its assets - not only the engineering elements, but also the people and processes essential to a complex railway system - to perform to their best ability.

2   THE VICTORIA LINE

The Victoria Line, like the original Underground, was built as a response to long-standing social and economic pressures. Traffic in London (both Underground and other modes) had grown substantially by the 1950s, and the new line was intended to reduce crowding in many stations and on certain sections of existing lines. It was also expected to relieve some street congestion. North-east London would gain a faster route to the centre of town and north-south journeys across the central area (especially between main line termini) would speed up considerably. Feasibility studies proved the case for building the line by showing that the initial cost would be more than justified by social benefits of reduced travelling time and reduced congestion.

Today the Victoria Line, approaching thirty years old, is the one of the busiest lines in London's Underground network, carrying 140 million people every year over a route length of some 21 kilometres from Walthamstow to Brixton (Figure 1). The line joins three of the Underground's busiest stations (Victoria, Oxford Circus and King's Cross) and all but one of the sixteen stations provide interchanges with other Underground lines, surface rail services and other surface transport modes. Principal traffic flows are in the central section between Victoria and Seven Sisters; the section between Green Park and Victoria has one of the heaviest peak inter-station loads anywhere on the Underground (Figure 2).

The average riding time for customers using the Victoria Line is about 8 minutes and at peak times the service intervals are approximately 135 seconds. The line was the first mass transit railway in the world to operate automatically and the original trains and signalling are still in place, yet today's traffic level is almost three times larger than the original forecast, indicating both the integrity of the original designs, and the substantial demand for the Line's services today. More large increases in demand (notably from the Channel Tunnel Rail Link) are forecast in the early years of the next century.

The Line has a vision of service with 90 second intervals at peak times, encapsulated by the mission statement:

'to be the 90 second railway, where every second counts and people matter'.

In the short to medium term, the line's goals are to operate the current service with sustained regular and reliable performance, and to maintain this while reducing service intervals to about 120 seconds - probably the optimum achievable with its existing assets - and this has been termed the '120 second railway'.

The strategic development of the line to meet its business goals is a crucial and complex process. It calls for a structured approach, attention to detail, and for teamwork across a wide range of disciplines and methods - operations, engineering, process analysis, modelling, economic appraisal, human resource development and project management. It is critical for success that everyone involved shares an understanding of the total railway system and how it delivers service to customers.

3   THE RAILWAY AS A SYSTEM

3.1   Components

Most Customers would recognise the assets - trains, stations, and information systems - as the railway system they use. These are the tools with which station staff and train operators, themselves key assets, deliver 'front line' customer service. However, the underground railway system is more complex, with many diverse elements not seen by customers, such as the energy supply, maintenance and management functions supporting the business. Others, such as tunnels, track and control centres, remain in the background of the operation, making only a fleeting impression on the traveller. One fundamental, and perhaps least visible, element of the railway system is its processes, ranging from work instructions for maintenance and testing, to emergency procedures, to train and crew schedules. Their role is to define people's methods of working and their use of assets, and thereby co-ordinate the operation of the system.

Finally, Customers are key elements of the system too. While travelling, their presence and behaviour affects the way the system operates - for example delaying a train about to depart by obstructing its doors. Customers' role in the system is most obvious when traffic is at a peak, when stations and trains have to cope with large demand for travel, and the service is most vulnerable to disruption.

3.2 Emergent Properties and system capacity

The emergent properties of the whole system are the performance measures of the business. It is the interaction between system elements that enables the business to operate to its timetable, incur and manage its costs, and attain its level of safety. Marketability is another key emergent property, and market research shows that Customers' valuation of the service, and their willingness to use and pay for it, is closely related to their experience of the system at each stage in their journey, with most emphasis on the speed of the point-to-point journey (street level to street level):

- Entering a station from the street
- Purchasing a ticket and transferring to the station's platforms
- Waiting for a train
- Riding on the train
- Interchanges from line to line at stations en route
- Leaving the system at the destination station

The qualities of the experience (cleanliness of the travel environment, levels of illumination, ventilation and air quality, noise and vibration, usefulness of information and additional design features) all have some effect on the perception of speed, and therefore value of the journey to the user. Nevertheless, the primary requirement remains the fastest journey, which includes waiting on the platform (which always seems to pass more slowly than travelling time!) and riding time, including any delays en route. Standing in crowded conditions, whether waiting or riding, reduces the value of the journey to customers and can be evaluated as a penalty factor on the journey time. The degree of crowding and loading is a significant consideration in mass transit systems where capacity (volume of station and train service) of a line is a critical emergent property of the system, setting limits on the level of service that can be provided.

The limiting capacity of a railway can be quantified in terms of the service frequency that minimises customers' perceived journey time. It is first determined by the design of the line - combining infrastructure constraints, train and signalling performance and station design. Ability to achieve and maintain capacity depends on the availability of the system, ease of recovering from disruptions and the flexibility and spare capacity of the line. Human factors also influence capacity (even in an automated railway like the Victoria Line) and generally, small variations in any system parameters can make significant impacts on performance.

3.3 Analysing the system

Computer modelling is a vital aid to the study of the complex behaviour of a railway system. London Underground has its own Train Service Model (TSM), which is an event driven simulation for Underground lines and their operation. Base information for the model includes performance and design parameters for trains, signalling and route, the volume of people waiting on platforms, and the impacts of boarding and alighting movements on station dwell times, all of which are calibrated from periodic surveys of the real service.

The model enables the service performance of a line to be assessed and quantified under various operating scenarios - timetables, disruptions, trains delayed or cancelled, and changing levels of demand - leading to a clearer understanding of options and their impact on journey times. Bottlenecks on the route can be studied and modifications tested, and the model also can play a role in developing appropriate control and recovery strategies. The TSM can help the Business to understand the economic value of improvements in journey time, and thereby weigh the benefits of running a better service against the cost of making the change and the time to achieve the result. For example, a previous timetable revision on the Victoria Line increased the peak service frequency from 24 to 26 trains per hour (reducing the service interval from 150 seconds to 138 seconds). By inspection of the effects on customer journey time, TSM analysis showed that existing customers would gain £5.1Mn social benefits over time for £0.69Mn p.a. increased operating costs (principally additional traincrew and traction supply charges). The impact of this change illustrates clearly the '...every second counts...' in the line's mission statement.

As well as assisting development of the existing line, the TSM (in conjunction with other economic models and appraisal techniques within London Transport) is a key input to feasibility studies for future upgrading of the line and the network.

4   IMPROVING THE PERFORMANCE OF THE VICTORIA LINE

4.1   Optimising capacity

The line's short to medium term goals have been developed and refined through a combination of practical experience, business analysis and use of the TSM. This has shown that, for any given service schedule, there is an optimum regularity and reliability of the line and therefore an optimum journey speed for the customer. Figure 3 illustrates how long a representative journey on the line could be with different service intervals and different levels of disruption. Existing infrastructure, trains and signalling are used and each point represents a service with a different average interval. Journey time of around 18 minutes is achieved with typical levels of disruption by operating a service interval of 135 seconds, whereas a reduction in disruption enables a journey time of under 17.5 minutes to be achieved with a 130 second interval. Results with more recent survey data and assumptions suggest journey time could be reduced further by running a service at around 120 second intervals, provided that low disruption could be maintained and minor improvements are made to the system. This illustrates the impact of better service reliability and consistency on customers' journey speed, with a difference of more than 30 seconds in point-to-point time made possible by reducing disruption.

In practice, it is essential to gain assurance of system reliability before committing to more demanding timetable, and the Victoria Line aims to change the emergent properties of its system by incremental improvements in capacity, service regularity and equipment reliability. These changes embrace people, process, assets and even customers, since no single area of change can effect the transformation alone. The approach also recognises the long lead times for engineering changes to take place, compared to the smaller changes in methods of working which could deliver benefits relatively rapidly.

The key improvement areas, identified by experience and analysis are:

 ♦ Station and train crowding
 ♦ Dwell times at stations
 ♦ Signalling capacity
 ♦ Service disruption
 ♦ People and organisation

The following sections (4.2 to 4.7) discuss projects and results in these areas - some already in place and others at the feasibility or experimental stage.

4.2 Station crowding

Modern modelling and planning aids have quantified station congestion in terms of delay to customers moving through various areas. Qualitatively, the effects of station congestion on the overall system are:

 ♦ Frequent closing of stations for short times at peak periods (delay to journey)
 ♦ Platform, station and train crowding (increased perceived journey time)
 ♦ Longer waiting, boarding and alighting times (increased journey time and delays to following trains)
 ♦ Trains not fully loaded (reduced actual service capacity)

The last item is a result of having platform access passages at the southern end of platforms in several of the Victoria Line's busier stations. This results in customers massing in particular places when entering or leaving trains, rather than distributing evenly.

The direct solution to station congestion, to change the layout, is a substantial investment with long lead times. It can be justified where there are benefits of improved safety and of improved journeys for the station's users. Victoria, which handles over 70 million passengers per year (more through traffic than London's three airports combined), is very congested at peak times and incoming customers may be held behind the station's gates several times every morning to ensure the platforms do not become overcrowded. A project is under way to install an additional escalator, enlarge circulating areas and construct an additional interchange route. As a result, passengers will have faster access to the Victoria Line platforms, and movement on platforms and station routeways will be easier. The train service will benefit, too, from speedier boarding and alighting.

4.3 Vehicle Design

Crowding can also be relieved by improvements in train design. At the conclusion of the recent fleet refurbishment programme, the opportunity was taken to modify four vehicles as a trial. Each had about eight seats removed in favour of extra standing space around doorways, and the 4-car unit is permanently marshalled at the south end of a train, coinciding with the location of greatest congestion. The train gains extra capacity, as well as allowing better freedom and speed of movement in and out of the vehicle during station stops.

So far, the modified train appears to improve the boarding and alighting speed and customer reaction is to be researched - a few comments have already been received on the reduction in seating space! There are no plans to modify the remaining 340 vehicles in the fleet, but the experiment is an important step in developing options for future operation, since station dwell times and train capacity are both critical properties of the 90 second railway service to which the line aspires.

4.4   Dwell times at stations

One consequence of crowding is excess dwell time (the time trains spend at rest in stations). Although the line's signalling was designed to operate with 30 second average dwell times, busy stations now average up to 45 seconds and a few trains in every peak hour may spend up to double the allotted time in a platform. This constraint on peak capacity takes most effect at Victoria, King's Cross and Oxford Circus, but other stations - Green Park, Stockwell and Finsbury Park - also suffer at times. Consistent dwell times at stations are essential - excess dwell times will slow or stop other trains behind, thereby slowing down customers' journeys elsewhere on the system. Variation in intervals between trains (caused by delays to other trains, or by inconsistencies in train performance) can also affect dwell times adversely. Figure 4 shows how dwell time varies on a busy morning at King's Cross.

Process analysis has been used to define the sequence of events from the arrival of one train to the arrival of the next. The process (summarised in figure 5) includes the functions of all assets involved and the parts played by train operator, platform staff and customers. The analysis was carried out by a team of experienced front-line staff, who identified a number of options for improvement, including technical measures, such as automatic departure warnings, removal of non-essential equipment on platforms, and changes to the process such as the revision of public address messages. Another practical suggestion was to educate regular Customers and guide them into more consistent boarding and alighting behaviour.

The team experimented with temporary clocks giving Train Operators a guide to the correct departure time and thereby reducing variance in dwell times. Figure 6 illustrates the cumulative distribution of dwell times achieved by the trial compared to 'as-is' performance. The shallow slope of the 'as is' distribution indicates a wide range of dwell times (up to 104 seconds), despite the average of 45 seconds. The clock achieved different results, and a steeper slope to the curve indicates a tighter distribution - the average dwell is reduced to 38 seconds and 20% more trains achieved this time. Better results are expected from a more permanent installation, and work is now extending to platform information and crowd control in order to compile an optimum package of engineering, people and process changes. Meanwhile, another team is studying the operation of terminal stations on the line using similar techniques.

4.5   Signalling capacity

Signalling, by controlling the safe passage of trains through a given area, sets limits on the frequency of the train service. In outline, the system has to allow first for the safe separation of trains running at their maximum speed, taking into account the train accelerating and braking performance, and track and traffic conditions. Next, allowance is made for stations and junctions, where the speed and separation of trains changes significantly. This allowance will include the time spent by trains at rest in stations. Finally, time is allowed for trains to recover from minor irregularities in running. The performance of signalling and layout at particular locations can determine the achievable level of service on the whole line.

The Victoria Line's Automatic Train Control (ATC) system was designed to achieve a minimum service interval of around two minutes (120 seconds). A signalled headway of under 90 seconds for full speed (up to 77 km/h) running was specified and average station dwell times of 30 seconds were assumed. With average dwell times now higher at many places and times, the underlying design assumptions appear compromised if trains are to run at their full speed. In practice, the signalling system tolerates extended dwell times by controlling the speed of trains approaching stations, or even stopping trains until the route ahead becomes clear. At the lower controlled speed of 35 km/h, closer intervals can be achieved, but fewer trains can run through the area. This reduces dramatically the line's carrying capacity and disbenefits the customer and the business by extending journey times.

Although radical improvements in line capacity can only be made by quantum change to the signalling system, modifications at particular places can complement changes in station capacity and dwell time management to improve overall performance. At the southbound approach to King's Cross, for example, trains were restricted to a controlled speed of 35 km/h if there was another train in the platform or in the process of leaving. The speed restriction applied for a long distance before the station and even if the train ahead moved clear, the ATC system did not permit the train to re-accelerate. Modifying the local signalling and ATC commands reduced the controlled speed run time by 33 seconds per train. The modification also made the local signalling more capable of tolerating excess dwell times without controlled speed running (though further measures on the station are still required).

There are similar bottlenecks at other locations where modification could improve line capacity. Oxford Circus (southbound) is typical, with dwell times averaging 45 seconds at present, and a full speed headway of 79 seconds, making possible a 124 second minumum interval. This implies that, with the current scheduled interval if 135 seconds, there is an 11 second recovery margin at this site. Re-acceleration, additional track circuits and improving tolerances on train stopping accuracy at signals, could together reduce the full speed headway to 61 seconds. When coupled with measures at the station to achieve consistent dwell times of around 40 seconds, this would enable an increase in the service frequency which could be operated reliably.

4.6   Service disruption

Although the small time losses and disturbances to process are not negligible, discrete incidents are the major causes of disruption on the Victoria Line today. For the financial year 1995/6, the breakdown of reported incidents causing delay to the service of two minutes or more was:

Cause of incident	Total incidents	Total Initial Delay (Minutes)
Rolling Stock	889 (41%)	5024 (33%)
Passenger Action	728 (26%)	4883 (32%)
Staff	528 (21%)	2351 (15%)
Signalling, track, infrastructure	188 (5%)	2228 (14%)
Stations and other causes	92 (4%)	832 (5%)
Grand Totals:	2340 (100%)	15 318 (100%)

Business cases for improvement are based on the value to Customers and the Business of eliminating the incidents and their effects on the system, taking into account both frequency of occurrence and amount of delay incurred. Incidents which involve Customers or Rolling Stock defects are clearly high on the list as a group, but the priorites for corrective action need to be determined through detailed analysis of incident data and the underlying trends and patterns.

Customer incidents are varied, including vandalism (such as graffiti), negligent acts (lost property) and people being taken ill. Reducing these incidents relies substantially on improved customer education, and recent campaigns have included raising awareness of security (lost property and suspicious objects) and indicating the hazards of obstructing train doors. The subject of customer behaviour is an important one, and best practice in this area is still being sought.

All causes of incidents are unwanted, but any contribution from the line's own assets, processes and people could be described as an 'own goal'. With this in mind the line is gradually reducing the delays to service from causes under its own control. In the engineering domain, a reliability improvement culture is growing, with work on rolling stock in the forefront. A train reliability improvement programme (TRIP) has now been in place for more than a year and has played a part in driving down train failure rates (Figure 7) by providing a common focus for the variety of improvements and modifications needed.

Reliability-centred Maintenance techniques are now being applied in London Underground to a variety of assets and this, too, is expected to benefit the Victoria Line, not only by reducing failure rates but also by ensuring that maintenance regimes are effective and appropriate for the demands of the service.

4.7   People and organisation

In the Line's mission statement, "...and people matter" emphasises not only the Underground's customers but also the people working for and with the Line, running it day-by-day and delivering improvements to the Customer in quality, cost and journey time.

In recent years, there have been several initiatives in London Underground, all designed to improve awareness of how individual staff play a part in the overall system and its processes. Service managers, traincrew, station staff and engineering maintenance staff were all involved and even here, the TSM played a key role by providing a graphic simulation of the line in operation to illustrate traffic scenarios, and the effects of disruption on the system. Opportunities were also provided for people to work in groups to solve real service problems. These programmes had positive effects on service performance, yet also had long-term value by starting to develop people's shared understanding of the system.

The line team is now implementing a locally-developed management system termed 'Process 90' which provides a consistent high-level process (Figure 8) to enable change activities at all levels to be structured and aligned with strategic goals. Everyone, not only those working day-to-day on the line, but also those in the supporting services, has the opportunity to be involved in the improvement work sponsored by the process. So far, the emphasis has been on tasks working towards the '120 second railway'; for example, a cross-disciplinary team within the line is carrying out the work on station dwell times described in section 4.4, and other teams are working on asset reliability, management systems, station crowd control and terminus operation.

Process 90 has been evolved with the experience of previous major projects and continuous improvement initiatives in the Underground and elsewhere, and seeks to balance the empowerment and openness of teamworking with the discipline and co-ordination of programme management techniques. It offers an opportunity for project teams to become part of the Line team, and thus identify with individual end users more closely, with mutual advantage and overall benefit to the end product.

5   DELIVERING IMPROVEMENTS

The paper has described how a systems approach, embracing people, processes and assets, is essential when planning and delivering improved performance on a complex mass transit railway. The practical examples of change in the paper have illustrated this approach, and show how journey speed and quality of service to the Customer are critical success factors for the Underground and for the Victoria Line in particular, where every second does count towards the ultimate goal of a '90 second railway'.

ACKNOWLEDGEMENTS
Everyone associated with the Victoria Line, both in day-to-day operation and maintenance and in improvement projects, has contributed to the story in some way, and many have also assisted in preparing this paper. In particular, I acknowledge the support and encouragement given to me by John Self, the outgoing General Manager, and his team. I would like to thank London Underground's Director of Passenger Services, Hugh Sumner, and Director of Engineering, David Hornby, for their permission to deliver this paper.

Figure 1 - The Victoria Line and its rail interchanges

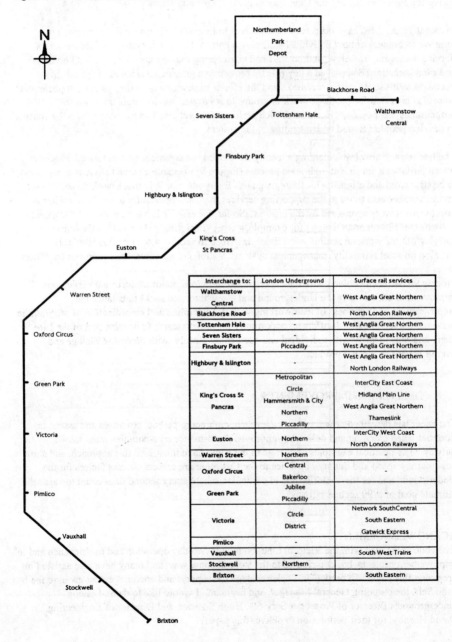

Interchange to:	London Underground	Surface rail services
Walthamstow Central	-	West Anglia Great Northern
Blackhorse Road	-	North London Railways
Tottenham Hale	-	West Anglia Great Northern
Seven Sisters	-	West Anglia Great Northern
Finsbury Park	Piccadilly	West Anglia Great Northern
Highbury & Islington	-	West Anglia Great Northern North London Railways
King's Cross St Pancras	Metropolitan Circle Hammersmith & City Northern Piccadilly	InterCity East Coast Midland Main Line West Anglia Great Northern Thameslink
Euston	Northern	InterCity West Coast North London Railways
Warren Street	Northern	-
Oxford Circus	Central Bakerloo	-
Green Park	Jubilee Piccadilly	-
Victoria	Circle District	Network SouthCentral South Eastern Gatwick Express
Pimlico	-	-
Vauxhall	-	South West Trains
Stockwell	Northern	-
Brixton	-	South Eastern

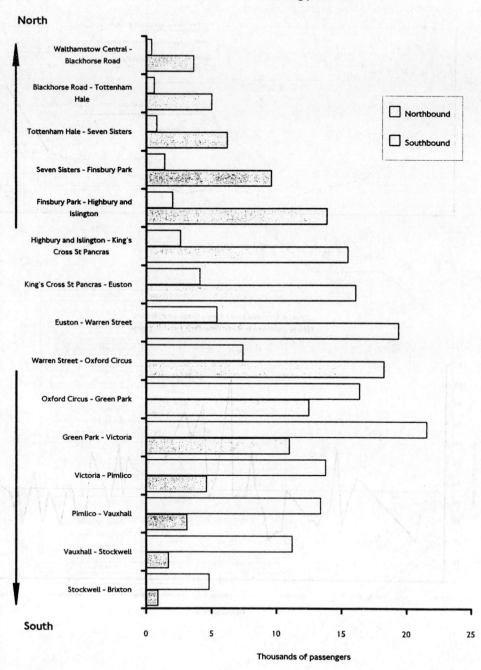

Figure 2 - Victoria Line inter-station loading 1994
Morning peak hour

North

South

Thousands of passengers

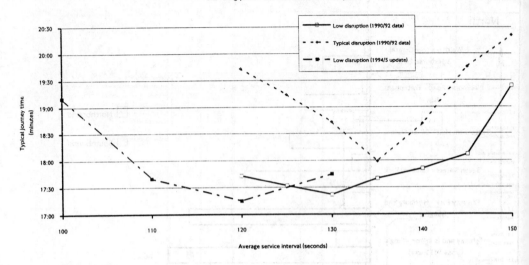

Figure 3 - Victoria Line - Simulated impact of service interval on journey time
Morning peak - various levels of disruption

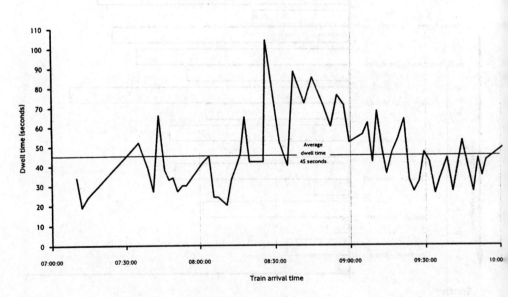

Figure 4 - Victoria Line - King's Cross southbound
Typical variations in dwell time during morning peak

© IMechE 1996 C514/068

Figure 5 - Victoria Line - Station Stop Process Summary

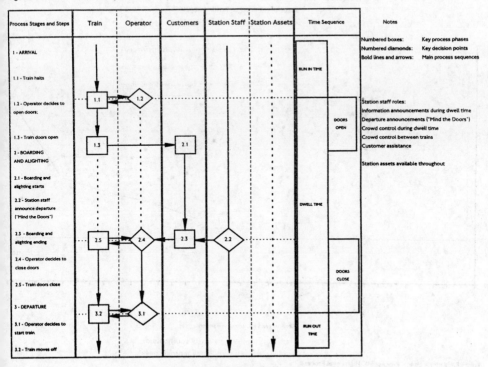

Figure 6 - Victoria Line - King's Cross southbound
Cumulative distribution of dwell times

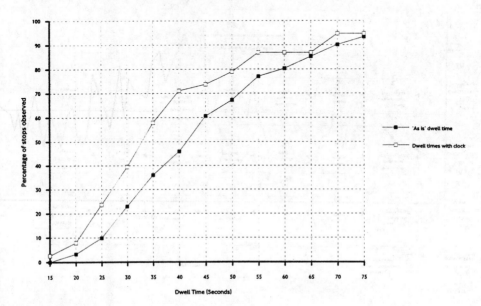

Figure 7 - Victoria Line - Incidents per 10,000 train kilometres
Causes attributed to Rolling Stock

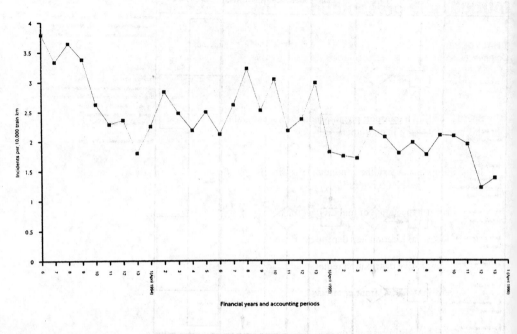

Financial years and accounting periods

Figure 8 - Victoria Line: "Process 90" High Level Process

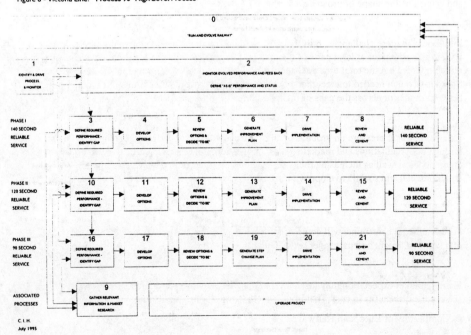

C. I. H.
July 1995

C514/061/96

Introductory presentation

T HALVORSEN
Community of European Railways, Brussels, Belgium

Speed is not just a transport phenomenon - it is a cultural icon. The presentation examines how this manifests itself in consumers' views and in competition between services.

In the European context, the continuing effects of the single market will present challenging demands for transport services:

rising expectation of mobility and flexibility;

industrial location less dependent on traditional patterns;

competitive distribution requirements;

erosion of past "frontier effects"

At national level the pressures on land use planning will influence transport policy and financing and determine the shape of urban development.

In response to these challenges, railways are confronted with the need to carve out a performance level which addresses at least parts of the market. The extent of the challenge covers not just the rail component but the total time package linked to inter-modality, information and systems. Speed is as much needed to meet market demand as to compensate for technical frontiers be they between modes or across the gaps of interoperability of rail equipment.

External political pressure on grounds of environment will seek to redraw the inter modal map through both regulations and market signals. Congestion will reduce the effective performance of other modes as well as increasing the cost. Speed will need to fit into an acceptable environmental and social envelope with increased scrutiny in areas such as noise emission and safety.

Further internal constraints will come from managing the mix of speed requirements - the trade offs between freight and passenger service quality.

Finally, market expectation may insist on spreading the advantage of higher speed without sharing the disbenefits in cost. That is a major challenge to the engineering community.

© Community of European Railways